**Anais da XXVIII
Reunião de Trabalho sobre
Física Nuclear no Brasil**
São Paulo, 2005

Anais da XXVIII
Reunião de Trabalho sobre
Física Nuclear no Brasil

Hotel Delphin no Guarujá-SP,
07–11 de setembro de 2005

Marcelo G. Munhoz – IFUSP (coord.)
Carlos Roberto Appoloni – UEL
Cibele Bugno Zamboni – IPEN-CNEN/SP
Fábio de Camargo – IPEN-CNEN/SP
Paulo Reginaldo Pascholati – IFUSP
Roberto Meigikos dos Anjos – UFF

Editora Livraria da Física
São Paulo

Copyright © 2007 Editora Livraria da Física

1a. Edição

Editor: José Roberto Marinho
Diagramação: Roberto Maluhy Jr & Mika Mitsui
Capa: Arte Ativa
Impressão: Gráfica Paym

Dados Internacionais de Catalogação e Publicação (CIP)
(Câmara Brasileira do Livro, SP, Brasil)

Reunião de Trabalho sobre Física Nuclear no Brasil (2005 : Guarujá, SP)

Anais da XXVIII Reunião de Trabalho sobre Física Nuclear no Brasil : Hotel Delphin no Guarujá – SP / Marcelo G. Munhoz, (coord.). – São Paulo : Editora Livraria da Física, 2007.

Vários autores.
Bibliografia

1. Física nuclear – Brasil – Congressos
I. Munhoz, Marcelo G.. II. Título.

07-4361   CDD-539.706081

Índices para catálogo sistemático:
1. Brasil : Física nuclear : Congressos 539.706081

ISBN: 978-85-88325-90-6

Todos os direitos reservados. Nenhuma parte desta obra poderá ser reproduzida sejam quais forem os meios empregados sem a permissão da Editora. Aos infratores aplicam-se as sanções previstas nos artigos 102, 104, 106 e 107 da Lei nº 9.610, de 19 de fevereiro de 1998.

Impresso no Brasil

Editora Livraria da Física
Telefone 55 11 3816 7599 / Fax 55 11 3815 8688

www.livrariadafisica.com.br

Article Index

The Brazilian Two-Pion Exchange Nucleon-Nucleon Potential 1
C. A. da Rocha, M. R. Robilotta and R. Higa

Measurement of Sr/Ca Ratio in Bones as a Temperature Indicator 9
P. R. dos Santos, N. Added, M. A. Rizzutto, J. H. Aburaya and M. D. L. Barbosa

The Relativistic Quasi-Particle Random Phase Approximation and Beta Decay 17
C. De Conti, B. V. Carlson and T. Frederico

Caracterização de Microporos produzidos por Rastos Iônicos em Polímeros 25
A. O. Delgado, M. A. Rizzutto, A. R. Lima, A. A. R. Da Silva, M. A. P. Carmignotto, M. H. Tabacniks, N. Added

Trace Elements in Blood Serum Measured by PIXE 37
S. Bernardes, M.H. Tabacniks, Marcel D.L. Barbosa and Márcia A. Rizzutto

Automatic Classification and Segmentation of Image Elements Using Computed Tomographic Data 45
J. M. de Oliveira Jr., E. C. Paizani, A. C. G. Martins and G. F. de Lima

PlanDose: A User-Friendly Program for Planning of Ablation of Thyroid Carcinoma 57
Lima, F. F., Benevides, C. A., Silveira, R., Stabin, M. G., Khoury, H. J.

Optimization of the ^{123}I Production in the Microtron MT-25, using the Code MCNP-X (Part 1) 63
B. Zaldívar-Montero, N. López-Pino, M. V. Manso-Guevara, K. D'Alessandro, J. N. Varona-Ayala

The Color Flavor Locked Phase in the Chromodielectric Model and Quark Stars 75
L. P. Linares and M. Malheiro

Investigation of Elemental Distribution in Lung Samples by X-Ray Fluorescence Microtomography 89

Gabriela R. Pereira, Ricardo T. Lopes, Henrique S. Rocha, Marcelino J. dos Anjos, Luis F. Oliveira and Carlos A. Pérez

ᴿᴶᴾ Rise and Fall of Pentaquarks in the QCD Sum Rules Approach 99

R. D. Matheus, F. S. Navarra and M. Nielsen

ᴿᴶᴾ Accumulation and Long-Term Behavior of Radiocaesium in Tropical Plants 127

C. Carvalho, B. Mosquera, R. M. Anjos, N. Sanches, J. Bastos, K. Macario, R. Veiga

Implications of the $\rho(770)^0$ Mass Shift Measured at RHIC 135

P. Fachini, R. Longacre, Z. Xu and H. Zhang

Real-Time Radiographic Imaging With Thermal Neutrons 143

Marcelo J. Gonçalves, Ricardo Tadeu Lopes, Maria Ines Silvani, Gevaldo L. de Almeida, Rosanne C.A.A. Furieri

A QTDA Model for Double Beta Decay 151

A. R. Samana, F. Krmpotić

ᴿᴶᴾ Quasiparticle-Rotor Model Description of Carbon Isotopes 161

T. Tarutina, A. R. Samana, F. Krmpotić, M. S. Hussein

Analytical Sensitivity Analysis and Optimization of a System Employing a Si-PIN Detector for the Study of Environmental Samples 171

Melquiades F. L., Parreira P. S., Appoloni C. R., Lopes F.

Caracterização Química de Minerais em um Suplemento Vitamínico Comercial por Fluorescência de raios X com Equipamento Portátil 179

Fabio Lopes, Fábio L. Melquiades, Paulo S. Parreira, Carlos R. Appoloni

Minimum Level of Component Discrimination (MLCD) for Characterization of Biphasic Mixtures (Oil / Brine or Gas) Using Gamma and X-Ray Transmission 187

Luiz Diego Marestoni, Filipe Sammarco, Carlos Roberto Appoloni, Jair Romeu Eichlt

Aplicação de um Detector de Telureto de Cádmio (CdTe) para Espectrometria Gama "IN SITU" 199

Luiz Diego Marestoni, Carlos Roberto Appoloni e Jaquiel Savi Fernandes

Calibração do Detector LR-115 II em Monitor de Placas Paralelas sob Condições de Ventilação Forçada 205

H. Javier Ccallata, E. M. Yoshimura

Study of DNA and Protein Structures Through Perturbed Angular Correlations 209

J. C. de Souza, N. Carlin, A. W. Carbonari, E. M. Szanto and F. J. de Oliveira Filho

Low-Energy Levels Calculation for [193]Ir 221

Guilherme Soares Zahn, Cibele Bugno Zamboni, Frederico Antonio Genezini, Joel Mesa-Hormaza and Manoel Tiago Freitas da Cruz

Evaluation of X-Ray Sensitive Image Plates as Detectors to Acquire Neutron Radiography Images 227

Maria Ines Silvani, Gevaldo L. de Almeida, Rosanne C.A.A. Furieri, Ricardo Tadeu Lopes, Marcelo J. Gonçalves and John Douglas Rogers

Development of Neutrongraphic Image-Acquisition Systems at the Instituto de Engenharia Nuclear 239

Maria Ines Silvani, Gevaldo L. de Almeida, Rosanne C.A.A. Furieri, Ricardo Tadeu Lopes and Marcelo J. Gonçalves

Systematics of Half-lives for Alpha-Emitter Nuclides 251

M. M. N. Rodrigues, S. B. Duarte, O. A. P. Tavares and E. L. Medeiros

Effects of the Mixed Thermal Neutron-Gamma Radiation on Plasmid DNA From IEA-R1 Reactor of IPEN 255

Maritza R. Gual, Airton Deppman, Felix Mas, Paulo R. P. Coelho, A. N. Gouveia, Oscar R. Hoyos and A. C. Schenberg

Simple Analytical Expression for Vector Hypernuclear Asymmetry in Nonmesonic Decay of $^5_\Lambda$He and $^{12}_\Lambda$C 263

César Barbero, Francisco Krmpotić, Alfredo P. Galeão

Atualização das Energias dos Padrões Gama para a Escala do CODATA2002: Efeito Sobre as Correlações 273

Zwinglio O. Guimarães-Filho, Otaviano Helene

Correlações Entre as Energias dos Raios Gama que Seguem o Decaimento do ^{56}Co e do ^{66}Ga 281

Zwinglio O. Guimarães-Filho, Otaviano Helene, Vito R. Vanin

Electric Field Interference in DNA Repair: Case Study and Perspectives to Improve Cancer Radiotherapy 289

J. D. T. Arruda-Neto, M. C. Bittencourt-Oliveira, E. C. Friedberg, E. C. Silva, A. C. G. Schenberg, M. C. C. Oliveira, T. C. Hereman, J. Mesa, T. E. Rodrigues, K. Shtejer, C. Garcia, F. Garcia

VIII ARTICLE INDEX

Single Neutron Pick-up on ^{104}Pd 299

*M. R. D. Rodrigues, J. P. A. M. de André, T. Borello-Lewin,
L. B. Horodynski-Matsushigue, J. L. M. Duarte, C. L. Rodrigues and G. M. Ukita*

The Nuclear Matter Effects in π^0 Photoproduction at High Energies .. 305

*T. E. Rodrigues, J. D. T. Arruda-Neto, J. Mesa, C. Garcia, K. Shtejer, D. Dale and
I. Nakagawa*

**Study of Different Configurations for a Subcritical Ensemble Using
MCNP-4C Code** 315

*A. O. Casanova, N. López-Pino, M. V. Manso Guevara, K D'Alessandro, E. Herrera
Peraza, A. Gelen, O. Díaz and M. N. Martins*

**Analysis of the Internal Structure in Bone From 3D Microtomography
Using Microfocus X-Ray Source** 325

Inaya C. B. Lima, Ricardo T. Lopes and Mônica Rocha

**Evaluation of entrance surface dose to chest x-rays in pediatric
radiography performed in a University Hospital in Curitiba, Brazil** ... 335

*N. A. Lunelli, H. R. Schelin, S. Paschuk, V. Denyak, J. G. Tilly Jr, E. Rogacheski,
H. J. Khoury and A. C. P. de Azevedo*

High-K Band in ^{140}Gd 343

*J.R.B. Oliveira, F. Falla-Sotelo, M.N. Rao, J.A. Alcántara–Núñez, E.W. Cybulska,
N.H. Medina, R.V. Ribas, M.A. Rizzutto, W.A. Seale, D. Bazzacco, F. Brandolini, S.
Lunardi, C. Rossi-Alvarez, C. Petrache, Zs. Podolyák, C.A. Ur, D. de Acuña, G.de
Angelis, E. Farnea, D. Foltescu, A. Gadea, D.R. Napoli, M. de Poli, P. Spolaore, M.
Ionescu-Bujor and A. Iordachescu*

Trajectory Effects in Coulomb Excitation 353

F. W. Fernandes and B. V. Carlson

Fissility of Actinide Nuclei by 60-130 MeV Photons 361

Viviane Morcelle and Odilon A. P. Tavares

**Analysis of Multi-Parametric Coincidence Measurements with n
Detectors** 373

*Frederico Antonio Genezini, Guilherme Soares Zahn, Marcus Paulo Raele, Cibele
Bugno Zamboni and Manoel Tiago Freitas da Cruz[3]*

**Compton Scattering and Absolute Measurement of Tagged Photon
Energies** 379

W. R. Carvalho Jr., Z. O. Guimarães-Filho and A. Deppman

Evaluation of Trace Elements in Whole Blood and its Possible use as Reference Values 387

C. B. Zamboni, L. C. Oliveira, M. R. Azevedo, E. C. Vilela, F. L. Farias, P. Loureiro, R. Mello, J. Mesa and L. Dalaqua Jr.

Active Photomultiplier Bases for the Saci Perere Spectrometer 391

T. F. Triumpho, D. L. Toufen, J. R. B. Oliveira, C. M. Figueiredo, R. Linares, K. T. Wiedemann, N. H. Medina, R. V. Ribas

Nuclear methodology for Al quantification in patients submitted to dialysis 399

N. Santos, C. B. Zamboni, F. F. Lima, E. C. Vilela

☞Artigos originalmente publicados no Brazilian Journal of Physics, Vol 36 – n° 4B, Special Issue: XXVIII Workshop on Nuclear Physics in Brazil, (2006).

Programa

- **Abertura** "Novas perspectivas para a atuação do físico nuclear: o papel da CNEN"
  Prof. Odair Dias Gonçalves (Presidente da CNEN)

- Mesa Redonda **"Pensando o Futuro"**
  Prof. Alejandro Szanto de Toledo (USP, Coordenador Projeto LINAC)
  Prof. Carlos Appoloni (UEL/PR, Coordenador do Laboratório de Física Nuclear Aplicada)
  Profa. Marina Nielsen (USP, Membro da Comissão Física para o Brasil-SBF)
  Prof. Raul Donângelo (UFRJ, Membro da Comissão Física para o Brasil-SBF)

  PA01 – Reações Nucleares/Núcleos Exóticos
  Prof. Joachin Gomez-Camacho
  "Understanding the Scattering of Exotic Nuclei"

  PA02 – Estrutura Nuclear
  Prof. Calin Ur
  "Modern Gamma-Ray Spectroscopy: AGATA and
  PRISMA-CLARA"

  PA03 – Estrutura Nuclear/Núcleos Exóticos
  Prof. Vladilen Z. Goldberg
  "Inverse Resonance Thick Target Scattering"

  PA04 – Reações Fotonucleares
  Prof. Mario Babilon
  "Bremsstrahlung Photons - An Ideal Tool in Nuclear Structure and
  Nuclear Astrophysics"

  PA05 – Física Hadrônica
  Prof. Fernando Navarra
  "The Pentaquark Physics"

  PA06 – Altas Energias
  Dra. Patrícia Fachini

Tabela 1 Programa – XXVIII RTFNB 2005

Horário	quarta-feira 07/09	quinta-feira 08/09	sexta-feira 09/09	sábado 10/09	domingo 11/09
09:00 - 09:45		Abertura	PA04 Babilon	PA08 Watanabe	Saída
09:45 - 10:30		PA01 Gomez-Camacho	PA05 Navarra	PA09 Rizzuto	
10:30 - 11:00			*Café*		
11:00 - 11:45		PA02 Calin Ur	PA06 Hallman	PA10 Leung	
11:45 - 12:30		PA03 Goldberg	PA07 Ferreira	PA11 Lopes	
12:30 - 14:00			*Almoço*		
14:00 - 15:40		Comunicações Orais	Comunicações Orais	Comunicações Orais	
15:40 - 17:00	Chegada	Café e Painéis	Café e Painéis	Café e Painéis	
17:00 - 17:30		PD1 Medina	Livre	PD3 Okuno	
17:30 - 18:00		PD2 Robilotta	Livre	PD4 Kodama	
18:00 - 19:00	Coquetel de Abertura	Livre	Assembléia	Premiação de Painéis	
19:00 - 21:00			*Jantar*		
21:00 - 22:00	Ano Mundial da Física	Mesa Redonda "Pensando o Futuro"	Colóquio "Divulgação Científica"		

PA07 – Astrofísica Nuclear/Altas Energias
Profa. Lidia Ferreira
"Physics at the drip-lines"

PA08 – Nuclear Aplicada
Prof. Shigueo Watanabe
"Aplicação de Física Nuclear em Dosimetria da Radiação,
Arqueologia e em Geologia"

PA09 – Nuclear Aplicada
Profa. Márcia Rizzuto
"Non-destructive analysis in art and archaeology"

PA10 – Instrumentação
Prof. Ka-Ngo Leung
"Compact Neutron Generators: New Facilities and Applications"

PA11 – Instrumentação
Prof. Ricardo Tadeu Lopes
"Sistemas de Detecção de Radiações para a Formação de Imagens"

PD1 – Estrutura Nuclear
Prof. Nilberto Medina

PD2 – Física Hadrônica
Prof. Manuel Robilotta

PD3 – Física Nuclear Aplicada
Profa. Emico Okuno

PD4 – Astrofísica Nuclear/Altas Energias
Prof. Takeshi Kodama

- **Colóquio** – Prof. Nelson Studart
"Einstein e a Divulgação Científica"

PA – Palestras Avançadas
PD – Palestras Didáticas

Introdução

XXXVIII Reunião de Trabalho sobre Física Nuclear no Brasil

A XXVIII Reunião de Trabalho sobre Física Nuclear no Brasil (RTFNB) realizou-se na cidade do Guarujá, litoral do estado de São Paulo, de 7 a 11 de Setembro de 2005. Participaram do evento 151 pessoas, que apresentaram um total de 138 trabalhos.

A Reunião de 2005 contou com algumas novidades cujo intuito era tornar a reunião mais dinâmica, aumentando a interação entre os participantes e possibilitando uma maior participação dos alunos de pós-graduação.

Em comparação aos eventos de outros anos, o número de participantes e trabalhos apresentados se manteve constante, como mostrado na figura 1. O evento de 2003 é uma exceção, pois nesse ano foi realizado simultaneamente o *V Latin American Symposium of Nuclear Physics and Applications*.

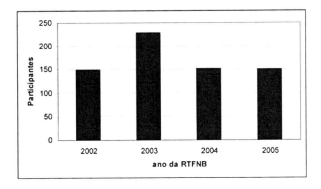

Figura 1 Número de participantes das RTFNB ao longo dos últimos anos

Na figura 2 é mostrada a distribuição de participantes da XXVIII RTFNB em termos da localização de sua instituição de origem, divididos em pesquisadores com doutorado (DR), alunos de pós-graduação (PG) e alunos de Iniciação Científica (IC), havendo três categorias: vindos de São Paulo, de outros estados e do exterior (Ext).

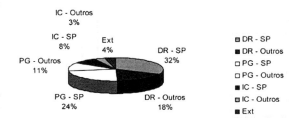

Figura 2 Distribuição de participantes da XXVIII RTFNB em termos da localização de sua instituição de origem.

Na figura 3 é mostrada a distribuição de áreas dos trabalhos apresentados. Nota-se a relevância de trabalhos de física nuclear aplicada nesta reunião.

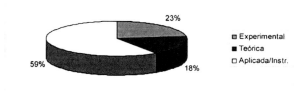

Figura 3 Distribuição das áreas dos trabalhos apresentados, divididos em: física nuclear básica experimental, física nuclear teórica e física nuclear aplicada e instrumentação.

O formato do programa da reunião seguiu o padrão estabelecido ao longo dos anos. A parte da manhã foi dedicada à palestras convidadas, contando com a presença de pesquisadores estrangeiros e brasileiros de reconhecida reputação internacional. No período da tarde havia as apresentações orais de 25 minutos e 90 minutos para a visita e apresentação dos painéis. O período da noite foi reservado para atividades extras como colóquios e discussões.

A comissão organizadora incentivou fortemente a apresentação oral de trabalhos inscritos por alunos de pós-graduação. Nas sessões de painéis, contamos com avaliadores informais, que tiveram a incumbência de discutir o trabalho exposto com seu apresentador principal (normalmente alunos de pós-graduação) e fazer um breve comentário sobre a forma como o trabalho foi apresentado no painel. Ao final da reunião, os painéis com as melhores avaliações foram premiados.

O programa também contou com uma sessão no período da tarde, onde pesquisadores de diferentes áreas da física nuclear expuseram de maneira didática o estado da arte de sua especialidade.

Para comemorar o Ano Internacional da Física, contamos com uma apresentação teatral da peça "Einstein" do grupo Arte e Ciência e com um colóquio do Prof. Nelson Studart (editor da Revista Brasileira de Ensino de Física) cujo título foi "Einstein e a Divulgação Científica", como uma forma de incentivar nossos colegas à importante prática da divulgação da ciência. Também organizamos um evento de divulgação de física em uma escola pública da cidade do Guarujá, em parceria com a Secretaria de Educação da cidade.

Finalmente, o programa contou com uma mesa redonda cujo tema foi "Pensando o Futuro". Esta atividade gerou discussões muito importantes para o futuro da física nuclear no Brasil.

Em suma, a reunião foi muito produtiva, com excepcional interação entre os participantes e excelentes trabalhos apresentados, o que certamente contribui muito para a vitalidade da física nuclear no Brasil.

A comissão organizadora agradece a todos os participantes e agências financiadoras, FAPESP, CNPq, CNEN, CAPES, FINEP e CLAF, sem os quais este evento não teria sido possível.

A Comissão Organizadora

Anais da XXVIII Reunião de Trabalho sobre Física Nuclear no Brasil
SP, Brasil, 2005

The Brazilian Two-Pion Exchange Nucleon-Nucleon Potential

C. A. da Rocha* (USJT), M. R. Robilotta† (IFUSP) and R. Higa‡ (JLAB)

1 Introduction

A considerable refinement in the description of nuclear interactions has occurred in the last decade, due to the systematic use of chiral symmetry, which provides the best conceptual framework for the construction of nuclear potentials. The works by Weinberg [1] restating the role of chiral symmetry in nuclear interactions were followed by an effort by many authors to construct the Two-Pion Exchange Potential (TPEP) under this symmetry [2, 3, 4, 5, 6, 7, 8, 9, 10, 11, 12, 13]. In the early 1990s, our group achieved a first version of this chiral TPEP, which we called "minimal model". The symmetry was then realized by means of non linear Lagrangians containing only pions and nucleons. This minimal chiral TPEP was consistent with the requirements of chiral symmetry and reproduced, at the nuclear level, the well known cancellations present in the intermediate pion-nucleon amplitude [14]. Afterward, other degrees of freedom were introduced from the empirical information of the low energy πN amplitude, which is normally summarized by means of subthreshold coefficients, that can be used either directly in the construction of the TPEP or to determine unknown coupling constants (low-energy constants-LEC's) in chiral Lagrangians. The inclusion of this information allowed satisfactory descriptions of the asymptotic NN data to be produced, with no need of free parameters [12], and gives rise to predictions, which are tested against phenomenological potentials. The dynamical structure of the eight leading non-relativistic components of the interaction is investigated and, in most cases, found to be clearly dominated by a well defined class of diagrams. A second step was to get the TPEP in configuration space [13], where we made the dependency on low-energy constants more explicit. In order to facilitate the use of the potential to fit data and get the LEC's values, we developed a parametrized version of our potential to $\mathcal{O}(q^4)$, which is present in this contribution. An important result is

---

* carlos.rocha@usjt.br
† mane@if.usp.br
‡ higa@jlab.org

obtained analyzing the functions representing covariant loop integrations. We show that the expansion of these functions in powers of the inverse of the nucleon mass reproduces most of the terms of the TPEP derived from the heavy baryon (HB) formalism, but such an expansion is ill defined and does not converge at large distances [13]. Our recent calculations has been devoted to perform a complete study of the phase shifts in nucleon-nucleon (NN) scattering for peripherical waves ($L \leqslant 3$) [15], which are sensitive to the tail of the potential.

2 Parametrized Potential

The $O(q^4)$ relativistic expansion of the TPEP produced in Refs. [12, 13] was based on the evaluation of three families of diagrams*. The first of them involves only pions and nucleon degrees of freedom into single loops and corresponds to the minimal realization of chiral symmetry [14]. It includes the subtraction of the iterated OPEP and yields the terms in the profile functions given below which are proportional to just g_A^4/f_π^4, g_A^2/f_π^4 or $1/f_\pi^4$. Terms proportional to $1/f_\pi^6$, on the other hand, come from two-loop processes, either in the form of t-channel contributions from the second family or s and u-channel terms embodied in the subthreshold coefficients of the third family. Finally, the third group of diagrams includes chiral corrections associated with other degrees of freedom, hidden within the LECs c_i and d_i, and gives rise to contributions which are proportional to either $(LEC)/f_\pi^4$ or $(LEC)^2/f_\pi^4$. In Ref. [13] we have expressed this last class of results in terms of the πN subthreshold coefficients. As these can be easily translated into LECs, in the present work we write the parametrized potential in terms of these constants, which appear directly into the effective Lagrangians. The following expressions correspond to the updated version as described in Ref. [15].

The configuration space potential has the isospin structure

$$V(\mathbf{r}) = V^+(\mathbf{r}) + \boldsymbol{\tau}^{(1)} \cdot \boldsymbol{\tau}^{(2)} \, V^-(\mathbf{r}), \tag{1}$$

with

$$V^\pm(\mathbf{r}) = V_C^\pm + V_{LS}^\pm \Omega_{LS} + V_T^\pm \Omega_T + V_{SS}^\pm \Omega_{SS} + V_Q^\pm \Omega_Q, \tag{2}$$

and $\Omega_{LS} = \mathbf{L} \cdot (\boldsymbol{\sigma}^{(1)} + \boldsymbol{\sigma}^{(2)})/2$, $\Omega_T = 3\,\boldsymbol{\sigma}^{(1)} \cdot \hat{\mathbf{r}}\,\boldsymbol{\sigma}^{(2)} \cdot \hat{\mathbf{r}} - \boldsymbol{\sigma}^{(1)} \cdot \boldsymbol{\sigma}^{(2)}$, $\Omega_{SS} = \boldsymbol{\sigma}^{(1)} \cdot \boldsymbol{\sigma}^{(2)}$. The form of the operator Ω_Q in configuration space is highly non-local and can be found in ref.[16].

The radial components of the potential are expressed in terms of the following dimensionless profile functions $U_I^\pm$ with $I = C, LS, T, SS$:

---

* Please see section II of Ref. [13], for a detailed discussion of the meaning and dynamical contents of these families of diagrams, which are given in its Fig.2.

$$V_C^\pm(r) = \tau^\pm \, U_C^\pm(x), \tag{3}$$

$$V_{LS}^\pm(r) = \tau^\pm \, \frac{\mu^2}{m^2} \, \frac{1}{x} \, \frac{d}{dx} \, U_{LS}^\pm(x), \tag{4}$$

$$V_T^\pm(r) = \tau^\pm \, \frac{\mu^2}{m^2} \left[\frac{d^2}{dx^2} - \frac{1}{x} \, \frac{d}{dx} \right] U_T^\pm(x), \tag{5}$$

$$V_{SS}^\pm(r) = -\tau^\pm \, \frac{\mu^2}{m^2} \left[\frac{d^2}{dx^2} + \frac{2}{x} \, \frac{d}{dx} \right] U_{SS}^\pm(x), \tag{6}$$

where $\tau^+ = 3$ and $\tau^- = 2$. Therefore, the configuration space potential is written in terms of numerical coefficients which multiply dimensionless functions arising form the Fourier transforms of Feynman loop integrals. The former are combinations of external parameters representing the pion and nucleon masses, μ and m, respectively, the pion decay constant f_π, the axial coupling constant g_A, and the LECs c_i and d_i. The latter, denoted by $U_I^\pm$, depend on just μ, m and $x \equiv \mu r$.

The final $V's$ functions were calculated and parametrized as follows: we keep the external quantities as free and parametrize the dimensionless Feynman loop integral functions and their derivatives in terms of $Z_i \equiv (G_i, H_i, I_i)$ functions, which have the following structure

$$Z_i = \frac{\mu}{(4\pi)^{5/2}} \left(\frac{\mu}{f_\pi} \right)^4 \left[\sum \gamma_i \, x^n \right] \frac{e^{-2x}}{x^2}. \tag{7}$$

which is a combination of an exponential function times a polynomial with coefficients γ_i, given in Tab. 1 to Tab. 4 at the end of this section. This parametrization is more than 1% accurate in the range $0.8 \text{ fm} \leqslant r \leqslant 10 \text{ fm}$.

Using the definition $\alpha \equiv \mu/m$, the profile functions are written as

$$V_C^+ = \frac{3g_A^4}{16} \, G_1 - \frac{3g_A^2 \alpha}{2} \left\{ 4m \, c_1 \left[2 \, I_2 - I_1 - 2\alpha(H_1 - H_2) \right] \right.$$

$$+ \frac{m \, c_2}{3} \, \alpha \left(3 \, H_1 - 2 \, H_3 \right) - 2m \, c_3 \left[I_1 - I_3 + \alpha \left(2 \, H_1 - 2 \, H_2 - H_3 \right) \right] \right\}$$

$$+ \frac{3\alpha^2}{2} \left[(4m \, c_1)^2 \, H_1 + \frac{1}{5} \, (m \, c_2)^2 \left(4 \, H_2 - H_3 \right) \right.$$

$$+ (2m \, c_3)^2 \left(H_1 - H_3 \right) - \frac{16}{3} \, m^2 c_1 \, c_2 \, H_2$$

$$\left. - 16m^2 c_1 \, c_3 \left(2 \, H_2 - H_1 \right) + \frac{4}{3} \, m^2 c_2 \, c_3 \left(2 \, H_2 - H_3 \right) \right]$$

$$+ \frac{3g_A^6 \mu^2}{16\pi \, f_\pi^2} \left(I_1 - I_3 \right) - \frac{3g_A^4 \mu^2}{256\pi^2 f_\pi^2} \left\{ \left[8 \left(I_6 - I_8 \right) - 7 \left(I_5 - I_7 \right) \right] + 4\pi \left[4 \, I_3 + 6 \, I_2 - 7 \, I_1 \right] \right\},$$

$$V_{LS}^+ = \frac{3g_A^4\,\alpha}{8}\,G_2 - 4g_A^2\,\alpha^2\,m\,c_2\,H_5\,,$$

$$V_T^+ = -\frac{g_A^4}{16}\,G_3 + \frac{g_A^2\,\alpha^2}{3}\,(m^2\,\tilde{d}_{14} - m^2\,\tilde{d}_{15})\left(H_3 - 3\,H_5\right) - \frac{g_A^6\,\mu^2}{96\pi^2 f_\pi^2}\left(H_3 - 3\,H_5\right),$$

$$V_{SS}^+ = \frac{g_A^4}{8}\,G_4 - \frac{2g_A^2\,\alpha^2}{3}\,(m^2\,\tilde{d}_{14} - m^2\,\tilde{d}_{15})\,H_3 + \frac{g_A^6\,\mu^2}{48\pi^2 f_\pi^2}\,H_3\,,$$

$$V_C^- = \frac{g_A^4}{8}\,G_5 + \frac{g_A^2}{12}\left\{\left[2\left(5\,H_2 - 3\,H_1\right) - 3\,\alpha\left(I_1 - I_3\right) - 3\,\alpha^2\left(2\,H1 - 2\,H2 - H_3\right)\right]\right.$$

$$+\,\alpha^2\left[2m\,c_4\left(5\,H_3 - 12\,H_1 + 12\,H_2\right) - 8(m^2 d_1 + m^2 d_2)\left(5\,H_3 + 2\,H_2 - 6\,H_1\right)\right.$$

$$\left.\left.-\,\frac{4m^2 d_3}{5}\left(7\,H_3 - 8\,H_2\right) + 32m^2 d_5\left(5\,H_2 - 3\,H_1\right)\right]\right\}$$

$$+\,\frac{1}{12}\left\{H_2 + \alpha^2\left[2m\,c_4\,H_3 + 8(m^2 d_1 + m^2 d_2)\left(2\,H_2 - H_3\right)\right.\right.$$

$$\left.\left.+\,\frac{12m^2 d_3}{5}\left(4\,H_2 - H_3\right) + 32m^2 d_5\,H_2\right]\right\}$$

$$-\,\frac{g_A^6\,\mu^2}{288\pi^2 f_\pi^2}\,\frac{1}{50}\,(201\,H_3 + 156\,H_2 - 300\,H_1)$$

$$-\,\frac{g_A^4\,\mu^2}{288\pi^2 f_\pi^2}\,\frac{1}{50}\left[1250\,H_8 - 1500\,H_7 + 450\,H_6 - 346\,H_3 + 1084\,H_2 - 300\,H_1\right]$$

$$-\,\frac{g_A^2\,\mu^2}{144\pi^2 f_\pi^2}\left\{5\,H_8 - 3\,H_7 - \frac{61}{20}\,H_3 + \frac{61}{5}\,H_2 - 3\,H_1\right\}$$

$$-\,\frac{\mu^2}{288\pi^2 f_\pi^2}\left[H_8 - \frac{1}{10}\left(14\,H_3 - 76\,H_2\right)\right],$$

$$V_{LS}^- = \frac{g_A^4\,\alpha}{16}\,G_6 - \frac{g_A^2\,\alpha}{12}\left[6\,I_4 + \alpha\,(3 - 40m\,c_4)\,H_5 + \alpha\,24m\,c_4\,H_4\right]$$

$$+\,\frac{\alpha^2}{24}\,(3 + 16m\,c_4)\,H_5\,,$$

$$V_T^- = \frac{g_A^4\,\alpha}{48}\,G_7 + \frac{g_A^2\,\alpha}{144}\,(1 + 4m\,c_4)\left\{6\left(I_3 - 3\,I_4\right) + \alpha\left[\left(8\,H_3 - 12\,H_1 + 12\,H_2\right)\right.\right.$$

$$\left.\left.-\,3\left(8\,H_5 - 3\,H_4\right)\right]\right\} - \frac{\alpha^2}{144}\,(1 + 4m\,c_4)^2\left(H_3 - 3\,H_5\right)$$

$$-\,\frac{g_A^6\,\mu^2}{96\pi\,f_\pi^2}\left(I_3 - 3\,I_4\right) + \frac{g_A^4\,\mu^2}{384\pi^2 f_\pi^2}\left[\left(I_8 - 3\,I_9\right) - 2\pi\left(I_3 - 3\,I_4\right)\right],$$

$$V_{SS}^- = -\frac{g_A^4\,\alpha}{24}\,G_8 - \frac{g_A^2\,\alpha}{72}\,(1+4m\,c_4)\left[6\,I_3 + \alpha\left(8\,H_3 - 12\,H_1 + 12\,H_2\right)\right]$$
$$+ \frac{\alpha^2}{72}\,(1+4m\,c_4)^2\,H_3 + \frac{g_A^6\,\mu^2}{48\pi\,f_\pi^2}\,I_3 - \frac{g_A^4\,\mu^2}{192\pi^2 f_\pi^2}\left(I_8 - 2\pi\,I_3\right).$$

The parametrized profile functions given above depend explicitly on four well known quantities, namely m, μ, g_A, f_π, and on the less known LECs c_i and d_i. Therefore, the latter may be extracted from fits to data. When doing this, however, one has to bear in mind that, as discussed in Ref. [13], the influence of the LECs over the profile functions is rather uneven. Indeed, their influence over V_C^+, V_{LS}^-, V_T^-, and V_{SS}^- is rather strong, but barely perceptible in V_C^-, V_{LS}^+, V_T^+ and V_{SS}^+.

Table 1 Values of the polynomial coefficients for the functions H_i.

γ_i^n	$-1/2$	$-3/2$	$-5/2$	$-7/2$	$-9/2$	$-11/2$
H_1	-1	-3/16	15/512	-105/8192	0.0069211	-0.002031054
H_2	—	3/2	45/32	315/1024	-0.050879	0.0105639
H_3	—	6	165/8	8715/256	27.45483	5.43256
H_4	—	2	23/8	153/256	-0.0723934	—
H_5	—	—	-3	-129/16	-3555/512	-1.33605
H_6	-3.89861	4.23305	-0.833136	—	—	—
H_7	—	5.78893	-7.63019	-2.69576	—	—
H_8	—	—	-14.3654	14.6375	39.3909	18.8729

Table 2 Values of the polynomial coefficients for the functions G_i.

γ_i^n	1	0	-1	-2	-3	-4	-5	-6
G_1	—	2.83823	-7.200711	38.9637	-55.5164	47.2443	-16.2395	—
G_2	—	—	-6.12315	-28.1422	-30.2813	0.023458	-15.8996	7.18869
G_3	—	0.5579	17.1039	16.8038	9.94755	3.40171	-2.7544	—
G_4	—	0.569624	15.9429	-4.26031	15.6445	-5.06641	—	—
G_5	-0.217221	-9.98415	-4.662	-36.9761	13.4087	-6.21047	—	—
G_6	—	—	7.90985	55.9568	86.3242	66.9540	-29.5680	11.8985
G_7	—	1.69219	25.5612	6.53589	160.459	-169.567	120.612	-36.7881
G_8	—	1.7661	21.2122	-9.87710	116.454	-144.344	103.063	-30.9265

Table 3 Values of the polynomial coefficients for the functions I_i, part A.

γ_i^n	2	1	0	-1	-2	-3	-4
I_1	0.000483761	-0.0226386	1.53346	0.0595627	-0.0913580	0.0291743	—
I_2	—	-0.000381934	0.0158372	-1.63009	-0.660019	0.0532419	—
I_3	—	—	0.242214	-8.87827	-6.47733	-30.5206	—
I_4	—	—	—	-0.00147946	2.99191	6.86185	1.82098

Table 4 Values of the polynomial coefficients for the functions I_i, part B.

γ_i^n	$3/2$	$1/2$	$-1/2$	$-3/2$	$-5/2$	$-7/2$	$-9/2$
I_5	0.168731	-6.00262	-2.18325	1.79108	—	—	—
I_6	—	-0.129344	5.46315	-2.54740	0.240122	—	—
I_7	—	-0.464737	20.2994	31.4762	-26.1396	—	—
I_8	—	—	—	-28.6452	-101.827	28.3771	99.2609
I_9	—	—	—	-0.454235	18.7535	24.4758	-31.2207

3 Conclusions

In taming the chiral TPEP over the last 15 years, one may say that considerable progress has been made, since several aspects of the problem have been clarified. The minimal realization of the symmetry is implemented by just nucleons and pions, but realistic potentials require other degrees of freedom, in the form of explicit deltas or hidden within πN subthreshold coefficients or low energy constants of effective Lagrangians (LECs). The asymptotic expressions for the potential have now the status of theorems and are written as sums of chiral layers, with little model dependence. The relation with πN scattering is understood and fixes free parameters and the various channels of the potential are clearly dominated by either nucleonic (minimal model) or non-nucleonic (LEC's) degrees of freedom, mainly because we found the role of intermediate $\pi\pi$ scattering (two-loops) to be small. Relativistic effects occurring inside loops are visible in the final form of the potential and, therefore, heavy baryon and relativistic derivations of the potential do not coincide, but differences are small in practical applications, which means that the chiral picture is well supported by empirical data.

Acknowledgements

The work of R. H. was supported by DOE Contract No.DE-AC05-84ER40150 under which SURA operates the Thomas Jefferson National Accelerator Facility, and M. R. R. and C. A. dR. were supported by FAPESP (Fundação de Amparo à Pesquisa do Estado de São Paulo).

Bibliography

[1] S. Weinberg, *Phys. Lett. B* **251**, 288 (1990); *Nucl. Phys.* **B363**, 3 (1991).

[2] C. Ordóñez and U. van Kolck, *Phys. Lett.* **B291**, 459 (1992).

[3] C. Ordóñez, L. Ray, and U. van Kolck, *Phys. Rev. Lett.* **72**, 1982 (1994); Phys.Rev. **C 53**, 2086 (1996).

[4] M. R. Robilotta, *Nucl. Phys.* **A595**, 171 (1995); *Phys. Rev. C* **63**, 044004 (2001).

[5] M. R. Robilotta and C. A. da Rocha, *Nucl. Phys. A* **615**,391 (1997).

[6] J-L. Ballot, M. R. Robilotta, and C. A. da Rocha, *Phys. Rev. C* **57**, 1574 (1998).

[7] N. Kaiser, R. Brockman and W. Weise, *Nucl. Phys.* **A625**, 758 (1997).

[8] N. Kaiser, S. Gerstendörfer, and W. Weise, *Nucl. Phys.* **A637**, 395 (1998); N. Kaiser, *Phys. Rev. C* **64**, 057001 (2001); *Phys. Rev. C* **65**, 017001 (2001).

[9] M. C. M. Rentmeester, R. G. E. Timmermans, J. L. Friar, and J. J. de Swart, *Phys. Rev. Lett* **82**, 4992 (1999); M. C. M. Rentmeester, R. G. E. Timmermans, and J. J. de Swart, *Phys. Rev. C* **67**, 044001 (2003).

[10] E. Epelbaum, W. Glöckle, and Ulf-G. Meißner, *Eur. Phys. Jour. A* **19**, 401 (2004); Nucl. Phys. **A747**, 362 (2005).

[11] D. R. Entem and R. Machleidt, *Phys. Rev. C* **66**, 014002 (2002).

[12] R. Higa and M. R. Robilotta, *Phys. Rev. C* **68**, 024004 (2003).

[13] R. Higa, M. R. Robilotta, and C. A. da Rocha, *Phys. Rev. C* **69**, 034009 (2004).

[14] C.A. da Rocha and M.R.Robilotta, *Phys. Rev.* **C49**, 1818 (1994).

[15] R. Higa, *JLAB-05-04-293*, Nov 2004. E-Print Archive: nucl-th/0411046.

[16] N. Hoshizaki and S. Machida, *Progr. Theor. Phys.* **24**, 1325 (1960).

Anais da XXVIII Reunião de Trabalho sobre Física Nuclear no Brasil
SP, Brasil, 2005
Artigo originalmente publicado no Brazilian Journal of Physics, Vol 36 – nº 4B,
Special Issue: XXVIII Workshop on Nuclear Physics in Brazil, pp.1388–1390 (2006).

Measurement of Sr/Ca Ratio in Bones as a Temperature Indicator

P. R. dos Santos*, N. Added, M. A. Rizzutto, J. H. Aburaya and M. D. L. Barbosa

Departamento de Física Nuclear, Instituto de Física,
Universidade de São Paulo, Caixa Postal 66318,
CEP 05315-970, São Paulo, SP, Brazil

Abstract. The purpose of this work is to correlate Sr/Ca ratio with internal body temperature from teeth and bones. Results obtained in exploratory measurements using human, bovine and swine teeth indicated some relation between temperature and Sr/Ca ratio, but no other parameters, as feeding habits that certainly has some influence over Sr/Ca ratio, were controlled. In this work, to eliminate feeding effects, we decided to compare Sr/Ca ratio of bones from some individual. The first bones irradiated were from a crocodile (*Caiman Yacare*), which regulates the internal body temperature by the temperature of its surroundings. The pieces irradiated were from the crocodile's tail, vertebral column and leg. To quantify Sr and Ca a 2.4 MeV proton beam was used in PIXE beam line at LAMFI - USP. Emitted X-rays were collected using a Si(Li) detector (150eV @ 6.4 KeV). First results show that the bones closer to the heart have a lower Sr/Ca ratio.

1 Introduction

Analysis of the Sr/Ca ratio in aragonite ($CaCO_3$) from sea shells and coral skeletons was used for precise measurement of sea water temperature [1, 2]. Coral skeletons grows in the same way as trees and the difference in Sr/Ca ratio along the growing direction is related to the temperature of the water surroundings at the respective time (figure 1).

In this way, Sr/Ca ratio from sea shells and coral skeletons can be used as historical data of sea water temperature. The results show great agreement if compared to satellite data, figure (2).

---

* psantos@if.usp.br

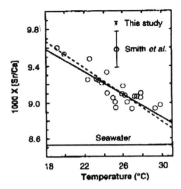

Figure 1 Sr/Ca relation of coral skeletons against sea temperature. Figure from [1].

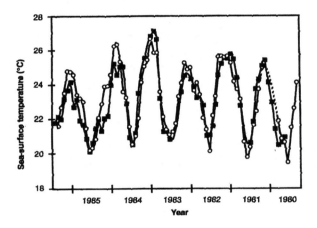

Figure 2 Sea water temperature from Sr/Ca ratio from coral sketeton results and satellite data along the years. Figure from [2].

So, the initial idea of our studies was to relate Sr/Ca ratio of human, bovine and swine teeth with body respective temperature. Teeth, as bones, are mainly composed by another calcium crystal called hydroxiapatite ($Ca_{10}(PO_4)_6(OH)_2$). Each species has a different body temperature. For humans, its average is 36,9(6) °C [3], for bovine, temperature average is 39,0(5) °C and for swine, its average is 39,0(10) °C [4]. The average values for Sr/Ca resulted of irradiation of 30 samples (enamel and dentine) can be seen in figures 3 and 4.

It can be observed that the angular coefficient from data prescuted in Sr/Ca × T plots from figure 1, 3 and 4 have opposite signs. These results show that probably strontium quantity in hydroxyapatite is dependent of some other parameters that can hide temperature effects.

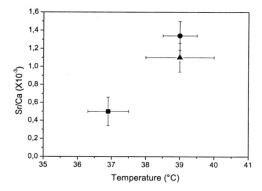

Figure 3 Average of Sr/Ca ($\times 10^{-3}$) ratio for dentine irradiation against body temperature. Human data is represented by a circle, bovine data by a square and swine data by a triangle.

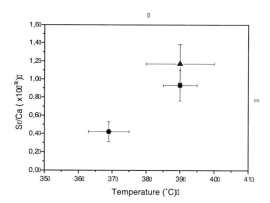

Figure 4 Average of Sr/Ca ($\times 10^{-3}$) ratio for enamel irradiation against body temperature. Human data is represented by a circle, bovine data by a square and swine data by a triangle.

One of these parameters is feeding habits [5, 6], that wasn't controlled in our samples. To eliminate feeding effects it was decided to compare Sr/Ca of bones from the same individual.

For every vertebrate, independent of its thermoregulation pattern (endothermic or ectothermic), there is a variation of temperature along the body. Usually the warmest parts are the parts closer to the heart and to the brain, and the coldest are the extremities, as feet [7].

It was chosen, for first irradiation, a crocodile (*Caiman Yacare*), which specie regulates its body temperature uptaken heat from the environment (ectothermy) and probably has a bigger range between the warmest and the coldest parts of the body than an endothermic animal.

2 Experimental Facility

Regular PIXE analysis were performed at the Laboratory for Material Analysis by Ion Beams at University of São Paulo (LAMFI-USP), using a 2.4 MeV proton beam. The pieces irradiated were a bone from one leg, three from the different points of vertebral column (see figure 5) and two from the tail.

Figure 5 One of the bones irradiated from the spinal column.

The samples were fixed in carbon targets and mounted in a multiposition target holder at the center of the PIXE vacuum chamber as it can be seen in figures 6 and 7.

Figure 6 Bones ready for irradiation in the multiposition target holder.

Figure 7 Multiposition target holder in the center of the PIXE chamber.

Figure 8 External view of the experimental setup showing Si(Li) detectors and vacuum chamber.

Two Si(Li) detectors with resolutions around 150 eV for 6.4 KeV were used to collect the X-rays emitted by the samples. An external view of the experimental setup is shown in figure 8.

3 Preliminary Results

All energy spectra were analyzed using ROOT code [8]. In figure 9 a typical X-ray energy spectrum can be seen.

The quantities of strontium and calcium were obtained integrating the respective peak areas and normalizing them by the number of incident particles (charge), thin film PIXE yields [9] and correction factors for thick target [10]. Uncertainties for Sr and Ca were evaluated using statistical considerations.

The results of preliminary irradiations can be seen in figure 10. It's not difficult to see that there is a tendency of a lower Sr/Ca ratio closer to the heart region and a higher Sr/Ca ratio in the extremities, as we expected.

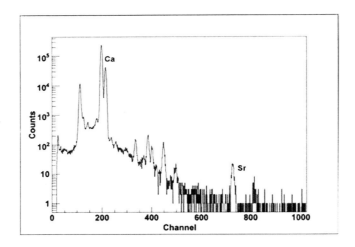

Figure 9 Energy espectrum from one of the irradiations using ROOT code.

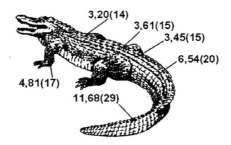

Figure 10 Schematic draw of Sr/Ca quantities in crocodile bones.

Acknowledgements

We would like to thank Dr. Manfredo H. Tabaknics for ideas and discussions. We also would like to thank FAPESP and CNPq for financial support.

Bibliography

[1] R. W. Buddemeier, R. C. Redale, J. E. Houce, Science **204**, 404 (1979);

[2] J. Warren Beck et al., Science **257**, 644 (1992);

[3] "Human Temperature" http://www.icb.ufmg.br/~neurofib/Engenharia/Temperatura/processos%20biologicos.htm (2005);

[4] "Animal Temperature"
http://www.saudeanimal.com.br/artigo97_print.htm (2005);

[5] M. Sponheimer et al., Journal of Human Evolutions **48**, 147 (2005);

[6] J. H. Burton and L. E.Wright, American Journal of Physical Anthropology **96**, 273 (1995);

[7] Reese E. Barrick and William J. Showers, Science **265**, 222 (1994);

[8] Root code source: http://root.cern.ch/ (2005);

[9] Tabacniks, M.H. CLCPIXE.xls (2005). Private communication (2005).

[10] Aburaya, J.H., Padronização de análises PIXE de amostras sólidas em alvos espessos., MS Dissertation, Instituto de Física, Universidade de São Paulo, Brasil (2005).

Anais da XXVIII Reunião de Trabalho sobre Física Nuclear no Brasil
SP, Brasil, 2005

The Relativistic Quasi-Particle Random Phase Approximation and Beta Decay

C. De Conti[1], B. V. Carlson[2,] and T. Frederico[2,*]*

[1] Unidade Diferenciada de Itapeva, UNESP - São Paulo State University,
18409-010, Itapeva, São Paulo, Brazil

[2] Departamento de Física, Instituto Tecnológico de Aeronáutica,
Centro Técnico Aeroespacial,
12228-900 São José dos Campos, SP, Brazil

Abstract. We have derived the relativistic quasi-particle random phase approximation (QRPA) equations in the particle-particle, particle-hole and hole-hole subspaces. We use the motion equation techniques to develop the QRPA, based on the Dirac-Hartree-Fock-Bogoliubov (DHFB) approximation to the single-particle motion. We reduce these equations by coupling them to well defined angular momentum. Our aim in this work is to obtain the relativistic QRPA equations appropriate to study the beta decay in medium to heavy nuclei.

1 Introduction

The random-phase approximation (RPA) is the simplest theory of excited states of a nucleus which admits the possibility that the ground state is not of a purely independent-particle character but may contain correlations. The nonrelativistic RPA furnishes a microscopic treatment of nuclear excitation in terms of the particle-hole modes of the system. It successfully describes many of the low-energy collective excitations of the nucleus.

When it is based on the Hartree-Fock-Bogoliubov approximation (HFB) to the single-particle motion, rather than the Hartree-Fock one, the RPA becomes the QRPA. It then also includes particle-particle and hole-hole contributions to the excited states and ground state correlations. The particle-particle ground state correlations, in particular, are important for the correct description of β^+ decay and double β decay in medium to heavy nuclei.[1]

---

* Partial support from FAPESP and CNPQ

In the second section of this work we present the Dirac HFB approximation. We then derive the Hamiltonian density that we will use to obtain the relativisitc QRPA equations. Next, we obtain the DHFB approximation in second quantization. In the fifth section, we derive with the equation of motion method the relativistic QRPA equations and couple them to well defined angular momentum. Finally in the last section we summarize and conclude this poster.

2 The DHFB Approximation

Starting from the following version of the QHD Lagrangian density

$$
\begin{aligned}
\mathcal{L} = {} & \bar{\psi}(i\gamma_\mu \partial^\mu - M)\psi \\
& + \frac{1}{2}\partial_\mu \sigma \partial^\mu \sigma - \frac{1}{2}m_\sigma^2 \sigma^2 - \frac{1}{3}g_2 \sigma^3 - \frac{1}{4}g_3 \sigma^4 - g_\sigma \bar{\psi}\psi\sigma \\
& - \frac{1}{4}\Omega_{\mu\nu}\Omega^{\mu\nu} + \frac{1}{2}m_\omega^2 \omega_\mu \omega^\mu - g_\omega \bar{\psi}\gamma_\mu \psi \omega^\mu \\
& + \frac{1}{2}\partial_\mu \vec{\pi} \cdot \partial^\mu \vec{\pi} - \frac{1}{2}m_\pi^2 \vec{\pi} \cdot \vec{\pi} - \frac{f_\pi}{m_\pi}\bar{\psi}\gamma_5 \gamma_\mu \vec{\tau}\psi \cdot \partial^\mu \vec{\pi} \\
& - \frac{1}{4}\vec{R}_{\mu\nu} \cdot \vec{R}^{\mu\nu} + \frac{1}{2}m_\rho^2 \vec{\rho}_\mu \cdot \vec{\rho}^{\,\mu} - g_\rho \bar{\psi}\gamma_\mu \vec{\tau}\psi \cdot \vec{\rho}^{\,\mu} \\
& \qquad\qquad\qquad\qquad - \frac{1}{4}F_{\mu\nu}F^{\mu\nu} - e\bar{\psi}\gamma_\mu \frac{1+\tau_3}{2}\psi A^\mu, \quad (1)
\end{aligned}
$$

and following Gorkov [2], we can obtain the self-consistency equation of Dirac-Hartree-Fock-Bogoliubov (DHFB) approximation. An alternative, more detailed account can be found in [3].

In the static limit we obtain the Dirac-Gorkov equation

$$
\int d^3 y M(\vec{x}, \vec{y}) \begin{pmatrix} U(\vec{y}) \\ \gamma_0 V(\vec{y}) \end{pmatrix} = 0 \tag{2}
$$

where

$$
M(\vec{x}, \vec{y}) = \begin{pmatrix} (\varepsilon + \mu)\delta(\vec{x} - \vec{y}) - h(\vec{x}, \vec{y}) & \bar{\Delta}^\dagger(\vec{x}, \vec{y}) \\ \bar{\Delta}(\vec{x}, \vec{y}) & (\varepsilon - \mu)\delta(\vec{x} - \vec{y}) + h_T(\vec{x}, \vec{y}) \end{pmatrix}
$$

The components U and V are Dirac spinors correspond to the normal and time-reversed components, of the positive-frequency and negative-frequency solution to the Dirac-Gorkov equation, and $h(\vec{x}, \vec{y}) = (-i\vec{\alpha} \cdot \vec{\nabla} + \beta M)\delta(\vec{x} - \vec{y}) + \beta\Sigma(\vec{x}, \vec{y})$. The self-energy $\Sigma(\vec{x}, \vec{y})$ and pairing fields $\Delta(\vec{x}, \vec{y})$ are

$$\Sigma(\vec{x}, \vec{y}) = \delta(\vec{x} - \vec{y}) \sum_{j} \Gamma_{j\alpha}(\vec{x}) \int d^3z D_j^{\alpha\beta}(\vec{x} - \vec{z}) \sum_{\varepsilon_\gamma < 0} \bar{U}_\gamma(\vec{z}) \Gamma_{j\beta}(\vec{z}) U_\gamma(\vec{z})$$
$$- \sum_{j} \Gamma_{j\alpha}(\vec{x}) D_j^{\alpha\beta}(\vec{x} - \vec{y}) \sum_{\varepsilon_\gamma < 0} U_\gamma(\vec{x}) \bar{U}_\gamma(\vec{y}) \Gamma_{j\beta}(\vec{y}) \quad (3)$$

and

$$\Delta(\vec{x}, \vec{y}) = - \sum_{j} \Gamma_{j\alpha}(\vec{x}) D_j^{\alpha\beta}(\vec{x} - \vec{y}) \sum_{\varepsilon_\gamma < 0} U_\gamma(\vec{x}) \bar{V}_\gamma(\vec{y}) A \Gamma_{j\beta}^{\mathsf{T}}(\vec{y}) A^\dagger. \quad (4)$$

The summation over negative frequency solutions takes into account the occupation of all states, both those in the Fermi sea (the surface of which is now diffuse, due to the pairing) and those in the Dirac sea. The index j indicates the meson of the model with the meson-nucleon coupling $\Gamma_{j\alpha}$.

3 Relativistic Hamiltonian

Our start point to derive the relativisitc QRPA equations is obtain the hamiltonian density from the Lagrangian density in (1). We obtain it with the Legendre transformation and introducing the instantaneous approximation to find

$$H = \int d^3x \psi^\dagger(\vec{x})(-i\vec{\alpha} \cdot \vec{\nabla} + \beta M)\psi(\vec{x})$$
$$+ \frac{1}{2} \int d^3x d^3x' \psi^\dagger(\vec{x})\psi^\dagger(\vec{x'})V(\vec{x} - \vec{x'})\psi(\vec{x'})\psi(\vec{x}) \quad (5)$$

where

$$V(\vec{x} - \vec{x'}) = -\frac{g_\sigma}{4\pi}Y(\vec{x} - \vec{x'}, m_\sigma)\gamma_0\gamma_0' + \frac{g_\omega}{4\pi}Y(\vec{x} - \vec{x'}, m_\omega)\gamma_0\gamma_0'\gamma^\mu\gamma_\mu'$$
$$+ \frac{g_\rho}{4\pi}Y(\vec{x} - \vec{x'}, m_\rho)\gamma_0\gamma_0'\gamma^\mu\gamma_\mu'\vec{\tau} \cdot \vec{\tau'}$$
$$- \frac{1}{4\pi}\left(\frac{g_\pi}{2M}\right)^2 \partial_i\partial_j'Y(\vec{x} - \vec{x'}, m_\pi)\gamma_0\gamma_0'\gamma_5\gamma_5'\gamma^i\gamma'^j\vec{\tau} \cdot \vec{\tau'}$$
$$+ \frac{1}{4\pi}\left(\frac{e}{2}\right)^2 Y(\vec{x} - \vec{x'}, m = 0)\gamma_0\gamma_0'\gamma^\mu\gamma_\mu'(1 + \tau_3)(1 + \tau_3')$$

and $Y(\vec{x} - \vec{x'}, m)$ is the Yukawa function. "Here, for simplicity we not have included the non-linear σ-meson terms." modificar para; "Here, for simplicity we not have included the non-linear σ-meson terms". For this purpose, we may formally include them in the "free" scalar meson propagator, and changing the form of the Yukawa function for this meson.

4 The DHFB Approximation in Second Quantization

We expand the wave function operator in an arbitrary basis as

$$\psi(\vec{x}) = \sum_a u_a(\vec{x})c_a, \tag{6}$$

where c_a is a particle annihilation operator and $u_a(\vec{x})$ the wave function in this basis. To simplify the notation, we do not consider the antiparticles distinctly. For the adjoint and time-reversed wave operators, we then have $\bar{\psi}(\vec{x}) = \sum_a \bar{u}_a(\vec{x})c_a^\dagger$ and $\psi_T(\vec{x}) = \gamma_0 B \sum_a u_a^*(\vec{x})c_a^\dagger$. The effective Lagrangian is [3]

$$L = \int d^3x d^3y \left(\ \bar{\psi}(\vec{x}), \ \ \bar{\psi}_T(\vec{x}) \ \right) \gamma_0 M(\vec{x},\vec{y}) \gamma_0 \left(\begin{array}{c} \psi(\vec{y}) \\ \psi_T(\vec{y}) \end{array} \right) \tag{7}$$

where $h_T(\vec{x},\vec{y}) = Bh^T(\vec{x},\vec{y})B^\dagger$ and we assume that $h(\vec{x},\vec{y})$ is hermitian. The nontrivial integrals in this expression yield

$$\int d^3x d^3y \bar{\psi}(\vec{x})\gamma_0 h(\vec{x},\vec{y})\psi(\vec{y}) = \sum_{a,b} h_{ab}c_a^\dagger c_b,$$

$$\int d^3x d^3y \bar{\psi}(\vec{x})\gamma_0 \bar{\Delta}^\dagger(\vec{x},\vec{y})\gamma_0 \psi_T(\vec{y}) = \sum_{a,b} \bar{\Delta}_{ab}^\dagger c_a^\dagger c_b^\dagger, \tag{8}$$

with

$$h_{ab} = \int d^3x d^3y u_a^\dagger(\vec{x})h(\vec{x},\vec{y})u_b(\vec{y})$$

$$\bar{\Delta}_{ab} = \int d^3x d^3y u_a^\dagger(\vec{x})B^\dagger \bar{\Delta}(\vec{x},\vec{y})u_b^*(\vec{y}).$$

The additional condition on the pairing field, $\bar{\Delta}(\vec{x},\vec{y}) = -B\bar{\Delta}^T(\vec{x},\vec{y})B^\dagger$, implies that the matrix elements are antisymmetric, $\bar{\Delta}_{ab} = -\bar{\Delta}_{ba}$, so that $\bar{\Delta}_{ab}^\dagger = -\bar{\Delta}_{ab}^*$.

We can now write the effective Langrangian as

$$L = \sum_{ab} \left(\ c_a^\dagger, \ \ c_a \ \right) \left(\begin{array}{cc} (\omega+\mu)\delta_{ab} - h_{ab} & -\bar{\Delta}_{ab}^* \\ \bar{\Delta}_{ab} & (\omega-\mu)\delta_{ab} + h_{ab}^* \end{array} \right) \left(\begin{array}{c} c_b \\ c_b^\dagger \end{array} \right). \tag{9}$$

To determine the quasi-particle operators, we look for eigenvectors of the matrix. That is, we now assume that we can find solutions of the form

$$\left(\begin{array}{cc} h-\mu 1 & \bar{\Delta}^* \\ -\bar{\Delta} & -h^* + \mu 1 \end{array} \right) \left(\begin{array}{c} U_\alpha \\ V_\alpha \end{array} \right) = \varepsilon_\alpha \left(\begin{array}{c} U_\alpha \\ V_\alpha \end{array} \right). \tag{10}$$

We observe that these matrix equations are exactly the same as that in (2) in the previous sections. We construct a matrix in which each column is one of these solutions as

$$W = \begin{pmatrix} U & V^* \\ V & U^* \end{pmatrix}. \tag{11}$$

This matrix diagonalizes the effective lagrangian-hamiltonian. We assume the matrix W to be unitary $W^\dagger W = WW^\dagger = 1$. It will be in any finite basis, since the diagonalized matrix is hermitian.

We now return to the effective lagrangian in the particle basis and use the unitary of the matrix W to rewrite it in the quasi-particle basis. We thus can associate the quasi-particle and particle operators as

$$\beta_\alpha = \sum_b (U^*_{b\alpha} c_b + V^*_{b\alpha} c_b^\dagger),$$

$$\beta_\alpha^\dagger = \sum_b (U_{b\alpha} c_b^\dagger + V_{b\alpha} c_b). \tag{12}$$

5 The Relativistic QRPA Equations

Using the expansion (6) in the Hamiltonian density (5) we have

$$H = \sum_{aa'} \int d^3x u_a^\dagger(\vec{x})(-i\vec{\alpha}\cdot\vec{\nabla} + \beta M)u_{a'}(\vec{x})c_a^\dagger c_{a'}$$

$$+ \frac{1}{2}\sum_{aa'bb'}\int d^3x d^3x' u_a^\dagger(\vec{x})u_b^\dagger(\vec{x'})V(\vec{x}-\vec{x'})u_{a'}(\vec{x'})u_{b'}(\vec{x})c_a^\dagger c_b^\dagger c_{a'}c_{b'}$$

After we replace the transformations (12) and we expand it in normal-ordered form with respect to the new quasi-particle vacuum and so

$$H = \sum_\alpha \tilde{N}(\varepsilon_\alpha \beta_\alpha^\dagger \beta_\alpha) + \frac{1}{2}\sum_{aa'bb'}\langle ab|V|b'a'\rangle \tilde{N}(c_a^\dagger c_b^\dagger c_{a'}c_{b'}) + U \tag{13}$$

where $\tilde{N}$ stands for the normal ordering with respect to the new operators and U is a unimportant constant.

The QRPA equations are obtained from equation of motion method. The phonon creation and destruction operators $Q_\lambda^\dagger$, Q_λ, of the Hamiltonian (13) can be obtained by solution of following equations of motion

$$[H, Q_\lambda^\dagger] = E_\lambda Q_\lambda^\dagger. \tag{14}$$

Since it is not possible find $Q_\lambda^\dagger$ exactly, we expand it in the form

$$Q_\lambda^\dagger = \sum_{\alpha\alpha'}\left[X_{\alpha\alpha'}^{(\lambda)}\beta_\alpha^\dagger\beta_{\alpha'}^\dagger - Y_{\alpha\alpha'}^{(\lambda)}\beta_\alpha\beta_{\alpha'}\right] \tag{15}$$

Linearizing the commutation relation $[H, Q_\lambda^\dagger]$ in $\beta_\alpha^\dagger \beta_{\alpha'}^\dagger$ and $\beta_\alpha \beta_{\alpha'}$, we obtain

$$\sum_{\alpha\alpha'} \left(A_{\alpha_1\alpha_2,\alpha\alpha'} X_{\alpha\alpha'}^{(\lambda)} + B_{\alpha_2\alpha_1,\alpha\alpha'} Y_{\alpha\alpha'}^{(\lambda)} \right) = E_\lambda X_{\alpha_1\alpha_2}^{(\lambda)}$$

$$\sum_{\alpha\alpha'} \left(B_{\alpha_1\alpha_2,\alpha\alpha'}^* X_{\alpha\alpha'}^{(\lambda)} + A_{\alpha_2\alpha_1,\alpha\alpha'}^* Y_{\alpha\alpha'}^{(\lambda)} \right) = -E_\lambda Y_{\alpha_1\alpha_2}^{(\lambda)} \qquad (16)$$

This is the usual form of the QRPA equations, the matrices $A_{\alpha_1\alpha_2,\alpha\alpha'}$ and $B_{\alpha_1\alpha_2,\alpha\alpha'}$ have the same form find in the literature, see [4, eq.(14.45)].

In the angular momentum coupled representation, $Q_\lambda^\dagger(JM)$ is

$$Q_\lambda^\dagger(JM) = \sum_{\alpha\alpha'} \left[X_{\alpha\alpha'}^{(\lambda J)} (1 + \delta_{\alpha\alpha'})^{-1/2} \langle j_\alpha m_\alpha j_{\alpha'} m_{\alpha'} | JM \rangle \beta_\alpha^\dagger \beta_{\alpha'}^\dagger \right.$$
$$\left. - Y_{\alpha\alpha'}^{(\lambda J)} (1 + \delta_{\alpha\alpha'})^{-1/2} \langle j_\alpha m_\alpha j_{\alpha'} m_{\alpha'} | J - M \rangle (-)^{J+M} \beta_\alpha \beta_{\alpha'} \right]$$

Comparing $Q_\lambda^\dagger(JM)$ above with (15), we observe that its possible to obtain uncoupled and coupled representation. Finally, the relativistic QRPA equations, for the angular momentum coupled representation can be written as

$$\sum_{d_\alpha d_{\alpha'}} \left(A_{d_{\alpha_1} d_{\alpha_2}, d_\alpha d_{\alpha'}}^{(J)} X_{d_\alpha d_{\alpha'}}^{(\lambda J)} + B_{d_{\alpha_2} d_{\alpha_1}, d_{\alpha'} d_\alpha}^{(J)} Y_{d_\alpha d_{\alpha'}}^{(\lambda J)} \right) = E_\lambda^{(J)} X_{d_{\alpha_1} d_{\alpha_2}}^{(\lambda J)}$$

$$\sum_{d_\alpha d_{\alpha'}} \left(B_{d_{\alpha_1} d_{\alpha_2}, d_\alpha d_{\alpha'}}^{(J)*} X_{d_\alpha d_{\alpha'}}^{(\lambda J)} + A_{d_{\alpha_2} d_{\alpha_1}, d_{\alpha'} d_\alpha}^{(J)*} Y_{d_\alpha d_{\alpha'}}^{(\lambda J)} \right) = -E_\lambda^{(J)} Y_{d_{\alpha_1} d_{\alpha_2}}^{(\lambda J)}$$

where

$$E_\lambda^{(J)} = \sum_{m_\alpha m_{\alpha'}} \epsilon_\lambda (1 + \delta_{\alpha\alpha'})^{-1/2} \langle j_\alpha m_\alpha j_{\alpha'} m_{\alpha'} | JM \rangle \langle j_\alpha m_\alpha j_{\alpha'} m_{\alpha'} | JM \rangle$$

and

$$A_{d_{\alpha_2} d_{\alpha_1}, d_{\alpha'} d_\alpha}^{(J)} = \sum_{m_{\alpha_1} m_{\alpha_2} m_\alpha m_{\alpha'}} (1 + \delta_{\alpha\alpha'})^{-1/2} \langle j_{\alpha_1} m_{\alpha_1} j_{\alpha_2} m_{\alpha_2} | JM \rangle$$
$$\times \langle j_\alpha m_\alpha j_{\alpha'} m_{\alpha'} | JM \rangle A_{\alpha_1\alpha_2,\alpha\alpha'}$$

$$B_{d_{\alpha_2} d_{\alpha_1}, d_{\alpha'} d_\alpha}^{(J)} = \sum_{m_{\alpha_1} m_{\alpha_2} m_\alpha m_{\alpha'}} (1 + \delta_{\alpha\alpha'})^{-1/2} \langle j_{\alpha_1} m_{\alpha_1} j_{\alpha_2} m_{\alpha_2} | JM \rangle$$
$$\times \langle j_\alpha m_\alpha j_{\alpha'} m_{\alpha'} | J - M \rangle (-)^{J+M} B_{\alpha_1\alpha_2,\alpha\alpha'}$$

6 Conclusions

We have derived the equations of the relativistic QRPA appropriate to describe β^+ and double β decay. We use the equation of motion techniques to develop the QRPA, based on the Dirac-Hartree-Bogoliubov (DHB) approximation to the single-particle motion. We show also the procedure utilized to obtain the coupled representation to well defined angular momentum for these equations. In this form the equations are reduced and are in a form appropriate for numerical calculations.

We have obtained the QRPA in matrix form and extended the particle-hole response, including particle-particle and hole-hole excitations. We intend to continue this work in order to solve the QRPA equations numerically to apply to the β decay of exotic nuclei.

Bibliography

[1] D. Cha, Phys. Rev. C **27** (1983) 2269; J. Engel, P. Vogel and M.R. Zirnbauer, Phys. Rev. C **37** (1988) 731.

[2] L. P. Gorkov, Sov. Phys. (JETP) **34(7)** (1958) 505.

[3] B. V. Carlson and D. Hirata, Phys. Rev. C**62** (2000) 054310.

[4] D. J. Rowe, *Nuclear Collective Motion* (Methuen, London, 1970).

Anais da XXVIII Reunião de Trabalho sobre Física Nuclear no Brasil
SP, Brasil, 2005

Caracterização de Microporos produzidos por Rastos Iônicos em Polímeros

A. O. Delgado*, M. A. Rizzutto, A. R. Lima, A. A. R. Da Silva,
M. A. P. Carmignotto, M. H. Tabacniks, N. Added

Instituto de Física, Universidade de São Paulo - IFUSP

Resumo. Um dos métodos de confecção de nanoporos em isolantes é a irradiação do isolante com feixe iônico produzido em um acelerador e sua posterior corrosão química. No presente trabalho apresenta-se a caracterização dos poros produzidos em amostras de CR-39 LANTRACK, irradiadas com feixe de ^{16}O de diferentes energias: 24, 32, 41, 49 MeV, e posteriormente corroídas em solução de NaOH 7,25 M / 70°C. Com a utilização de um microscópio óptico, foi possível verificar a dependência linear do diâmetro dos poros em função do tempo de corrosão e o aumento do comprimento dos poros até a obtenção de um patamar de saturação. Dos ajustes de dados obteve-se a velocidade de corrosão volumétrica V_B do CR-39: 1,620(12), 1876(6), 1,646(9), 1,412(9) μ m/h, para íons das respectivas energias de 24, 32, 41 e 49 MeV. A comparação dos resultados obtidos com valores simulados no programa TRIM 98, mostrou que os resultados experimentais para alcance do íon foram inferiores ao esperado.

---

1 Introdução

Partículas energeticamente carregadas, ao atravessar um meio material, interagem com os constituintes atômicos do meio, transferindo-lhes energia e resultando em um rasto com moléculas danificadas. Esses rastos criados em membranas podem ser revelados para serem visualizados com o auxílio de microscópios ópticos. O processo de revelação consiste no uso de soluções químicas ácidas ou básicas que aumentam e tornam visíveis os danos gerados, técnica utilizada em longa data como forma de detecção de partículas [Knoll, 1989]. Além de tornar os danos visíveis é possível corroer a membrana, atingida pelas partículas, ao longo do rasto criado

---

* adelgado@dfn.if.usp.br

pelas mesmas, gerando poros de diferentes formas e tamanhos e visando diversos usos e aplicações.

Em especial, uma das áreas que se desenvolveu nos últimos anos é a produção de poros em polímeros, devido ao grande potencial econômico. A aplicação desses poros na fabricação de microfiltros já foi bastante desenvolvida [Yamasaki e Geraldo, 1990; Yamasaki e Outi, 1994]. Porém, novas idéias de aplicações em diferentes áreas como eletrônica, óptica e medicina, entre outras, têm surgido nos últimos anos [Mitrofanov et al, 1995; Fink et al., 2003] e motivado o desenvolvimento avançado da técnica de produção de poros por feixes iônicos.

As diferentes aplicações dependem da fabricação de membranas porosas, nas quais o tamanho e a densidade dos poros sejam características muito bem controladas. Essas características, por sua vez, dependem de parâmetros como os íons e polímeros utilizados, bem como condições de irradiação e corrosão [Tripathy et al., 2003].

O processo de abertura dos poros é descrito por dois parâmetros característicos: velocidade de corrosão ao longo do rasto V_T e velocidade de corrosão volumétrica V_B ou V_G. [Apel, 2001]. A figura 1 mostra de forma simples como se comportam as velocidades de corrosão para formação do poro no local onde se encontra o traço latente (rasto iônico).

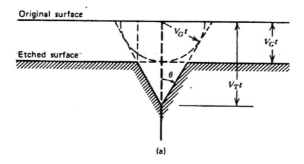

Figura 1 Formação dos poros a partir das velocidades de corrosão V_G e V_T para incidência normal do feixe (Knoll, 1989)

Como se observa, V_B atua igualmente em todas as direções e em toda superfície do plástico. Essa atuação leva a diminuição da espessura da amostra irradiada. V_T, por sua vez atua apenas na direção do rasto formado na irradiação.

2 Irradiação das Amostras

As amostras de CR-39 do tipo LANTRACK, de 0,9 mm de espessura, foram irradiadas no acelerador Pelletron do IFUSP com feixes de ^{16}O, com diversas energias:

24, 32, 41, 49 MeV. Antes de atingirem as amostras, os feixes foram espalhados em uma folha de ouro de 70(1) µg/cm², visando a redução da intensidade dos íons. Em duas irradiações, a energia do feixe espalhado foi, ainda, atenuada em uma folha de Al de 2,88(1) mg/cm². O tempo de irradiação foi definido para a obtenção de uma densidade de cerca de 10^6 íons/cm².

Utilizando um detector de barreira de superfície (Si), foi possível determinar o valor da energia após atenuação no Al. Para tanto, foi necessário, primeiramente, calibrar o ganho do detector a partir das irradiações sem atenuação. Nessa calibração foi considerada desprezível a perda de energia na folha de Au.

Os valores de energia, bem como o tempo de irradiação são dados na Tabela 1.

Tabela 1 Energia de incidência e tempo de irradiação do CR-39

E_{in} (MeV)	Atenuação (Al) (MeV)	t_{ir} (min)
41,0		12
49,0		19
41,0	23,8(3)	10
49,0	32,4(2)	26

As amostras de CR-39 foram dispostas em um suporte montado sobre um braço giratório, composto por 9 posições, com separação angular de 5° entre cada posição, conforme ilustrado na figura 2. A cada irradiação, duas dessas posições eram expostas ao feixe, posicionadas nos ângulos de espalhamento $\theta_1 = 42,5°$ e $\theta_2 = 47,5°$. Para vedar o espalhamento nas demais direções foi utilizado um colimador composto por um *Place de Cobre* montado sobre um caneco de aço inox, disposto no centro da câmara de irradiação.

Em cada posição do braço, foram dispostas de 2 a 3 amostras de CR-39, sendo que para cada valor de energia foram irradiadas 5 amostras. A posição das amostras variou de 1 a 3 no eixo z, sendo 3 a posição mais elevada. Além disso, foram colocados detectores de controle nas posições 1 e 5 do braço, conforme visualizados na figura 3.

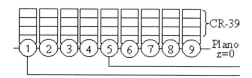

Figura 2 Vista do braço giratório na câmara de irradiação

Figura 3 Disposição das amostras de CR-39 e dos detectores de controle

Após a irradiação as amostras foram armazenadas em ambiente seco, para preservação dos rastos latentes.

3 Corrosão das Amostras

As amostras foram submetidas à solução de corrosão NaOH 7,25 M a temperatura de 70°C. O aquecimento foi efetuado em banho térmico de glicerina. Todo o sistema estava sobre uma placa de agitação magnética, para garantir a uniformidade da temperatura e da concentração da solução corrosiva, durante a reação. No aquecimento do banho utilizou-se um aquecedor, conectado a um termostato de precisão 1°C.

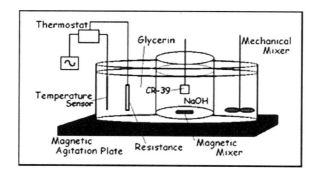

Figura 4 Arranjo Experimental

Para acelerar o equilíbrio térmico da glicerina, utilizou-se um agitador mecânico, conforme se observa na figura 4, que apresenta o arranjo experimental completo.

O tempo de imersão de cada amostra variou entre 0,3 e 2,6 horas. As amostras retiradas da solução eram lavadas imediatamente com água destilada para interromper o processo químico.

4 Análise das Amostras

O sistema de análise era composto pelo microscópio óptico Leitz Diaplan. Para as análises efetuadas nesse trabalho, utilizou-se uma ampliação de 63x da lente objetiva.

As figuras 5 e 6 mostram a vista superficial de algumas amostras de CR-39, onde é possível perceber a distribuição e a área dos poros produzidos. Nessas figuras observa-se que as amostras apresentam diferentes densidades.

A figura 7, por sua vez, mostra a vista lateral de um detector de CR-39 após a corrosão. Na figura é possível observar a profundidade e as bordas laterais dos poros, que não se apresentam completamente nítidas, dificultando a definição do tamanho do poro. Essa dificuldade acabou, inclusive, prejudicando a boa determinação dos valores de profundidade.

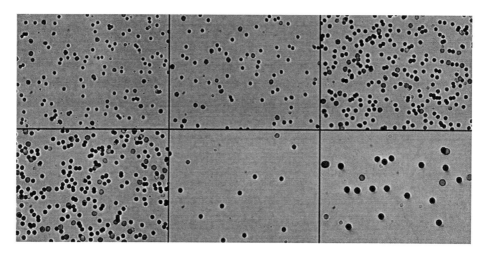

Figura 5 Vista superficial dos poros produzido em CR-39 com ^{16}O de 49 MeV, para diferentes tempos de corrosão em horas: a) 0,8; b)0,9; c)1,0; d)1,1; e)1,2; f) 1,3.

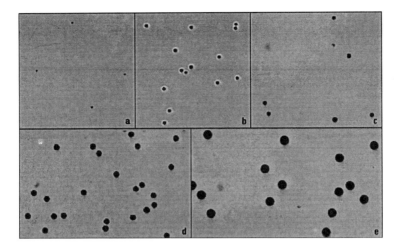

Figura 6 Vista superficial dos poros produzido em CR-39 com ^{16}O de 24 MeV, para diferentes tempos de corrosão em horas: a) 0,3; b)0,4; c)0,5; d)0,6; e)1,0

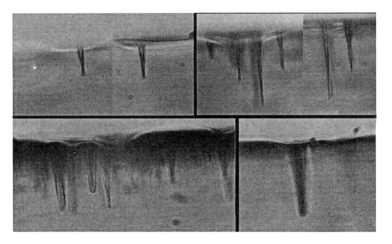

Figura 7 Vista lateral dos poros produzido em CR-39 com ^{16}O de 49 MeV para diferentes tempos de corrosão em horas: a) 0,7; b)0,9; c)1,1; d)1,3.

O programa de análise determina, automaticamente, várias características dos poros, tais como: densidade, diâmetro, área, excentricidade, perímetro*. Além disso, ele também calcula, estatisticamente, os valores médios dessas grandezas nas áreas da amostra definidas pelo usuário.

---

* Essa determinação é baseada na calibração do sistema de pixels para μm.

A figura 8 apresenta um gráfico típico dos valores médios de diâmetro ϕ dos poros em função do tempo t de corrosão, para irradiações em CR39. As incertezas correspondem ao desvio padrão da média, para todos os poros observados e analisados em cada amostra. O número total de poros analisados variou entre 200 e 500, conforme a densidade da amostra.

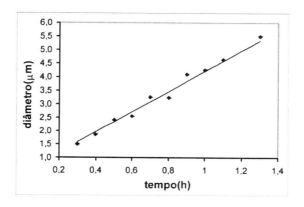

Figura 8 Diâmetro médio dos poros em função do tempo de corrosão para ^{16}O de 32,4 MeV.

O aumento do diâmetro do poro ocorre à taxa de variação $2V_B$, que corresponde ao dobro da velocidade de corrosão volumétrica do material. Do ajuste dos dados para cada valor de energia dos íons, obteve-se os valores dos coeficientes lineares e angulares, sendo possível calcular as velocidades de corrosão volumétrica V_B do CR-39. Os valores obtidos são apresentados na tabela 2.

A figura 9 apresenta o gráfico típico da profundidade L dos poros em função do tempo de corrosão. As incertezas correspondem ao desvio padrão da média, para um número de cerca de 30 poros para cada amostra. É possível observar que a profundidade dos poros aumenta até atingir um patamar de saturação.

Tabela 2 Valores de V_B e coeficiente linear b, obtidos dos ajustes de diâmetro do poro em função do tempo de corrosão.

$E(MeV)$	$b(\mu m)$	$V_B(\mu m/h)$
23,8	0,148(23)	1,620(12)
32,4	0,461(9)	1,876(6)
41,0	0,403(16)	1,646(9)
49,0	0,653(17)	1,412(9)

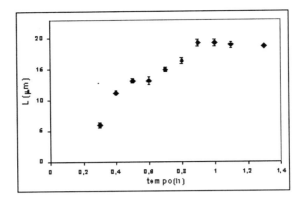

Figura 9 Profundidade dos poros em função do tempo de corrosão para ^{16}O de 32,4 MeV.

5 Discussão

Durante a análise superficial das amostras de CR-39 corroídas, observou-se uma diferente densidade de poros. Essa diferença foi atribuída às diferentes posições durante a irradiação. Para verificar essa hipótese, construiu-se o gráfico da densidade de poros em função da posição durante as irradiações, que pode ser visualizado na figura 10.

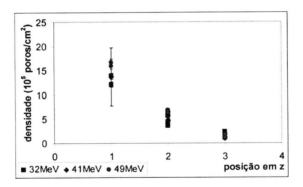

Figura 10 Densidade de poros em função da posição de irradiação.

Imediatamente, observa-se no gráfico a dependência da densidade com a posição. Nas amostras posicionadas na posição 1 (mais próxima ao plano de espalhamento) a densidade de poros foi da ordem de 10^6 poros/cm^2. Esse resultado é bastante significativo, pois indica a validade de todos os parâmetros e modelos utilizados para controlar o experimento. Além disso, indica que a eficiência do processo de corrosão dos traços latentes foi próxima de 100 %. Para as posições 3 e 2 a densidade de poros foi da ordem de $1,5.10^5$ e 5.10^5 poros/cm², respectivamente. Esses valores são facilmente explicados ao verificar a dependência da seção de choque com o ângulo de espalhamento. Da expressão de Rutherford [Knoll, 1989], verifica-se que o número de partículas espalhadas no ângulo θ_{CM} depende de $1/\operatorname{sen}^4 \theta_{CM}$.

Dos valores obtidos para as velocidades de corrosão volumétrica V_B, apresentados na tabela 2, não é possível observar nenhuma variação sistemática com o aumento da energia do íon. Os valores obtidos são incompatíveis entre si e apresentam uma variação percentual de aproximadamente 30 %.

Com os valores de V_B e do patamar de saturação da profundidade do poro, foi possível determinar o valor experimental do alcance dos íons nas amostras, para as diferentes energias irradiadas, supondo que o máximo comprimento obtido corresponde ao alcance máximo do íon. De acordo com o modelo teórico para a corrosão ao longo do rasto descrito em [Dörschel, 1999], a distância R alcançada pelo íon e corroída nessa direção, para incidência normal, é dada por:

$$R = L(E, t) + V_B t \tag{1}$$

onde L é o valor de profundidade medido e t é o tempo de corrosão.

Para comparar o valor experimental obtido para o alcance do íon, efetuou-se uma simulação com o programa TRIM 98. Esse programa calcula vários parâmetros físicos de interação de íons com a matéria, inclusive a trajetória na penetração. A tabela 3 apresenta os valores dos patamares de saturação de profundidade, o valor calculado experimentalmente para o alcance do íon e o valor simulado pelo programa TRIM.

Tabela 3 Valor do comprimento L máximo dos poros e alcance experimental R_{exp} e simulado R_{sim} dos íons, para diferentes energias em CR-39.

E_{in} (MEV)	$L(\mu m)$	$R_{exp}(\mu m)$	$R_{sim}(\mu m)$
23,8	11,5(5)	13,1(5)	19,4
32,4	19,5(5)	20,9(5)	27,2
41	25,5(5)	27,9(5)	36,9
49	29(1)	31(1)	46,4

Da tabela observa-se que o valor experimental obtido para o alcance dos íons de ^{16}O, nas diferentes energias, foi cerca de 70 % do valor de alcance simulado, o que não era esperado, pois em trabalho análogo [Dörschel et al., 2002; Dörschel et al., 2003] os valores experimentais de alcance em CR-39, obtidos, eram compatíveis com os valores de alcance simulado, para mesmo íon incidente. Uma hipótese é que durante o espalhamento na folha de Au, a energia tenha sido atenuada e, portanto, os valores de energia utilizados na simulação computacional estejam superestimados. Estando superestimados levariam a um valor de alcance simulado superior ao experimental.

6 Conclusões

Foi possível produzir poros em CR-39, utilizando a corrosão de rastos iônicos latentes criados por feixe de ^{16}O de diferentes energias. Também foi possível caracterizar esses poros com a utilização de um microscópio óptico acoplado a um sistema de análise computacional.

Da caracterização dos poros, verificou-se que a densidade de rastos apresenta forte dependência com a posição de irradiação.

Além disso, foi possível checar o aumento linear do diâmetro dos poros com o tempo de corrosão e o aumento da profundidade dos mesmos até a ocorrência de um patamar de saturação correspondente ao alcance máximo do íon de ^{16}O no CR-39. Os valores de alcance experimental do íon em comparação com os alcances simulados através do programa TRIM 98 mostraram-se cerca de 30 % inferiores ao esperado.

Referências Bibliográficas

[1] Apel, P. Yu, *Radiation Measurements* **34**, (2001) 559-566

[2] Da Silva, A.A.R., Yoshimura, E. M., *Radiation Measurements* **39**, (2005) 621-625

[3] Dörschel,B., Bretschneider,R., Hermsdorf,D., Kadner, K., Kühne, H., *Radiation Measurements* **31**, (1999) 103-108.

[4] Dörschel, B., Hermsdorf, D., Kadner, K., Starke, S., *Radiation Measurements* **35**, (2002) 177-182.

[5] Dörschel, B., Hermsdorf, D., Starke, S., *Radiation Measurements* **37**, (2003) 583-587.

[6] Fink, D., Alegaonkar, P.S., Petrov, A.V., Berdinsky, A.S., Rao, V., Muller, M., Dwivedi, K.K., Chadderton, L.T., *Radiation Measurements* **36**, (2003) 605-609.

[7] Knoll, G. F., *Radiation Detection and Measurement*, John Wiley & Sons, 1989, 2nd Edition.

[8] Mitrofanov, A.V., Tokarchuk, D.N., Gromova, T.I., Apel, P. Yu, Didyk, A. Yu, *Radiation Measurements* **25**, (1995) 733-734.

[9] Tripathy, S.P., Mishra, R., Dwivedi, K.K., Khathing, D.T., Ghosh, S., Fink, D., *Radiation Measurements* **36**, (2003) 107-110.

[10] Yamazaki, I.M., Geraldo, L. P., *Desenvolvimento de microfiltros nucleares utilizando a técnica do registro de traços de fissão*, Publicação IPEN 319, 1990.

[11] Yamazaki, I.M., Outi, E. Y., *Radiation Measurements* **23**, (1994) 203-207.

Anais da XXVIII Reunião de Trabalho sobre Física Nuclear no Brasil
SP, Brasil, 2005

Trace Elements in Blood Serum Measured by PIXE

S. Bernardes, M. H. Tabacniks, Marcel D. L. Barbosa and Márcia A. Rizzutto

Department of Applied Physics, Institute of Physics
University of S. Paulo, S. Paulo, SP
Rua do Matão, travessa R 187, 05508-900 - suene@if.usp.br

Abstract. The quantitative measurement of trace elements in blood serum and its comparison with data from literature are shown. To verify and to optimize the analytical conditions, and the detection limits of the elementary PIXE (*Particle Induced X-ray Emission*) analysis, 7 µl blood-serum samples were pipetted on paper filter substrates without any previous treatment and left to dry in a clean bench at room temperature. PIXE analysis was done using 2.4 MeV proton beam and two Si(Li) detectors respectively with 65µm Be and 290 µm thick *Mylar* filters. Elementary concentrations were calculated relative to an internal Gallium standard, previously added to the liquid sample. A total of 15 trace elements were detected with concentrations ranging from 10 ng/g to 1 mg/g .

1 Introduction

The precise determination of trace elements in biological fluids and tissues is of vital importance when investigating the role that such elements play in several biochemical processes, to prevent and to recognize the possible accumulation of toxic elements in living bodies, and to reveal low levels of essential elements in patients. Nevertheless, it must be stressed that trace element concentrations in blood serum are always influenced by physical and pathological conditions, by environmental conditions, age, sex and diet, which makes a direct comparison of data a troublesome work [1, 2].

Since 1970, the PIXE (*Particle Induced X-ray Emission*) analysis has become an important and widespread technique for the determination of the trace elements in biological and environmental materials. The principal advantages of a PIXE analysis is its non-destructiveness, its intrinsic multielementary nature, and its high sensitivity (ppm), hence requiring very small sample quantities.

The objective of this work was to develop a reliable measuring protocol for trace elements in sanguineous plasma detected by PIXE (*Particle Induced X-ray Emission*) and to compare the results with data from literature attempting to guide the answers of two basic questions: a) Which are the elements of biological interest? b) What is the necessary detection limit? PIXE is an elementary spectrometric method that relies on the detection of characteristic X-rays emitted by bombarding a sample with an energetic ($\sim$ MeV) ion beam (H^+, He^+, ...). PIXE is sensitive to all the elements of the periodic table above the Aluminum, with a primary detection limit for solid samples of the order of $1\mu g/g$ which can be extended to the ng/g level, for liquid samples, after a delicate pre-concentration procedure [3].

2 Earlier Measurements

The development of many element sensitive techniques and the systematic measurement of trace elements in materials and more specifically in biological materials can be tracked down to early 1950 [4, 5]. Initially, and probably aiming toxicological studies, many investigations dealt with the detection of poisonous contaminants at relatively high concentrations (like Pb, Hg, Cd etc.) [6]. Willing to understand the role of trace and ultra-trace elements in the environment and in the living bodies, the search for trace elements required sequentially lower detection limits, while also reducing the amount of sample material needed. The multi-elementary and almost non-destructive physical spectrometric methods are specially suited for this kind of analysis. Starting around the 50ties, the X-ray Fluorescence analysis and the Neutron Activation Analysis, extended and complemented the data of the widespread "mono-elementary" Atomic Absorption (and Emission) Spectrometry, a common equipment available in many chemical laboratories almost since the beginning of the 20th century. With the development and general dissemination of the PIXE analysis in the 70ties [7], the search for trace elements in the environment and biology got a remarkable push. Thanks to its low sensitivity (in the ppm range), speediness and allowing yet smaller samples, when compared to the XRF, PIXE analysis grew in less than 10 years to a mature and powerful micro-analytical method [3]. Yet, there was still the need to lower the detection limits down to sub *ng/g* (ppb) level. The Inductive Coupled Plasma Atomic Emission (ICP-AES) and its successor, the ICP - Mass Spectrometer, developed in the 80ties [8, 9], achieved the ultimate trace element detection limits at a reasonable cost. Hence, the question nowadays changed from *"How many elements were detected in the sample?"* to *"What are the necessary detection limits?"* or *"Which are the elements that need to be detected?"*.

A survey on 10 of the most accessible journals and publications supplied data for elementary concentration data for 48 elements in blood serum. Crude data

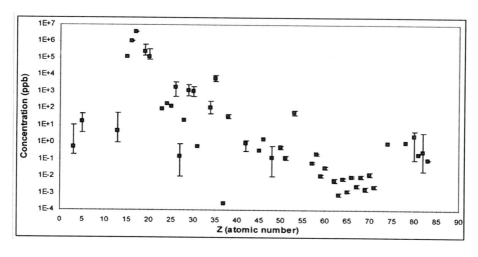

Figure 1 Medians of trace element concentration (μ g/l) in blood serum of literature valuesv 5. Straight lines indicate the maximum-minimum range of published data.

were compiled in simple scatter-plots of their mean and median values for blood serum, as displayed in **Figure 1,** ordered accordingly to their atomic number. The maximum-minimum ranges, when available, were also plotted as straight lines. Concentration data range over 11 orders of magnitude, and include samples from different healthy populations, most measured by PIXE and by ICP [10-21].

3 Material and Methods

To access our own sampling-PIXE analysis protocol, human blood samples were collected from 2 healthy persons. After coagulation, the serum was separated after a centrifugation at 3000 rpm for 10min. The samples were doped with Gallium, used as an internal standard, resulting in a concentration of 40ppmg of Ga in the final solution. The doped serum samples were pipetted (7μ L) on paper filter (Wattman 40) and let to dry at room temperature in a clean bench with laminar air flow (HEPA filtered).

PIXE analysis was done using a 2.4 MeV proton beam collimated to a diameter of 5mm. X-rays were measured with two Si(Li) detectors, respectively with a 65μm Be filter and a 290μm Mylar^{TM} filter, to record the low energy and the high energy X-ray spectrum. The samples were irradiated during 20 minutes with a beam current of 20 nA. Since Whatman 40 (9,2 mg/cm^2) paper cannot be considered a thin sample for PIXE analysis, and due to probable chromatographic effects during sample pipetting and diffusion from the front side to the back side of the paper, both sides where analyzed by PIXE. The complete PIXE setup at LAMFI [22] is described elsewhere [23]. The X-ray spectra were analyzed using the well-known non-linear multiple gaussian fitting program AXIL [24]. Quantitative calibration of the PIXE setup was made using a set of mono-elementary evaporated thin film standards (mostly from Micromatter Co.) and a semi-empirical fit.

4 Results

For each sample, a set of two PIXE spectra of blood serum, were collected, irradiating the front side (solid line) and the back side (dotted line) of a Whatman paper filter, as shown in **Figure 3**. The Ga peak is due to the internal standard used for elementary mass quantification.

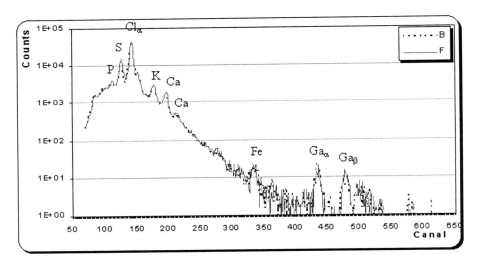

Figure 2 A set of two PIXE spectra of blood serum, irradiating the front side (solid line) and the back side (dotted line) of a Whatman 40 paper filter.

5 Discussion

As expected, due to chromatographic effects, there are significant differences in the concentration of several elements when measured from the front or from the back side of the filter. After self absorption corrections, and supposing an homogeneous mass distribution in the paper filter, S, Cu and Br concentrations tend to be higher on the front side while V and Cr show an opposite behavior. As known, minerals in blood can be found in a free state, or combined to proteins and other components. Potassium use to be free, in the form of an ionic salt. Calcium can be partially free and partially linked to proteins. Phosphorus exists in the form of a sodium or potassium salt. Metals, like iron and zinc, are in their majority attached to proteins [25]. This element-compound combination in the blood (the organic molecules are not detected by PIXE), associated to chromatographic effects of the paper filter, may explain the differences of the front and back side X-ray spectra. This effect has to be further investigated and may provide an interesting lever helping to discriminate the function and state of combination of the different trace elements.

Measured elementary concentrations were visually compared to serum values from literature in Figure 3. The PIXE data follow the general trend. Cl, K, Ca, Mn, Fe and Zn concentrations agree within one standard deviation of published data, though there is no statistical significance of any of these data. The samples were taken from only two individuals and there are none data from São Paulo region to compare to. The agreement of Cl, K, Ca, Mn, Fe and Zn to published data may be fortuitous, but may also indicate the correctness of our sampling-measurement protocol. A more rigorous test, using certified serum samples is under way.

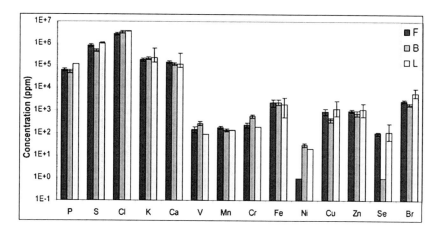

Figure 3 The average of the elementary concentrations in serum measured by PIXE compared to data from literature (L). The (F) data refer to filter front side mass concentrations while (B) indicates the back side measurements. The error bars indicate one standard deviation for this work data and maximum-minimum range literature values.

6 Conclusions

PIXE analysis of serum samples collected on paper filter (ore some other/better ashless absorbing substrate) seems to be a promising technique for elementary analysis above the 10 ppb wgt concentration. The chromatographic effects of blood serum in paper filter and its consequences in PIXE analysis need to be better understood, and may be an interesting method to access trace element and organic compound combination.

Studies using larger sampling sizes are necessary to draw more definitive conclusions on trace element concentrations in blood serum for donors that live in São Paulo.

Bibliography

[1] E.A. Hernández-Caraballo, L.M. Marcó-Parra, *Spectrochimica Acta Part B*, **58**, (2003) 2205-2213.

[2] J. Dombovári, Zs. Varga, J.S. Becker, J. Mátyus, Gy. Kakuk, L. Papp, *Atomic Spectroscopy*, **22(4)**, July/August (2001), 330-335.

[3] S.A.E. Johansson and J.L. Campbell, *PIXE, A Novel Technique for Elemental Analysis*. John Wiley and Sons, (1998).

[4] W. Maenhaut, B. Ottar and J.M. Pacyna eds. *Control and Fate of Atmospheric Heavy Metals*, NATO ARW Series, Kluver Ac. Pub. Amsterdam. (1988) 259-301.

[5] R. Jenkins, R.W. Gould and D. Gedke, *Quantitative X-ray Spectrometry*, Marcel Dekker, New York, USA. (1981)

[6] W. Maenhaut, L. de Reu, H.A. Van Rinsvelt, G.S. Roessler, J.W. Swanson and M.D. Williams, *IEEE Trans. on Nucl. Science*. NS-28-2 (1981) 1386-91.

[7] T.B. Johansson, K.R. Akselsson and S.A.E. Johansson, *Nucl. Instr. Meth.*, **84**, (1970) 141.

[8] P.J. Potts, *A Handbook of Silicate Rock Analysis*. Chapman & Hall, (1987) 575-584.

[9] M.F. Giné-Rosias, ed. *Espectrometria de Massas com Fonte de Plasma* (ICP-MS), CGP/CENA. Piracicaba, Brasil. (1999)

[10] Ch. Zarkadas, A.G. Karydas, T. Paradellis, *Spectrochimica Acta Part B*, **56**, (2001) 2219-2228.

[11] H. Afarideh, A. Amirabadi, S.M. Hadji-Saeid, N. Mansourian, K. Kaviani, E. Zibafar, *Nuclear Instruments and Methods in Physics Research B*, **109/110**, (1996) 270-277.

[12] H. Vanhoe, R. Dams, C. Vandecasteele, J. Versieck, *Analytica Chimica Acta*, **281**, (1993) 401-411.

[13] H. Vanhoe, C. Vandecasteele, J. Versieck, R. Dams, *Analytica Chimica Acta*, **244**, (1991) 259-267.

[14] K. Inagaki, H. Haraguchi, *Analyst*, **125**, (2000) 191-196.

[15] H.B. Lim, M.S. Han, K.J. Lee, *Analytica Chimica Acta*, **320**, (1996) 185-189.

[16] H. Vanhoe, F. Van Allemeersch, J. Versieck, R. Dams, *Analyst*, **118**, August, (1993) 1015-1019.

[17] W. Maenhaut, L. De Reu, H.A. Van Rinsvelt, J. Cafmeyer, *Nuclear Instruments and Methods*, **168**, (1980) 557-562.

[18] E. Bárány, I.A. Bergdahl, L.E. Bratteby, T. Lundh, G. Samuelson, A. Schütz, S. Skerfving, A. Oskarsson., *Toxicology Letters*, **134**, (2002) 177-184.

[19] R.A. Kumar, V. John Kennedy, K. Sasikala, A.L.C. Jude, M. Ashok, Ph. Moretto, *Nuclear Instruments and Methods in Physics Research B*, **190**, (2002) 449-452.

[20] E. Bárány, I.A. Bergdahl, L.E. Bratteby, T. Lundh, G. Samuelson, A. Schütz, S. Skerfving, A. Oskarsson, *Environmental Research*, **A89**, (2002) 72-84.

[21] H. Vanhoe, R. Dams, J. Versieck, *Journal of Analytical Atomic Spectrometry*, **9**, (1994) 23-31.

[22] www.if.usp.br/lamfi/

[23] M.H. Tabacniks, in *Nuclear Physics*, Eds. S.R. Souza, O.D. Gonçalves, C.L. Lima, L. Tomio, V.R. Vanin, World Scientific, Singapore, (1998).

[24] P. Van Espen, K. Janssens, & I. Swenters, *AXIL X-Ray Analysis software*. Camberra Packard, Benelux (1986).

[25] Villela, G.G., Bacila, M. and Tastaldi, H., *Bioquímica*, Capítulo XV.

Anais da XXVIII Reunião de Trabalho sobre Física Nuclear no Brasil
SP, Brasil, 2005

Automatic Classification and Segmentation of Image Elements Using Computed Tomographic Data

J. M. de Oliveira Jr.[1,2], E. C. Paizani[1], A. C. G. Martins[3] and G. F. de Lima[1]

[1] Universidade de Sorocaba - UNISO,
Cidade Universitária, C.P. 578, Sorocaba, SP, 18023-000, Brazil

[2] Faculdade de Engenharia de Sorocaba - FACENS,
Rod. Sen. José Ermirio de Moraes, Km 1.5, Alto da Boa Vista,
Sorocaba, SP, 18087-090, Brazil

[3] Universidade Estadual Paulista Julio de Mesquita Filho - Unesp, LAPI
Av. 3 de Março, 511, Alto da Boa Vista, Sorocaba, SP, 18087-180, Brazil

Abstract. Computerized Tomography (CT) imaging was originally developed to produce images of cross-sections of the human body. Nevertheless, its industrial application may ultimately far exceed its medical applications. Computed tomography methods have been used in many other areas in recent years such as application in soil science, monitoring of multi-phase flow phenomena in pipes, studies of living trees in order to measure growing rings, detection of structural defects and other heterogeneities in concrete samples. In this work, using data obtained by the Gamma-ray CT Miniscanner at University of Sorocaba (MTCU), we implemented a new computer tool in the program "**TOMOGRAFIA TRI-DIMENSIONAL - TTD**", that allows automatic classification and segmentation of elements presented in the tomography data of concrete samples, such as: i) identification of the number of stones presented in the mortar as well as the calculation of its volume and surface area; ii) identification of fractures, voids and cracks inside the concrete and perform its evaluation, etc. The information obtained from this implementation haves been used as an important instrument for tomographic data analysis. Using data obtained by MTCU, we apply this new tool in the study of concrete samples. The image processing techniques used to do the classification and segmentation are discussed and some results of concrete sample analysis are presented.

1 Introduction

The increasing availability of powerful processing computer has allowed the development of new methods for the analysis of discrete three dimensional data. Our implementation was developed to be used with data obtained by the Gamma-ray CT Miniscanner at University of Sorocaba (MTCU) [1]. The implementation was based in the visualization method called ray casting volume rendering, in which, the data array is displayed directly, i.e. without pre-fitting of geometric primitives. The use of such technique for nondestructive evaluation is a powerful tool that permits a visual trip inside an object without physically opening or cutting it. This technique also permits the extraction of some important parameters, performing classification and segmentation of elements presented in the tomographic data. In this study, our team used concrete samples and showed that it is possible the identification of the number of stones presented in the mortar, as well as, the calculation of its volume, surface area and localization, and the identification of fractures, air voids and cracks inside the concrete along with its evaluation, such as, calculus of volume and extension, etc. The segmentation of elements presented are obtained using a modified version of the program "**TOMOGRAFIA TRI-DIMENSIONAL - TTD**" [2]. This package runs on PCs, using Windows platform. The information obtained from this implementation have been used as an important instrument for the detection, quantification and further analysis of structural changes in cement-based material. CT offers the potential for more rational approach of designing, testing, repairing, or replacement of concrete structures [3, 4, 5].

2 Materials and Methods

The gamma-ray CT miniscanner employed in this experiment was built at the Experimental Nuclear Physics Laboratory at University of Sorocaba (LAFINAU) [2]. This scanner allows the use of nondestructive techniques for visualizing features in the interior of opaque solid objects and permits the analysis of digital information on their 3D geometry and topology. In the case of concrete, the coarse aggregates, air voids and cracks can be visualized inside a sample. The images to be shown in this paper, were reconstructed from the projection obtained by the miniscanner, a gamma-ray first-generation tomographic system. The tomograph operate with a gamma-ray source of ^{241}Am (photons of 60 keV and 100 mCi or 3.7×10^9 Bq of activity) and a NaI(Tl) detector. A PC controls the data acquisition system as well as the movement of translation and rotation of the sample. A tomographic image can be reconstructed, after a number of profiles of narrow-beam transmission are made at different orientations around the sample. These measurements represent the total linear attenuation coefficient of γ-rays along the path of the ray. Let N_i denotes the number of γ-rays photons incident on the body per unit of time for a particular ray path. Let N be the corresponding number of photons exiting the body. Then, for a mono-energetic γ-rays beam we can write:

$$\int \mu(x, y) ds = \ln\left(\frac{N_i}{N}\right), \tag{1}$$

where N_i and N are Poisson variables, $\mu(x, y)$ represents the linear attenuation coefficient at (x, y) position, ds represents the elemental distance along the ray, and the integral is calculated along the ray path from the γ-rays source to the detector. A measurement of the incident and the exiting γ-rays give only an the integral of $\mu(x, y)$ over the ray path. The details of the body along the ray path are then summed on to a single measurement. The image reconstruction problem is therefore to determine the distribution of the linear attenuation coefficient $\mu(x, y)$ through the section of the specimen crossed by each ray. This problem is solved in our case, using two different algorithms: i) Algebraic Reconstruction Techniques (ART) [6] and ii) Discrete Filtered Backprojection (FBP) [7, 8]. Visual criterion was used to choose which algorithm produces the best image. With several 2D images and a stack procedure, 3D volume data is produced. The concrete samples used in this study were obtained using cement type Portland-CP-II, fine natural sand, stones (coarse aggregate) and water. The cylindrical concrete samples measured 30 mm of diameter and 40 mm of height.

3 Theoretical discussions: Volume Rendering, Classification and Segmentation

Volume rendering is a family of techniques for visualizing sampled scalar or vector fields of three spatial dimensions, without fitting geometric primitives to the data. The techniques used in the process of visualization can be resumed in four steps: data acquisition, classification, shading and projection. The first step includes data preparation, such as interpolation between 2D acquire images, contrast enhancement and use of specific filters to improve the data quality. These 2D data are then used as input for the formation of the 3D volume data, that are divided into volume elements called voxel. This voxels array is then used in the classification procedure, where one opacity or transparency is associated to each voxel by making use of transfer functions like the one showed in equation 2, proposed by Levoy [9] and modified by us, in which we introduced the parameters δ_1 and δ_2. In this equation an opacity α_v is associate to all voxels with color or intensity value f_v, and assigning an opacity zero to all other voxels. To avoid aliasing artifacts, it is desired that the voxels with values close to f_v are assigned opacities close to α_v. The parameters δ_1 and δ_2 introduce some tolerance in equation 2, facilitate the choice of regions of interest inside the volume data and the assignment of specific opacity α_v.

$$\alpha(x_i) = \alpha_v \begin{cases} 1 & \text{if } |\nabla f(x_i)| \leqslant \delta_1 \text{ and } |f(x_i) - f_v| \leqslant \delta_2 \\ 1 - \frac{1}{r}\left|\frac{(f_v - f(x_i))}{|\nabla f(x_i)|}\right| & \text{if } |\nabla f(x_i)| > 0 \text{ and } f(x_i) - r \\ & |\nabla f(x_i)| \leqslant f_v \leqslant f(x_i) + r|\nabla f(x_i)| \\ 0 & \text{otherwise} \end{cases} \quad (2)$$

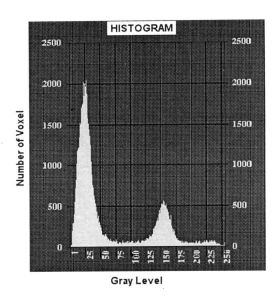

Figure 1 Image histogram. Using this it is possible to choose the ROI and start the classification process.

In Eq. (2) r is a voxel thickness and $|\nabla f(x_i)|$ is the gradient vector. In a separate step, the voxel array is used as the input to a shading model like the one proposed by Phong [10, 11]. The shaded color volume provides a satisfactory illusion of smooth, 3D shape cues and represents the sum of light emitted by the volume and scattered by the surfaces. Finally, rays are then cast into these two arrays from the observer eyepoint. For each ray a vector of sample color and opacities is computed by resampling the voxel database at some evenly spaced locations along the ray. Using these tools implemented in the TTD program it is possible to perform classification of tomographic data. Classification is the process that divides an image into its constituents parts, objects or Regions of Interest (ROI) [12]. This process is done by TTD with the following steps: i) using the image histogram showed in

Fig. 1, the intensity or gray level (ROIs) f_v is selected; ii) using Levoy's transfer function showed in Eq. 2, one opacity is associate to all voxel with the selected value of gray level f_v, (see step one), and assigning an opacity zero to all other voxel. This procedure enables us to view some special structures or ROI, contained within the volume data. The result of classification is showed in Fig. 2.

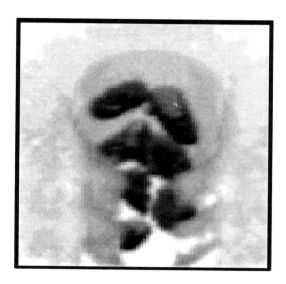

Figure 2 Portion of volume data displayed after the choice of ROI in the image histogram. This image represents a concrete block, with coarse aggregates appearing in black. Using Levoy's transfer function, it was possible to remove the cement material around the coarse aggregates and showed the air (gray portion of figure) around the concrete block.

After the classification of an image, segmentation and recognition can be accomplished. This step is done with the use of a new tool implemented in program TTD denominate "**SEGMENTA**". The implemented method in the SEGMENTA tool uses the flowchart showed in Fig. 3.

The SEGMENTA algorithm starts with the information generated by TTD program. After the choice of a ROI in the image histogram and with a parameter adjustments in Levoy's equation, it is possible to separate elements or voxels of interest (visual criterion was used to define if the correct elements were selected). Then using some threshold, the volume data is discretized and an opacity value zero ($\alpha_v = 0$) or value one ($\alpha_v = 1$) is associate to all voxels of the volume data. SEGMENTA algorithm is then started and when one ROI is encountered (voxel with $\alpha_v = 1$) a region growing process is initiate. This process is implemented with the following steps:

a) the first voxel is computed and its position in the volume data (rectangular coordinates x, y and z) is written in a file denominated Volume-File;

b) all neighbor voxels are computed and their position in the volume data are written in another file denominated Neighbor-File;

c) a change on the opacity value of the first voxel from $\alpha_v = 1$ to $\alpha_v = 0$ is done;

d) using now a voxel belonging to the Neighbor-File, its position in the volume data (coordinates x, y and z) is written in the Volume-File, and their neighbors are computed and their position is written in the same Neighbor-File as defined in step b;

e) a opacity change is done as defined in step c;

f) all described step above are repeated to all voxels belonging to the Neighbor-File.

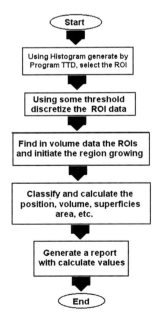

Figure 3 Flowchart used by SEGMENTA tool.

After, voxels that were computed more than once are eliminated by the analysis of their rectangular coordinates. The procedure described above permits the selection of the position and calculus of the volume of the ROI. In a separated step if the voxels with a six neighbor is removed, the remaining voxels are equal the surface area of selected ROI. SEGMENTA algorithm stops and a report is generated with: number of objects or ROIs found, position, volume, surface area and compactness factor (F_C) of each ROI. This compactness factor, defined by Eq. 3, provides some idea of the shape or appearance of the ROI.

$$F_C = \left(\frac{V^2}{A^3}\right), \qquad (3)$$

where V is the ROI volume and A is the ROI surface area.

4 Experimental Results and Discussion

A special phantom was fabricated in order to test the SEGMENTA algorithm. This phantom was constructed with known formats. Using the MTCU a tomographic volume data was obtained, see Fig. 4.

This phantom has three object whose shapes are: i) a cylinder with 8 mm of height and 10 mm of diameter approximate; ii) a sphere with 11.5 mm of diameter approximate and iii) a pyramid with area of base equals 121 mm^2 and 11 mm of height approximate. Results obtained with segmentation of this phantom were used to test the program SEGMENTA. The results are showed in Table I.

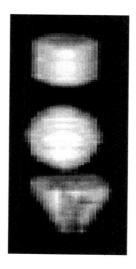

Figure 4 Tomographic image phantom used to test the algorithm.

Table 1 Results of phantom analysis. In the calculation of the surface area, an average was obtained using internal and external radiis of the cylinder and its height, internal and external radii of the sphere and internal and external heights of the pyramid. The uncertainty was estimated using the number of voxels selected by the program SEGMENTA in each of the selected ROI. The coordinates x = 0, y = 0 and z = 0 are in the center of sphere.

ROI	Real phantom volume (mm³)	Tomographic Phantom volume (mm³)	Real phantom average surface area (mm²)	Tomographic phantom surface area (mm²)	Real phantom F_C	Tomographic phantom F_C	Tomographic phantom rectangular coordinates of ROI (mm)
Cylinder	749.52	665 ± 26	374.11	353 ± 19	0.0107	0.014 ± 0.003	x = 0.12 y = -0.08 z = -15.00
Sphere	904.32	901 ± 30	383.08	361 ± 19	0.0145	0.019 ± 0.003	x = -0.01 y = -0.25 z = -0.85
Pyramid	732.33	679 ± 26	732.33	371 ± 20	0.0065	0.013 ± 0.004	x = 1.19 y = -0.82 z = -10.23

Another test of the program SEGMENTA was performed using a specific concrete sample, fabricated with cement type Portland-CP-II, fine natural sand, coarse aggregates (stones) and water. By using the MTCU a 3D volume rendered image was obtained, combining 81 2D CT scan images into a volume data. Each 2D image contains 81 per 81 pixels. The resolution of the image is 1 mm per pixel. Fig. 5 shows the image, where only the distribution of coarse aggregates inside the concrete was selected to be visible.

Figure 5 Tomographic image of a concrete block with 81^3 voxels with coarse aggregates (high density) shown in white. Cement was removed using Levoy's transfer function.

This concrete sample was analyzed by algorithm SEGMENTA and the final result is shown in the report generated by algorithm in Table II.

In Table II we present the results obtained in the segmentation of coarse aggregates found inside the concrete sample shown in Fig. 5. The first selected object (object N^0 1 on top) is an artifact generated by program TTD, caused by the density change in the transition between air cement region and interpreted by program as a stone. It can be seen that it isn't a stone because the surface area is equal to the volume. The other objects (N^0 2, 3, 4, 6 and 7) are stones that were correctly segmented and recognized. The base of the concrete sample was incorrectly added in the volume and surface area of object number 6. It is possible to identify this fact because its enormous volume and surface area values presented in the report. Objects 5 and 8 are only small stones presented inside the concrete

Table 2 Report generate by algorithm SEGMENTA.

REPORT OF SELECTED OBJECTS

OBJECT N: 1 Compactness Factor = 0.00100000005
Volume = 1000 mm^3 Surface area = 1000 mm^2
COORDINATES: X = 0.50 Y = -0.74 Z = -40.00

OBJECT N: 2 Compactness factor = 0.0118007241
Volume = 1490 mm^3 Surface area = 573 mm^2
COORDINATES: X = 11.71 Y = -3.97 Z = -27.63

OBJECT N: 3 Compactness factor = 0.00313178496
Volume = 1360 mm^3 Surface area = 839 mm^2
COORDINATES: X = -6.31 Y = -3.59 Z = -19.88

OBJECT N: 4 Compactness factor = 0.00796011928
Volume = 2145 mm^3 Surface area = 833 mm^2
COORDINATES: X = 7.16 Y = -0.40 Z = -0.52

OBJECT N: 5 Compactness factor = 0.111111112
Volume = 9 mm^3 Surface area = 9 mm^2
COORDINATES: X = 1.11 Y = -15.56 Z = 12.00

OBJECT N: 6 Compactness factor = 0.0013636565
Volume = 4180 mm^3 Surface area = 2340 mm^2
COORDINATES: X = -3.97 Y = -5.61 Z = 32.13

OBJECT N: 7 Compactness factor = 0.0084967399
Volume = 1112 mm^3 Surface area = 526 mm^2
COORDINATES: X = 15.63 Y = 0.65 Z = 32.49

OBJECT N: 8 Compactness factor = 0.0358518511
Volume = 11 mm^3 Surface area = 15 mm^2
COORDINATES: X = 10.55 Y = -15.09 Z = 35.91

that were also segmented and recognized by the program. All other objects with volume smaller than 5 voxel were reject by the program.

5 Conclusion

Computerized Tomography associated with volume rendering technique is a powerful tool to analyze concrete structures. The results obtained in this work, using a phantom and samples of concrete, confirm that our algorithm called SEGMENTA, is able to segment different objects and showed the high potential of the method for qualitative and quantitative analysis, opening possibilities for the use of this technique in others areas. TTD program along with SEGMENTA algorithm were develop using image processing techniques, to analyze any three dimensional volumetric data, and generating a complete report of selects objects. The quantitative characterization done in this study using a concrete sample shows the strength of the technique in analysing cement-based material. In future works, we intend to extend this analysis to the detection of voids (cracks and air hole) and determination of their surface area and volume.

Acknowledgements

The authors thank to Universidade de Sorocaba - UNISO and Fundação de Amparo à Pesquisa do Estado de São Paulo - FAPESP, for the financial support. The authors would like to thank the SCHAEFFLER BRASIL LTDA industry, by the donation of a linear guide used in the construction of the Mini Computerized Tomograph of UNISO.

Bibliography

[1] J. M. Oliveira Jr., Brazilian Journal of Physics, **33**(2), 273 (2003).

[2] J. M. Oliveira Jr., Brazilian Journal of Physics, **35**(3B), 789 (2005).

[3] L. B. Wang, J. D. Frost, G. Z. Voyiadjis, T. P. Harman, Mechanics of Materials, **35**, 777 (2003)

[4] E. J. Garboczi, Cement and Concrete Research **32**, 1621 (2002)

[5] S. R. Stock, N. K. Naik, A. P. Wilkinson, K. E. Kurtis, Cement and Concrete Research **32**, 1673 (2002)

[6] R. Gordon, R. Bender and G. T. Herman, J. Theor. Biol. **29**, 471 (1970).

[7] R. N. Bracewell and A. C. Riddle, Astrophys. J. **150**, 427 (1967).

[8] G. N. Ramachandran and A. V. Lakshminarayanan, Proc. Nat. Acad. Sci. USA **68**, 2236 (1971).

[9] M. Levoy, IEEE Computer Graphics & Applications, **8**(3), 29 (1988).

[10] B. T. Phong, Communications of the ACM, **18**(6), 311 (1975).

[11] J. Blinn, Computer Graphics, **11**(2), 192 (1977).

[12] Rangayyan, R. M., Biomedical Image Analysis. Boca Raton, FL: CRC Press, 2005

Anais da XXVIII Reunião de Trabalho sobre Física Nuclear no Brasil
SP, Brasil, 2005

PlanDose: A User-Friendly Program for Planning of Ablation of Thyroid Carcinoma

Lima, F. F.[1], Benevides, C. A.[1], Silveira, R.[1], Stabin, M. G.[2], Khoury, H. J.[3]

[1] Centro Regional de Ciencias Nucleares - CRCN/CNEN, Rua Prof. Luiz Freire, 200, Cidade Universitaria, Recife-PE, Brazil,CEP:50740-540

[2] Department of Radiology and Radiological Sciences, Vanderbilt University, Nashville-TN, USA

[3] Departamento de Energia Nuclear, Universidade Federal de Pernambuco, Recife-PE, Brazil

Abstract. Thyroid cancer therapy using ^{131}I is based on the strategy of concentrating radioactive iodine in thyroid tissue, to completely eliminate thyroid tissue and functioning thyroid cancer metastases remaining after thyroidectomy. In Brazil, fixed activities of ^{131}I generally are given, with insufficient activities to ablate all of the remnants sometimes delivered, or with unnecessarily high activities given, and patients thus remaining in the hospital unnecessarily. Some studies have considered protocols for planning and individualized dose calculations, considering the metabolism of the patient and the mass of thyroid remnants. In this study, we present a computer program for planning and calculation of the therapeutic activity of ^{131}I to be given to individual patients. Users must enter data from the biokinetic study for the subject, and the program fits the data to a two component exponential function using a Figure of Merit analysis. The parameters are obtained through fitting and minimization of this function. To treat a nonlinear function, the minimization is done through an iterative (Levenberg-Marquardt) model. The mass of the thyroid tissue remnants and the activity given in the metabolic study also must be provided. The program then calculates the maximum thyroid uptake and therapeutic activity recommended for the subject. Data from the biokinetic and anatomical studies in nine patients after thyroidectomy are shown for examples. The fittings of the curves supplied from the PlanDose has been compared with those obtained from the Grace 5.0 program; good agreement was found for all the patients. Although this work considered only thyroid ablation, the principles could be extended to the treatment of hyperthyroidism.

1 Introduction

Thyroid cancer therapy using ^{131}I is based on the strategy of concentrating radioactive iodine in thyroid tissue, to completely eliminate thyroid tissue and functioning thyroid cancer metastases remaining after thyroidectomy. The most common method used to ablate thyroid remnants is the use of fixed levels of activity for all patients, which may result in the administration of insufficient levels of activity to cause complete ablation or overdosage of iodine to the subject. With large administrations of radioactive iodine, this may cause subjects to require unnecessary hospital stays, with their attendant costs, as the optimum amount of activity was not given, based on individual patient anatomical and biokinetic data.

An alternative to the use of fixed activities is the administration of tracer levels of ^{131}I to determine the characteristics of each patient, including the thyroid uptake and elimination half-times for radioactive iodine in the remaining thyroid tissue [1], and the volume of thyroid remnants. Such data may be used to establish more exactly the activity needed for ablation.

Lima et al [2, 3] suggested a method for determining volume and activity concentration values using quantification of SPECT images for dose planning of thyroid ablation therapy in patients with differentiated thyroid cancer. Good agreement between the measured and expected volumes was found for the majority of patients.

As such techniques may be difficult to implement in routine clinical use, mainly the calculations, this work describes a user-friendly computer program, called PlanDose, which provides the optimum therapeutic activity of ^{131}I to be administered to a patient, given the input of appropriate data.

2 Materials and Method

2.1 Calculation of Therapeutic Activity

From Lima et al [2, 3], the anatomical and bioknetic data from 9 patients with differentiated thyroid cancer were used. Activity values determined for each patient were plotted, and the uptake and elimination phases were fit to a two component exponential function using the Grace - 5.0 code. The fitted function was thus of the form:

$$y(t) = a_0 \times \left(e^{-a_1 \times t} - e^{-a_2 \times t}\right), \qquad (1)$$

where $y(t)$ is the activity in the thyroid at any time, and a_0, a_1, and a_2 are fitted parameters. The cumulated activity $(\tilde{A}_h)$ is obtained by direct integration of equation (1), and the maximum activity in the thyroid is given as:

$$A_{MAX} = f \times A_0 = a_0 \times C, \qquad (2)$$

where f is the maximum fraction of administered activity (A_0) in the thyroid and

$$C = \left[\frac{\left(\frac{a_2}{a_1} \right) - a_1}{a_2 - a_1} - \frac{\left(\frac{a_2}{a_1} \right) - a_2}{a_2 - a_1} \right], \tag{3}$$

The standard dose equation for average dose in an organ is given as [4]:

$$D = \frac{\tilde{A}_h}{m} \times \sum \Delta_i \phi_i, \tag{4}$$

where m is the mass of the organ (here the remnant thyroid tissue), Δ_i is the mean energyof the ith radiation emitted per unit cumulated activity and ϕ_i is the absorbed fraction of energy of the ith radiation component.

The chosen values of the absorbed fractions are based on the energy values of the emitted radiations and from ^{131}I, that are, respectively, 192 keV (90 %) and 364 keV (82 %), and assuming that the thyroid remnant masses are less than 10g. According to Stabin and Konijnenberg [5], the recommended values of the fractions absorbed for electrons and photons with these energy characteristics and for source regions with these mass values are 0,982 and 0,032, respectively. Thus, it can be considered that the fraction of absorbed energy proceeding from particles is total ($\phi_\beta = 1$) and that the contribution of the radiation is negligible to the total absorbed dose ($\phi_\gamma = 0$).

Thus, the activity that should be administered to any subject is calculated as:

$$A_0 = \frac{C \times a_1 \times a_2 \times D \times m}{f \times (a_2 - a_1)}, \tag{5}$$

where Δ_i for ^{131}I is 3.07×10^{-14} Gy.kg/Bq.s [4].

2.2 Computer Program for Dose Planning

To facilitate the implementation of the above considerations in routine nuclear medicine practice, we created a compute program, called PlanDose, which provides the optimum therapeutic activity of ^{131}I to be administered to a patient, given the input of appropriate data, as discussed above. The program was written in C, which allows for flexibility and portability on many platforms when compiled. Users must enter data from the biokinetic study for a subject, and the program fits the data to the function shown in Eq. (1), using a Figure of Merit analysis (χ_i^2):

$$\chi_i^2 = \sum_i [y_i - f(x_i)]^2, \tag{6}$$

where, y_i is a measurement-dependent variable and $f(x_i)$ is the value of the variable which depends on the point x in the model. The variables a_0, a_1 and a_2

are obtained through fitting and minimization of this function. To treat a nonlinear function, the minimization is done through an iterative (Levenberg-Marquardt) model. The minimization criteria are that $(\chi_i^2 - \chi_{i+1}^2)^2 < 0,001$ or N = 40 where N is the number of iterations. Then, once the fitted coefficients of the time-activity curve are obtained, the maximum thyroid uptake and therapeutic activity needed for the subject are calculated according to equations (2) and (5).

3 Results and Discussion

The program requires as input the results of the study of the individual subject's biokinetics, i.e. the activity measured at various times after administration of the tracer quantity of ^{131}I. The user must also provide the mass of the thyroid remnant tissue and the amount of ^{131}I administered in the tracer study. The user may also enter some limited data for patient identification, as shown in Figure 1. The program uses units for input and output as are commonly used in most nuclear medicine centers currently. The program fits the entered data to the functional form of Eq. (1), and calculates the maximum thyroid uptake and therapeutic activity recommended for the subject. The program also shows the figure of merit observed for the data that aims to give more confidence in the results (Figure 2).

The fitting results found with this program are the same of those obtained by Grace 5.0 program for all the patients.

The PlanDose program facilitates the application of this technique for the clinical staff. Expertise in radiation dosimetry is not required, only the accurate entry of the subject's biokinetic data. The target radiation dose value of 300 Gy is also an adjustable parameter in the program, so that the physician can vary this as desired. As with any computer program, however, care must be taken in the evaluation of the output, with both a physicist and the attending physician carefully

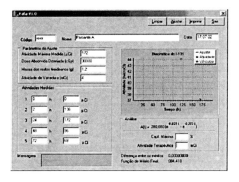

Figure 1 Form for the PlanDose Program, showing the input data for one thyroidectomy patient.

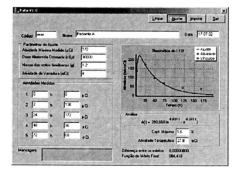

Figure 2 Form for the PlanDose Program, the results of the curve fit to the entered data and the calculation of suggested therapeutic activity.

studying the results and making the final recommendation on the desired level of therapeutic activity to be administered.

In our experience, the PlanDose converged rapidly on an optimum fit in all cases. In one case, however, the fit was interrupted as it reached the maximum number of iterations. Computer fitting of measured data must also be reviewed critically by the professional staff in all cases to ensure that reasonable values are used in the final analysis.

Although this work considered only thyroid ablation, the principles could be extended to the treatment of hyperthyroidism since that metabolic data are collected and the thyroid mass is determined. As patients with hyperthyroidism have all of their thyroid glands, the mass can be finding by ultrasonography.

4 Conclusion

PlanDose is a useful, user-friendly program for planning of ablative dose for patient with thyroid carcinoma, allowing calculation of patient-specific ^{131}I therapeutic activity to be administered. The program has been set up to evaluate thyroid remnant ablation, but it can also be used for the calculation of the activity to be administered for treatment of hyperthyroidism.

Bibliography

[1] BECKER, D.V., HURLER, J.R., MOTAZEDI, A., et al. Ablation of postsurgical thyroid remnants in patients with differentiated thyroid cancer can be achieved with less whole body radiation", J Nucl Med, vol. 23, pp. 43, 1982.

[2] LIMA, F.F., STABIN, M., KHOURY, H.J. Volume and Activity Concentration Estimation in Phantom of Thyroid Tissue Remnants Using SPECT, Radioproteccao, vol.2 (4 e 5), pp.44-53, Dez/2003 a Mai/2004.

[3] LIMA, F.F., KHOURY, H.J., STABIN, M. Estimativa do Volume de Restos Tiroideanos, por meio do SPECT, em Pacientes Tiroidectomizados, Rev Imagem, vol.25(4), pp. 239-246, 2003.

[4] LOEVINGER, R.; BUDINGER, T.F.; WATSON, E.E. in collaboration with MIRD Committee. MIRD PRIMER for absorbed dose calculation, 1991.

[5] STABIN, M.G.; KONIJNENBERG, M. Re-evaluation of Absorbed Fractions for Photons and Electrons in Small Spheres. J Nucl Med, 41:149-160, 2000.

Optimization of the ^{123}I Production in the Microtron MT-25, using the Code MCNP-X (Part 1)

B. Zaldívar-Montero, N. López-Pino, M. V. Manso-Guevara,
K. D'Alessandro, J. N. Varona-Ayala

Instituto Superior de Tecnologías y Ciencias Aplicadas (InSTEC),
Ave. Salvador Allende y Luaces. Quinta de los Molinos. Habana 10600.
A.P. 6163, InSTEC, La Habana, Cuba. email: susanms@cubarte.cult.cu

Abstract. Iodine 123 is a radionuclide with a relatively short half-life of appproximately 13 hours. Its emitted gamma photon has an anergy of 159 KeV compatible with current imaging systems. I-123 has been labeled to several ligands that have been applied clinically. Because of this, we reproduce the iodine production in an electron linear accelerator, carried out by Nordell B. et al in Sweden (Int. J. Appl. Radiat. Isot. 33, 1982), using the code MCNP-X. The first step was to check the experimental results obtained by this group, using the proposed geometry, and ^{124}Xe as target with 0.01 % of radionuclide purity. The second step was to analyze possible configurations and other parameters in order to get better or similar values of activity, using an electron cyclic accelerator (Microtron MT-25).

1 Introduction

The production of radioisotopes in particle accelerators is nowadays a great important aspect, due to the increased use of them in medical treatments. In particular, the obtaining of I-123 through a nuclear reaction, using an electron accelerator have became a very efficient way [1-3].

The final objective of this work is to check the feasibility of the production of I-123 through the reaction

$$^{124}Xe(\gamma, n)^{123}Xe \rightarrow_{\beta^+} {}^{123}I$$

using the electron cyclic accelerator Microtron MT-25, which is planned to be installed at Nuclear Physics Department of InSTEC, Havana. In order to reach this, first we will compare the experimental results reported by Nordell B.[1] with those results obtained by Monte Carlo simulation of the same experiment, using MCNPX-Code [4]. Then, we proceed to optimize the experiment taking into account the parameters of MT-25, the materials and geometry used, to obtain enough ^{123}I activity , suitable to be used for clinical pourposes.

2 Production and Decay Equations

The above reaction can be represented briefly as the process

$$A \to B \to C$$

Where the elements A, B and C are ^{124}Xe, ^{123}Xe, ^{123}I respectively.

The activity of B can be calculated using the known differential equation:

$$\frac{dN_B}{dt} = N_A R - N_B \lambda_B$$

where:

- N_B: number of atoms of B
  (with $N_B = 0$ for $t = 0$)

- N_A: number of atoms of A that can react with the incident particles

- R: probability for occuring the reaction, per unit time (reaction rate)

- λ_B: decay constant of B

If we consider that the probability per unit time for occuring the reaction is much less than the probability per unit time for the B decay, i. e. $R << \lambda_B$ [5], then the activity A_B of B is

$$A_B = A_A (1 - e^{-\lambda_B t})$$

With a similar computation we can then obtain an expression for the activity A_C of C:

$$A_C = A_A \left[1 + \frac{\lambda_C}{\lambda_B - \lambda_C} e^{-\lambda_B t} - \frac{\lambda_B}{\lambda_B - \lambda_C} e^{-\lambda_C t} \right]$$

After an irradiation time t, we let the sample to be cooled for a time t_c, then the differential equation to be solved is:

$$\frac{dN_C^*}{dt_c} = N_B^* \lambda_B - N_C^* \lambda_C$$

with the conditions:

$$N_C^*(t = 0) = N_C = \frac{A_C}{\lambda_C}$$

$$N_B^* = \left(\frac{A_B}{\lambda_B}\right)e^{-\lambda_B t_c}$$

Doing some algebraic work, and replacing the actual elements, we can finally get an expression for the ^{123}I activity:

$$A(^{123}I) = A(^{124}Xe)f(t, t_c)E_{brem}E_\mu \tag{1}$$

Here,

$$f(t, t_c) = \frac{\lambda_{123\,I}}{\lambda_{123\,I} - \lambda_{123\,Xe}} \times \left[e^{-\lambda_{123\,Xe}\,t_c}\left(1 - e^{-\lambda_{123\,Xe}\,t}\right) - \frac{\lambda_{123\,Xe}}{\lambda_{123\,I}}e^{-\lambda_{123\,I}\,t_c}\left(1 - e^{-\lambda_{123\,I}\,t}\right)\right]$$

also,

$$E_\mu = \frac{(1 - e^{\mu X})}{\mu X}$$

where μ is in units of cm^2/g and X in g/cm^2

$$E_{brem} = \begin{cases} S_b/S_t & \text{if } S_b \leqslant S_t \\ S_t/S_b & \text{if } S_t \leqslant S_b \end{cases}$$

$$A(^{124}Xe) = R \cdot N_{124\,Xe}$$

where

$$R = \int \Phi(E)\sigma(E)dE$$

$$N_{124\,Xe} = \frac{m(^{124}Xe)N_o}{124} = \frac{m_{sample}\Theta N_o}{124}$$

where:

- Θ: isotopic abundance of ^{124}Xe in the sample

- N_o: Avogadro Number

- $124 \approx$ atomic mass of ^{124}Xe

- E_μ: target utilization factor allowing for photon absorption and scattering in the irradiated sample [2]

- E_{brem}: bremsstrahlung beam utilization factor

- S_b, S_t: cross-sectional areas of the bremsstrahlung beam and target

- $\lambda_{123\,Xe} = 0.33324h^{-1}$

- $\lambda_{123\mathrm{I}} = 0.052116 \mathrm{h}^{-1}$

The procedure we followed was to simulate, with MCNPX, the selected experiment [1], and the magnitude that we will obtain from this simulation is the reaction rate R. Then, substituting this value in the equation (1), and taking into account the rest of the parameters, such as sample mass, and irradiation (and cooling) time, etc; we can get the final value of Iodine activity.

3 Testing Simulation Procedure

Comparing experimental results obtained by Nordell [1] is made as follows:

An electron beam from linear accelerator with 40 MeV of energy and an electron current of 100 µA, reaches a gold target (3mm thickness) producing bremsstrahlung. Aluminum collimators are used to improve the irradiation geometry. A cylinder of graphite was used as electron stop, to prevent heating of the cylindrical Xe-target by electrons transmitted through the gold target. Finally, gamma rays from gold could reach Xe-124 (0.01 % of radionuclide purity) and to induce the reaction mentioned above. Figure 1 shows a cross-sectional view of the geometry.

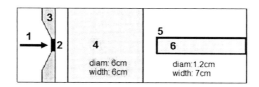

Figure 1 MCNP-X Irradiation Plot. 1-Electron beam; 2-Gold target; 3-Aluminum; 4-Graphite; 5-Copper cover; 6-Xe Target.

Bremsstrahlung flux in different components of the irradiation geometry was estimated using tally 4 of the MCNP-X Code [4]. Figure 2 shows us the result of this simulation.

Now we are able to compare the experimental results obtained by Nordell, with the one obtained by us using the MCNP-X Code.

They repeated 5 times the measurements, varying the electron current in ±20µA. The values of Iodine activity, once they have cooled the sample, lies between 133 kBq and 629 kBq. Taking an average value this gives:
$A_{\mathrm{Nordell}} = 0.115 \ \mathrm{mCi} \cdot \mathrm{\mu A}^{-1} \mathrm{h}^{-1} \mathrm{g}^{-1}$.

In our simulation, using the average electron current used by Nordell, and taking into account all the needed parameters of the experiment, we get, after estimating the reaction rate R, a value of specific activity of ^{123}I:

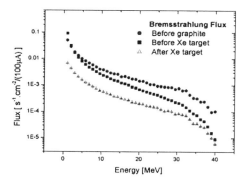

Figure 2 Bremsstrahlung flux behavior in geometry elements

$$A_{\text{MCNPX}}(^{123}\text{I}) = 0.116 \pm 0.005 \ \text{mCi} \cdot {-A}^{-1}h^{-1}g^{-1}$$

As we can see, the simulation value is agree with the experimental value. An ilustrative view of the distribution of the gamma flux in the target region is shown in Figure 3:

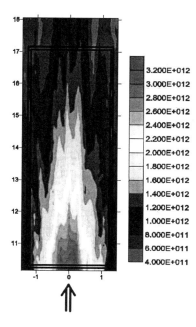

Figure 3 Mesh type 1 of MCNPX in the target region

4 Application to the Microtron MT-25

A second step of this work consists in the analysis of different configurations, in order to optimize the irradiation geometry and compound materials, to evaluate the capability of radioactivity iodine production in the electron cyclic accelerator Microtron MT-25, with the following parameters:

- Maximum Electron beam energy of 25 MeV

- Maximum Electron current of 20 µA

- Average bremsstrahlung Flux of 10^{14} photons $cm^{-2}s^{-1}$

4.1 Analysis of Bremsstrahlung target

Tantalum and Tunsgten are frequently used as bremsstrahlung target materials. We analyzed for each of them the optimum thickness, to produce maximum bremsstrahlung. The results are displayed in Figures 4, 5 and 6:

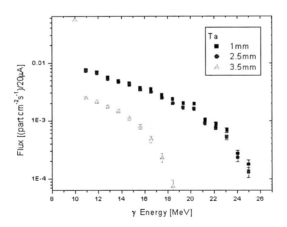

Figure 4 Bremsstrahlung flux from Ta target

As we can see, for both materials, the optimum thickness is around 1mm, but the bremsstrahlung produced by Tungsten is even bigger. The energy range was 10-25 MeV because the threshold of the reaction $^{124}Xe(\gamma, n)^{123}Xe$ is 10.5 MeV.

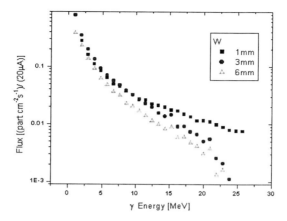

Figure 5 Bremsstrahlung flux from Tungsten target

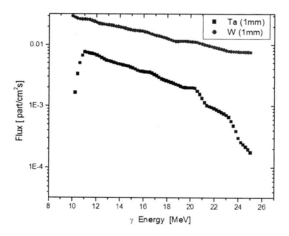

Figure 6 Bremsstrahlung fluxes from Tantalum (1mm) and Tungsten (1mm)

4.2 Analysis of the dimensions of the Xe-target

The main difference between the geometry used by Nordell and the one used by us, is that we took off the electron stop, in order to get as much gamma flux as possible at the target. This variant has been analyzed by Malinin [2] and others. We proceeded then with the analysis of the Xe target.

The behavior of the reaction rate R if we vary the width of the Xe cell is displayed in Figure 7:

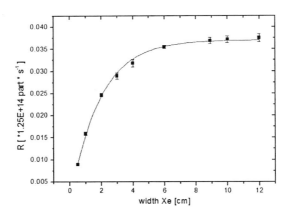

Figure 7 Reaction rate R vs. Width of Xe target

A similar analysis was made with the diameter of the Xe target (see below, Figure 8):

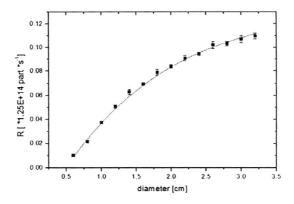

Figure 8 Reaction rate R vs. Diameter of Xe target

The results of the exponential fit of this two curves are shown in Table 1:

From this curves we can see a saturation of R with the increasing of width and diameter. Then we choose as optimum dimensions of the Xe target: width-10cm; diameter-3cm. The Figure 9 shows that choosed like this, the Xe cell practically has absorb all the gamma flux.

Table 1 Exponential fit of the curves in Fig.7 and Fig.8. R-reaction rate; x_1-width of Xe cell; x_2-diameter of the Xe cell

Models: $R = A_1 \exp(-x_1/t_1) + y_0$			$R = A_2 \exp(-x_2/t_2) + z_0$		
coeff.	value	error	coeff.	value	error
A_1	-0.0366	0.0005	A_2	-0.186	0.002
t_1	1.87	0.08	t_2	1.60	0.08
y_0	0.0370	0.0004	z_0	0.137	0.004
Chi square: 0.999			Chi square: 0.997		

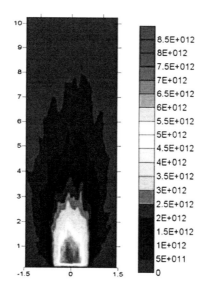

Figure 9 Distribution of the gamma flux inside the Xe target

4.3 Irradiation time and Cooling time

Now we look for the combination of irradiation time and cooling time that gives the maximum possible value of Iodine activity, through the dependence of the function $f(t, t_c)$ seen before. It's behavior is shown below (see Fig. 10)

If we increase the irradiation time for too many hours, we will obtain a lower value of the function; however, if we decrease the irradiation time for a few hours, we will get only a few amount of ^{123}Xe, and consequently, less activity of Iodine is produced. Then, we can say that a factible irradiation time could be 10 hours, and for that time, we get the maximun activity with 3 hours of cooling time.

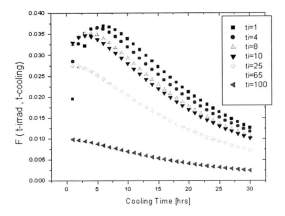

Figure 10 Function $f(t, t_c)$ vs. Cooling time, for differents Irradiation times

4.4 Iodine activity

From the equation (1) we can calculate the Iodine activity, substituting the following values of the parameters:

- Mass of the sample: 5.8g (99% purity of ^{124}Xe)
- Dimensions of Xenon vessel: Diameter 3cm; Width 10cm
- Irradiation and Cooling times: 10 hours and 3 hours, respectively
- Current of electrons from the Microtron MT-25: 20μA
- Maximum energy of electrons: 25 MeV

All this gives us an Iodine activity of **50.9 ± 1.4mCi**.

5 Conclusions

The indirect way of producing ^{123}I via ^{123}Xe is free of the disadvantage of having the ^{124}I impurity within the sample, which means that we are able to get a high quality product [2]. Also, as we found in the literature [2]-[3], the Xe target can be used hundred of times, which makes cheaper the production costs. On the other hand, the diagnostic for thyroid requires an individual dose of 0.1 mCi, and of 1mCi for heart[2]. With the iodine activity we have obtained, we are able to treat more than one hundred patients from one production of this radioisotope. This is a good result that finally shows us the capability of producing ^{123}I in the Microtron MT-25.

Bibliography

[1] Nordell, et al. Production of ^{123}I by Photonuclear Reactions on Xenon. Int J. Appl. Radiat. Isot. Vol 33 (1982).

[2] Malinin, A. B., Kurenkov N. V.. Production of Radionuclides by Photonuclear Reactions. On the Possibility of Iodine-123 Production. Radiochem. Radioanal. Letters (1982).

[3] Vognar M., Simane C., Laboratory apparatus for preparation of ^{123}I from ^{124}Xe by photonuclear reaction. Faculty of Nuclear Sciences and Physical Engineering, Czech Technical University, Prague.

[4] Los Alamos National Laboratory, CCC-700. MCNP-4C, USA, 2001.

[5] Evans R., The Atomic Nuclei. USA, 1967.

Anais da XXVIII Reunião de Trabalho sobre Física Nuclear no Brasil
SP, Brasil, 2005
Artigo originalmente publicado no Brazilian Journal of Physics, Vol 36 – nº 4B,
Special Issue: XXVIII Workshop on Nuclear Physics in Brazil, pp.1391–1396 (2006).

The Color Flavor Locked Phase in the Chromodielectric Model and Quark Stars

L. P. Linares[1], M. Malheiro[1,2], A. R. Taurines[3] and M. Fiolhais[4]

[1] Instituto de Física, Universidade Federal Fluminense, Niterói 24210-340, RJ, Brazil

[2] Dep. de Física, Instituto Tecnológico de Aeronáutica, CTA,
12.228-900, São José dos Campos, Brazil

[3] Department of Physics, University of Wales Swansea,
Singleton Park, Swansea, SA2 8PP, United Kingdom

[4] Departamento de Física and Centro de Física Computacional,
Universidade de Coimbra, P-3004-516 Coimbra, Portugal

Abstract. Recent results obtained in the Chromodielectric Model (CDM) have shown that strange quark matter at very high densities may appear in two phases, namely a chiral broken and a chiral symmetric phase, which may not be absolutely stable. In the chiral symmetric phase, the abundance of the quarks flavors u, d and s is the same and there are no electrons. In this paper we study an extended version of the Chromodielectric model (CDM) with a BCS quark pairing implemented, and analyze the superconducting color flavor locked phase. We show that the inclusion in the free energy density of a negative term of the diquark condensate guarantees the stability of quark matter. We also analyze the phase transition between matter described by different equations of state and only find a first order transition, at a very low pressure, from the CFL phase to the unpaired strange quark matter, which opens the possibility for a quark-hadron phase transition. Our study has implications in astrophysics, in particular regarding the formation and the structure of compact quark stars. We explicitly show that CFL stars can be absolutely stable and more compact than strange stars.

1 Introduction

Twenty years ago, Bailin and Love [1, 2] conjectured that the attractive channel of the one gluon exchange of QCD could lead to quark pairs of equal momenta and opposite directions near the Fermi surface, in analogy with the electron pairs of the BCS theory [3] of superconductivity in solid state physics. For these pairs to be created (BCS ground state), an attractive interaction, even very small, is enough. The QCD attractive interaction of quarks close to the Fermi surface may lead to the formation of Cooper pairs which condense since they are bosons. The condensates generate gaps in the fermionic spectra which can be understood as the masses of these quasi-quarks. Since quark pairs in QCD cannot be color singlets, the color symmetry must be broken and this phenomenon is known as color superconductivity. In the case of three flavors and massless quarks (including the strange quark), a new phenomenon occurs when a condensate of quark pairs is formed: since the Cooper pairs cannot be flavor isosinglets, both the chiral symmetry and the color symmetry are broken. The condensate formed through the attractive channel of the one gluon exchange exhibits what it is known in literature as Color Flavour Locking (CFL) [4]. The rotation transformations in the flavor space $SU(3)_L$ are "locked" with the color rotations $SU(3)_{color}$, in the sense that the condensate is not invariant for none of these transformations separately, but it is invariant for the simultaneous transformations of the color-flavor $SU(3)_{L+color}$ (the same happens for the condensates formed by the right hand helicity spinores which are invariant under $SU(3)_{R+color}$ rotations). Thus, the chiral symmetry is broken in the CFL phase due to this new phenomenon and not because of the formation of any quark-antiquark condensate which, at these densities, is simply absent. In this work we are going to investigate, using the chromodielectric model (CDM) [5, 6, 7], high density strange quark matter in the CFL phase. We choose this model, where quarks interact with effective mesons fields, since its chiral symmetry is spontaneously broken, generating dynamical masses for quarks (a common feature with NJL models), and since it confines at low energies. The confinement results from an effective field, χ, that can be regarded as an integrated gluon field [8]. At low energies the model yields a good account of the nucleon phenomenology [9, 10]. Therefore, it is tempting to investigated in the same model whether the confinement information still remains at high densities where we expect the CFL quark matter to be formed.

Moreover, in the framework of the same model, a study of the unpaired three flavor quark matter in β equilibrium has been carried out on [11, 12] and problems with the stability of this strange matter at high densities have been found [13]. These problems are solved through the introduction of quark pairing. since the strange matter in the superconducting phase has a lower energy per particle than in the unpaired quark phase. Preliminary results have already been presented in [14, 15] and, in this work, we will add more results and also analyze the importance of the CFL phase as far as the structure of compact stars is concerned [16].

The CDM model will be used with a quartic potential. This version of the model yields two self-consistent equations of state (EOS) for strange quark matter in β equilibrium [11, 12]. One of them, for low densities, shows up a chiral symmetry breaking, a symmetry that is restored in the other one, for high densities. In this one, the Fermi momenta are the same for all three flavors and, therefore, there are no electrons in the neutral quark matter. Similarly, quark matter in the CFL phase also does not contain any electrons, but in the unpaired phase the gap energy is zero [11, 12].

We investigated the phase transition between the equations of state and concluded that, at intermediate densities, the color superconducting phase may undergo a transition to the strange unpaired quark matter, if pressure is low enough. However, at sufficiently high densities the transition does not occur. Hence, one should conclude that the CFL state is the true ground state of strange quark matter in the CDM model in that regime.

The superconducting color phase of quark matter may have interesting consequences in astrophysics. Inside a neutron star, densities ten times larger than the nuclear matter density can be reached, with chemical potentials of the order $\mu \sim 400$ to 500 MeV. This exceeds the strange quark mass and, for that density, the CFL phase is favored. The existence of a core in a neutron star made up of strange matter, or even the possibility that stars made up entirely of quark matter may exist, was also a motivation for our study. In the normal unpaired quark matter, it was shown that very small compact stars with radius less than 8 km are metastable [11, 12]. We will show by means of $M \times R$ diagrams obtained for CFL stars (pure quark stars with the quark matter in the superconducting phase), that it is possible to obtain well compact stars, with maximum mass $M/M_\odot = 1.43$ and radius $R \sim 7.5$ km, where the quark matter is stable ($\varepsilon < \rho M$). This result seems to indicate that pure quark stars, if they really exist in nature, must be in the CFL phase.

2 The CFL phase in the Chromodielectric Model

The Lagrangian density of the CDM can be written as [5, 6, 7]

$$\mathcal{L} = i\bar{\psi}\gamma^\mu\partial_\mu\psi + \frac{1}{2}(\partial_\mu\sigma\partial^\mu\sigma + \partial_\mu\vec{\pi}\cdot\partial^\mu\vec{\pi}) - W(\sigma, \vec{\pi})$$
$$+ \frac{g}{\chi}\bar{\psi}(\sigma + i\vec{\tau}\cdot\vec{\pi}\gamma_5)\psi + \frac{g_s}{\chi}\bar{\psi}_s\psi_s + \frac{1}{2}\partial_\mu\chi\partial^\mu\chi - U(\chi). \quad (1)$$

The first and second terms describe the quark and meson kinetic energies, respectively, and the third one the chiral meson self-interaction (Mexican hat potential for the scalar σ and the pseudoscalar π mesons):

$$W(\sigma, \vec{\pi}) = \frac{m_\sigma^2}{8f_\pi^2}(\sigma^2 + \pi^2 - f_\pi^2)^2, \quad (2)$$

where f_π is the pion decay constant.

The fourth and fifth terms in (1) describe the meson-quark interaction: the former refers only to the two light quark flavors (u and d) and the latter to the strange quark s, so the model is extended to the strange sector. The last two terms in the Lagrangian density (1) refer to the dynamical confining χ field, namely to its kinetic (sixth term) and potential, $U(\chi)$, energies. We use a CDM version with the following quartic potential for the χ field:

$$U(\chi) = \frac{1}{2} m_\chi^2 \chi^2 \times \left[1 + \left(\frac{8\eta^4}{\gamma^2} - 2 \right) \left(\frac{\chi}{\gamma m_\chi} \right) + \left(1 - \frac{6\eta^4}{\gamma^2} \right) \left(\frac{\chi}{\gamma m_\chi} \right)^2 \right] \quad (3)$$

where m_χ is the mass of the χ field. The potential has an absolute minimum at $\chi = 0$. The other two constants acquire a simple meaning, when the potential is written as in Eq. (3): γ defines the position of the second (local) minimum, located at $\chi = \gamma m_\chi$, and η the value of the potential at this minimum: $U(\gamma m_\chi) = (\eta m_\chi)^4$. The range for η in this work is the same as in [11, 13].

In the mean field approximation, where the meson fields are constant classical fields, the energy density for an homogeneous system of u, d and s quarks, interacting with the χ and σ fields is given by [11, 12, 13]:

$$\varepsilon = \alpha \sum_{i=u,d,s} \int_0^{k_i} \frac{d^3k}{(2\pi)^3} \sqrt{k^2 + m_i(\sigma, \chi)^2} + U(\chi) + \frac{m_\sigma^2}{8f_\pi^2} (\sigma^2 - f_\pi^2)^2 \quad (4)$$

(due to parity, the terms involving the pion field vanish in this approximation). The degeneracy factor is $\alpha = 2$ (spin) $\times$ 3 (color) $= 6$, and $f_\pi = 93$ MeV and $m_\sigma = 1.2$ GeV. Since we are dealing with an infinite and homogeneous system, the fermions are described by plain waves. The last two terms correspond to the χ and σ potential energies, respectively. The first term of Eq. (4) is the relativistic kinetic energy of a fermion gas, corresponding to the three quark flavors.

The Fermi momentum of each quark, k_i, is relate to the corresponding density, ρ_i, through:

$$\rho_i = \alpha \frac{k_i^3}{6\pi^2} = \frac{k_i^3}{\pi^2}, \quad (5)$$

where $i = u$, d and s.

The meson-quark interaction terms generate dynamical quark masses that are all different [18], namely

$$m_u(\sigma, \chi) = \frac{g_u \sigma}{\chi f_\pi}, \qquad m_d(\sigma, \chi) = \frac{g_d \sigma}{\chi f_\pi}, \qquad m_s(\chi) = \frac{g_s}{\chi}, \quad (6)$$

and the coupling constants for the u, d and s quarks are given, respectively, by

$$g_u = g(f_\pi + \xi_3), \qquad g_d = g(f_\pi - \xi_3), \qquad g_s = g(2f_K - f_\pi) \quad (7)$$

where $\xi_3 = -0.75$ MeV and the kaon decay constant $f_K = 113$ MeV. The parameters that better reproduce the nucleon properties in this model are the coupling constant, $g = 0.023$ GeV, the χ field mass, $m_\chi = 1.7$ GeV and $\gamma = 0.2$, which we keep fixed at this value as in [17, 11]. Finally, since we are interested in the study of strange quark matter in compact stars, we have to take in account the β equilibrium. Thus a relativistic free electron gas should be added [11, 13] to the energy density, Eq. (4).

We now extend the chromodieletric model presented so far in order to include the diquark condensate. The superconducting CFL phase appears when an attractive term responsible for the quark pairing is introduced. This term reads $-3(\frac{\Delta\mu}{\pi})^2$ and its effect is to decrease the free energy of the system. The gap energy, Δ, in principle, depends on the quark chemical potential, μ, in the range of neutron star densities (μ between 400 to 500 MeV), but we are going to take it as a constant [19]. This new term has not been derived in the CDM but rather in the framework of QCD, in the one gluon approximation. That interaction can be modelled by a four fermion interaction that is linearized, in the spirit of NJL models [19, 20]. In the CDM, a similar interaction could be generated through the introduction of an effective meson field that would represent the diquark pair. In this case, it would be possible to derive a self-consistent equation for the gap as a function of the density. Such a study is going to be carried on in a future work, but here, we will be working in the most simple BCS approximation, assuming the gap energy to be constant.

Moreover, as noted in [11, 12], the CDM model at high density predicts an EOS for the unpaired strange quark matter that is essentially the same as the one obtained by a perturbative QCD expansion [21]. Since, the CDM model incorporates the main QCD dynamics at high densities, we would expect this to be also the case for the diquark condensate, making it reasonable the introduction of the QCD pairing term in the CDM.

For systems at zero temperature, the grand-potential divided by the system volume is given by

$$\Omega = \varepsilon - \mu\rho \tag{8}$$

where ε is the energy density, μ is the chemical potential and ρ the particle density.

The grand-potential density of CDM in the superconducting color phase (CFL), is given by [14, 15]

$$\Omega_{\mathrm{CFL}} = \alpha \sum_{i=u,d,s} \int_0^{k_i} \frac{d^3k}{(2\pi)^3} \left(\sqrt{k^2 + m_i^2} - \mu_i \right) - 3 \left(\frac{\Delta\mu}{\pi} \right)^2 + U(\chi) + W(\sigma, \vec{\pi}) \tag{9}$$

where the quark masses m_i are given by Eqs. (6), and the meson potentials by Eqs. (2) and (3). Using spherical symmetry and $\alpha = 6$, Eq. (9) can be written as:

$$\Omega_{\mathrm{CFL}} = \varepsilon_k - \frac{1}{\pi^2} \sum_{i=u,d,s} \mu_i k_i^3 - 3 \left(\frac{\Delta\mu}{\pi} \right)^2 + U(\chi) + W(\sigma, \vec{\pi}) \tag{10}$$

where

$$\varepsilon_k = \frac{3}{\pi^2} \sum_{i=u,d,s} \int_0^{k_i} \left(k^2 \sqrt{(k^2 + m_i^2)} \right) dk \tag{11}$$

is the kinetic energy for a relativistic quark gas [22, 14, 15].

Inserting (10) into (8), we obtain the following quark matter energy density:

$$\varepsilon = \varepsilon_k - \frac{1}{\pi^2} \sum_{i=u,d,s} \mu_i k_i^3 - 3 \left(\frac{\Delta\mu}{\pi} \right)^2 + \sum_{i=u,d,s} \mu_i \rho_i + U(\chi) + W(\sigma, \vec{\pi}) \tag{12}$$

In the CFL phase, in order to guarantee the local charge neutrality [22], the Fermi momenta ($k_i = \sqrt{\mu_i^2 - m_i^2}$) of all quarks are equal, as explained previously:

$$k_f = k_u = k_d = k_s = 2\mu - \sqrt{\mu^2 + \frac{m_s^2}{3}} \tag{13}$$

where we define the quark chemical potential μ by

$$3\mu = \sum_{i=u,d,s} \mu_i \tag{14}$$

and admit that the u and d quark masses are very small [22].

When we derive the grand-potential, given by Eq. (10), to respect the chemical potential of each quark, μ_i, we obtain the same quark density for u, d and s if we take in account, as explained before, that all quarks have the same Fermi momentum in the CFL phase. Thus, if we use the definition of the average quark chemical potential, μ, expressed by Eq. (14), the baryonic density ρ is given by

$$\rho = \rho_u = \rho_d = \rho_s = \frac{1}{\pi^2} (k_f^3 + 2\Delta^2 \mu). \tag{15}$$

As we can see from this expression, the baryonic density in the superconducting phase also depends on the value of the pairing constant, Δ, and not only on the Fermi momentum, k_f. This result reflects the well known fact that BCS theory violates the particle number conservation since now there is a dependence on the number of diquark pairs. Thus, for a given density, the larger the gap, Δ, the smaller the Fermi momenta. As we will see, this will affect the contribution to the energy density of the quark kinetic energy term given in Eq. (12).

Inserting (13), (14) and (15) back into Eq. (12), one gets

$$\varepsilon = \varepsilon_k - \frac{3\mu k_f^3}{\pi^2} - 3 \left(\frac{\Delta\mu}{\pi} \right)^2 + U(\chi) + W(\sigma, \pi) + \frac{3\mu}{\pi^2} (k_f^3 + 2\Delta^2 \mu). \tag{16}$$

This expression can still be re-arranged so that the energy density of the CFL phase of strange quark matter can be cast in the form:

$$\varepsilon = \varepsilon_k + 3 \left(\frac{\Delta\mu}{\pi}\right)^2 + U(\chi) + W(\sigma, \pi), \tag{17}$$

where ε_k is given by Eq. (11).

3 Results

As for the strange quark matter without pairing [11, 12], the model accommodates two solutions: one for small values of χ (solution I) and the other one for large values of this field, near to the second minimum of the confining potential $\chi \sim \gamma m_\chi$ (solution II). Keeping the masses of the mesons and of the χ field as fixed parameters, the free parameters of the model are the coupling constant g, and the γ and η parameters of the χ potential energy. These two parameters matter for solution II since they determine the second minimum $U(\chi)$ potential (the other minimum is always at $\chi = 0$). However, the sensitivity of the EOS for solution II on the parameter γ is negligible, since all three quark masses are very small because they are determined by the inverse of the χ field, which is always large for solution II [11]. On the other hand, since the dependence of both EOS on the coupling constant g is only through the quark dynamical masses, which is almost zero in solution II, the EOS for this solution only depends on the parameter η which fixes the value of the potential energy at the local minimum. Moreover, for both solutions, the self-consistent scalar field is $\sigma \sim f_\pi$, so the contribution of the Mexican hat potential almost vanishes. The dependence of solution II on just one parameter explains why the results for this EOS in the superconducting phase are similar to those obtained using the MIT bag model [22]. Increasing η, which means increasing the potential energy at the local minimum, is equivalent to increase the bag constant, B, of the MIT bag model. So, as in [22], the CFL strange quark matter of the solution II is more stable than the unpaired one.

In previous works, the stability of the unpaired strange quark matter in β equilibrium [11] for solution II ($\varepsilon < \rho M$) was studied as a function of the model parameters. It was shown that, for large confining potential energies (large values of η), the quark matter was unstable. However, as it can seen from Eq. (10), the presence of the attractive pairing term in the grand-potential may balance the positive contribution coming from the confining potential energy, so the supercon-dutivity of quark matter turns out to be the way to solve the meta-stabilidade of the unpaired quark matter. From Eqs. (8) and (17), and taking into account that the grand-potential vanishes when the energy per particle has a minimum (zero pressure, $\mu = \varepsilon/\rho$), one readily obtains the relationship between the maximum value of η and the gap constant Δ, for which the CFL quark matter in the solution II (where the quarks are almost massless) is still stable ($\varepsilon/\rho = M$):

$$
\begin{aligned}
\eta_{max} &\sim \frac{M}{m_\chi} \left(\frac{1}{108\pi^2}\right)^{1/4} \left(1 + \frac{6\Delta^2}{M}\right)^{1/4} \\
&\sim 0.0966 \left(1 + \frac{6\Delta^2}{M}\right)^{1/4}.
\end{aligned}
\tag{18}
$$

This expression shows that, for unpaired quark matter ($\Delta = 0$), $\eta > 0.0966$ will make the quark matter unstable, corroborating what was obtained in Ref. [11]. For the CFL phase with a strong pairing interaction, large values of η are still allowed (larger potential energy) by the stability condition (18).

In Fig. 1-a, the energy per particle for the solution II is shown as a function of the baryonic density, for various pairing parameters. As we can observe, increasing of the pairing interaction Δ lowers the energy per particle, leading to a more robust quark matter stability. In the case of solution I ($U(\chi) \sim 0$), the strange matter is stable only for values of $\Delta \lesssim 65$ MeV. This happens because only for these values of Δ, the pressure vanishes producing a minimum in Fig. 1-b. For larger values of the pairing, the energy per particle does not show a minimum, and the matter collapses for low densities. In solution I, the values for the χ field are very small and the dynamical quark masses large. This collapse indicates that the superconducting phase of massive quarks is not favored.

After analyzing the effect of the strong pairing in the energy density, it is worth investigating its effect on the pressure for the two solutions. First, let us take $\eta = 0.0966$ and increase the other parameter, starting at $\Delta = 0$. In Fig. 2-a we present the EOS for solution II, while the EOS for solution I is shown in Fig. 2-b. In these two figures we can observe that, for a given energy density, increasing the pairing strength leads to a corresponding increase in the pressure. This result shows that the pairing interaction makes the EOS more stiff in the CFL phase.

In order to study the possible phase transitions between different EOS (paired and unpaired strange quark matter) we present, in Fig. 3 a plot of the pressure versus the quark chemical potential, μ. These curves indicate a phase transition from the CFL quark matter (solution II) to the unpaired quark matter in β equilibrium (solution I). We remind that the unpaired strange matter appears in the chiral broken phase with massive quarks [11]. This is a first order phase transition, indicating the existence of a mixing phase between these two phases. The system changes from the CFL phase, with a strong pairing interaction ($\Delta = 150$ MeV), at a density around 0.67 fm$^{-3} \sim 4.5\rho_0$ to a smaller density of around 0.41 fm$^{-3} \sim 3.0\rho_0$ ($\rho_0 = 0.15$ fm^{-3} is the normal nuclear matter density), with the quark matter in the unpaired phase and in β equilibrium. This result shows that, in the CDM model,

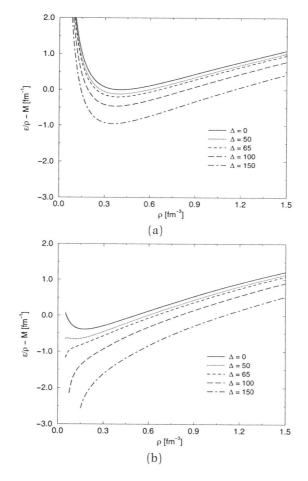

Figure 1 Energy per particle as a function of the baryonic density with $\eta = 0.0966$ and different Δ pairing interaction for the solution II (a) and for solution I (b).

there is a parameter range for which the CFL phase coexists with the chiral broken phase of unpaired strange quark matter. Since this transition occurs at a much too low pressure, even close to zero, it is possible that a quark-hadron transition also takes place. To investigate this possibility some hadronic EOS would be required. This is certainly a topic for future work; here we rather concentrated on the strange quark matter transitions.

One motivation for the present work, already pointed out in the Introduction, is to investigate whether bare quark stars in the CFL phase are absolutely stable, and how compact they are in comparison with strange stars made up of unpaired quark

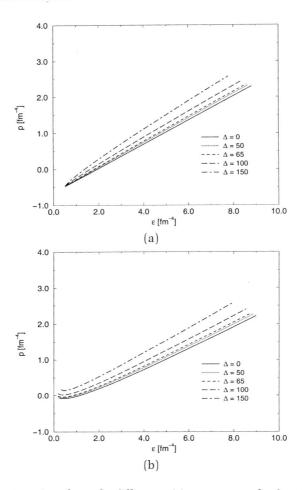

Figure 2 Equation of state for different pairing parameters for the solution I (a) and solution II (b).

matter in β equilibrium. To address this issue, we solved the Oppenheimer-Volkov equation for relativistic stars and obtained the mass *vs.* radius diagram for the bare quark stars, shown in Fig. 4. As the results show, keeping the pairing interaction fixed and increasing the potential energy (*i.e.* increasing η) we get CFL stars that are very compact and still stable in contrast with the strange stars made up of unpaired quark matter, which are metastable [11].

We present in Table 1 the results for the maximum mass, M of CFL strange stars (in units of the solar mass, $M_\odot$), the radius, R, and the central energy density, ε_c, for different values of η (all yielding a stable quark matter). As we can read from this table, relatively large maximum masses for CFL stars can be obtained.

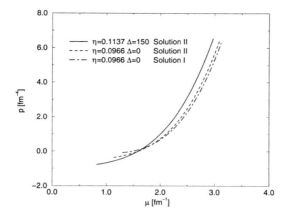

Figure 3 Pressure as a function of the quark chemical potential for different quark EOS as explained in the text.

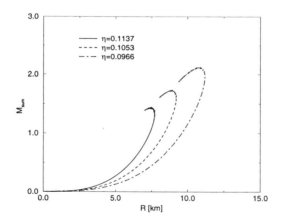

Figure 4 Mass - radius diagram with $\Delta = 150$ MeV and different values of η.

Most of the of neutrons stars have a mass $M \sim 1.4\, M_\odot$. It is interesting that in the maximum limit of quark matter stability (large η, for $\Delta = 150$ MeV) the maximum mass is around this value but with a much smaller radius – only $R \leqslant 8$ km – in comparison with the radius range from 10 to 14 km of a typical neutron star.

With the parameters considered in our model, the maximum limit for the potential energy and for the gap energy to have stable quark matter are $\eta = 0.1137$ and $\Delta = 150$ MeV respectively. In this case we get very compact CFL stars as shown in Fig. 4, which are stable, something that is not possible to obtain in the case of the unpaired quark matter without facing the problem of the metastability [11]. Thus, we can conclude that the color superconducting matter is stable and CFL stars can be rather small.

Table 1 Values of the maximum mass M, radius R and central density energy ε_c for the stable CFL star with different values of η and Δ.

η	Δ [MeV]	M [$M_\odot$]	R [km]	ε_c [fm^{-4}]
0.0966	0	1.56	8.48	9.62
	100	1.81	9.61	7.18
	150	2.11	10.81	5.72
0.1053	100	1.49	7.94	10.62
	150	1.72	8.89	8.46
0.1137	150	1.43	7.45	11.87

4 Conclusions

Through the introduction of a QCD inspired diquark interaction in the CDM model, we have studied the superconducting color flavor locked phase of strange quark matter at high densities, and compare with the previous results obtained in this model for the unpaired quark matter in β equilibrium. Our findings can be summarized as follow:

1. The CFL quark matter is more stable that the unpaired phase; when pairing is allowed, the range of the η parameter (which mimics the bag constant) is wider;

2. For standard CDM parameters, the CFL phase is absolutely stable at high densities, showing a first order phase transition to the β equilibrium unpaired quark phase at very small pressures;

3. CFL stars are stable and can be very compact with a mass similar to a neutron star mass but a smaller radius of the order of 8 km.

Acknowledgements

This work was supported by Capes and CNPq, Brazil, by FCT (POCTI/FEDER program), Portugal, and by CNPq/GRICES Brazilian-Portuguese scientific exchange program.

Bibliography

[1] D. Bailin and A. Love, Phys. Rept. **107**, 325 (1984)

[2] D. Bailin and A. Love, *Supersymmetric Gauge Field Theory and String Theory*, (IOP, London, 1994).

[3] J. Bardeen, L. N. Cooper and J. R. Schrieffer, Phys. Rev. **106**, 162 (1957); **108**, 1175 (1957).

[4] M. Alford, K. Rajagopal and F. Wilczek, Nucl. Phys. B **537**, 433 (1999).

[5] M. K. Banerjee, Prog. Part. Nucl. Phys. **31**, 77 (1993).

[6] M. C. Birse, Prog. Part. Nucl. Phys. **25**, 1 (1990).

[7] H. J. Pirner, Prog. Part. Nucl. Phys. **29**, 33 (1992).

[8] H. B. Nielsen and A. Patkós, Nucl. Phys. B **195**, 137 (1982).

[9] T. Neuber, M. Fiolhais, K. Goeke and J. N. Urbano, Nucl. Phys. A **560**, 909 (1993).

[10] A. Drago, M. Fiolhais and U. Tambini, Nucl. Phys. A **609**, 488 (1996).

[11] M. Malheiro, M. Fiolhais and A. R. Taurines, J. Phys. G **29**, 1045 (2003).

[12] M. Malheiro, E. O. Azevedo, L. G. Nuss, M. Fiolhais and A. R. Taurines, AIP Conference Proceedings **631** (1) 658 (2002), hep-ph/0111148

[13] M. Fiolhais, M. Malheiro and A. R. Taurines, *Stability of Quark Matter and Quark Stars*, Proceedings of the 4th international conference "New Worlds in Astroparticle Physics", Faro 2002, A. Krasnitz et al. (Eds.), World Scientific, Singapore, 123 (2003).

[14] M. Fiolhais, L. P. Linares, M. Malheiro and A. Taurines, Bled Workshops in Physics, **5** 18 (2004).

[15] L. P. Linares, M. Malheiro, M. Fiolhais and A. Taurines, AIP Conference Proceedings **739** 488 (2004).

[16] L. P. Linares *Color superconducting phase in the chromodielectric model and applications to quark stars*, Thesis (in Portuguese), Universidade Federal Fluminense, Niterói, 2005 (unpublished).

[17] A. Drago, M. Fiolhais and U. Tambini, Nucl. Phys. A **588**, 801 (1995).

[18] J. A. McGovern and M. Birse, Nucl. Phys. A **506**, 367 (1990); Nucl. Phys. **506**, 392 (1990).

[19] K. Rajagopal and F. Wilczek, *The Condensed Matter Physics of QCD*, Chapter 35 'At the Frontier of Particle Physics / Handbook of QCD', M. Shifman, ed., (World Scientific), vol. 3 2061-2151.(2001), hep-ph/0011333

[20] M. Alford, *Color Superconducting Quark Matter* Ann. Rev. Nucl. Part. Sci. **51**, 131-160 (2001), hep-ph/0102047

[21] E. S. Fraga, R. D. Pisarski and J. Schaffner-Bielich, Phys. Rev. D **63**, 121702 (2002).

[22] G. Lugones and J. E. Horvath, Phys. Rev. D **66**, 074017 (2002).

Anais da XXVIII Reunião de Trabalho sobre Física Nuclear no Brasil
SP, Brasil, 2005

Investigation of Elemental Distribution in Lung Samples by X-Ray Fluorescence Microtomography

Gabriela R. Pereira[1,], Ricardo T. Lopes[1], Henrique S. Rocha[1], Marcelino J. dos Anjos[2], Luis F. Oliveira[2] and Carlos A. Pérez*

[1] Laboratório de Instrumentação Nuclear - COPPE/UFRJ
C.P. 68509, Ilha do Fundão
21941-972 Rio de Janeiro, RJ, Brazil

[2] Instituto de Termodinâmica e Física Aplicada - UERJ - Rio de Janeiro, RJ, Brazil

[3] Laboratório Nacional de Luz Síncrotron - Campinas, SP, Brazil

Abstract. X-Ray Fluorescence Microtomography (XRFCT) is a suitable technique to find elemental distributions in heterogeneous samples. While X-ray transmission microtomography provides information about the linear attenuation coefficient distribution, XRFCT allows one to map the most important elements in the sample. The X-ray fluorescence tomography is based on the use of the X- ray fluorescence emitted from the elements contained in a sample so as to give additional information to characterize the object under study. In this work several tissues of rat lung and human lung have been investigated in order to verify the efficiency of the system for the determination of the internal distribution of detected elements in these kinds of samples and to compare the elemental distribution in the lung tissue of an old human and a fetus.The experiments were performed at the X-Ray Fluorescence beamline (XRF) of the Brazilian Synchrotron Light Source (LNLS), Campinas, Brazil. A white beam was used for the excitation of the elements and the fluorescence photons have been detected by a HPGe detector. The incident beam was monitored by an ionization chamber and a fast scintillator detector was used to detect the transmitted radiation. All the tomographies have been reconstructed using a filtered-back projection algorithm. In these samples the elements of higher concentration were Zn, Cu and Fe.

---

* gabriela@lin.ufrj.br

1 Introduction

Since tomographic techniques were developed, transmission tomography has been used in nondestructive testing for investigating the internal structure of samples [1] and with the advent of intense synchrotron radiation sources the resolution of the tomography was improved into μm regime [2]. Although the transmission tomography provides information about the distribution of the attenuation coefficients, the technique cannot give any information about elemental distribution.

Other complementary tomographic techniques have been developed based on the detection of the scattered [3] and fluorescence photons [4] in other to get some properties that also depend upon the distribution of individual elements within the sample. The X-ray fluorescence associated with tomographic techniques can supply important information of the sample chemical properties and produce high contrast in conditions where transmission tomography is not adjusted.

One of the drawbacks of fluorescence tomography is the reconstruction calculation that is more complex than transmission tomography's algorithm [5]. Hogan et al. proposed adapting one of the algorithms used in X-ray transmission tomography [6]. The simplest algorithm is based on the classical back projection algorithm used in transmission tomography. An algorithm more accurate applies corrections for absorption before and after the fluorescence point.

This paper presents the development of a system to study fluorescence microtomography at the X-Ray Fluorescence Facility (D09B-XRF) at the Brazilian Synchrotron Light Laboratory (LNLS), Campinas, Brazil, in order to study biological samples.

In this paper some tissues of rat lung and human lung have been investigated in order to verify the efficiency of the system for the determination of the internal distribution of detected elements in these kinds of samples and to compare the elemental distribution in the lung of an old human and a fetus.

2 Theory

X-ray transmission computed tomography is a well established diagnostic technique in Medicine. It is a non-destructive testing and allows reconstructing the image of the cross-sections of a sample starting from a set of X-ray "radiographic" measurements taken at different angles. It is performed by placing an X-ray source and an X-ray detector aligned at opposite sides of the sample. The detector records the part of the radiation emitted by the X-ray source that crosses the sample without interacting with it. The reconstructed image represents a map of the cross section in terms of linear attenuation coefficient of the material, which is a function of the mean atomic number (Z) and/or density in each single volume element (voxel) [7].

X-ray fluorescence tomography is based on detection of photons from fluorescent emission from the elements in the sample. These photons are acquired by an energy dispersive detector, placed at 90 degrees to the incident beam direction, and they are used as additional information for sample characterization [4].

For a particular element i and an atomic level v, the fluorescence radiation hitting the energy dispersive detector can be obtained through integration over y' [5,6],

$$I_{i,v}(\theta, x) = I_0 \int_{-\infty}^{\infty} f(\theta, x, y) g(\theta, x, y) p(x, y) \, dy'$$

Where I_0 is the intensity of the incident beam; $f(\theta, x, y)$ is the survival probability of a photon from the source to the point y', which is related to the sample absorption coefficient, μ_B, at the beam energy:

$$f(\theta, x, y) = \exp\left(-\int_{-\infty}^{y'} \mu_B(x, y'') \, dy''\right)$$

and $g(\theta, x, y)$ is the probability that a fluoresce photon emitted from y' reaches the detector:

$$g(\theta, x, y) = 1/4\pi \int_{\Omega_D} d\Omega \exp\left(-\int_{x,y'}^{Det} \mu_F(l) \, dl\right)$$

where μ_F is the absorption coefficient of the sample at the fluorescence energy and Ω_D is the solid angle from the fluorescence point to the detector surface. The function $p(x, y)$ is the probability that a fluorescence photon is produced from an incident photon along the path dy':

$$p(x, y) dy' \propto Nelem$$

where, the concentration of the element at the point (x, y) is Nelem. If the solid angle defined by the detector surface is almost constant and the attenuation is small then:

$$I_{i,v}(\theta, x) \propto I_0 \int_{-\infty}^{\infty} Nelem(x, y) \, dy'$$

That means that the concentration of the element is proportional to the experimental projections and the usual algorithms of transmission tomography can be used for fluorescence tomography. Otherwise, beam attenuation must be taken into account and new algorithms for tomographic reconstruction are indispensable [5,8]

3 Experiment

The experiments were performed at the X-Ray Fluorescence (XRF) beamline of the Brazilian Synchrotron Light Source (LNLS). A white beam (4-23 keV), collimated to a 100 x 100 µm² area with a set of slits, were used to sample excitation. The intensity of the incident beam was monitored with an ionization chamber placed in front of it, before the sample. A schematic of the experimental setup for an X-ray fluorescence microtomography measurement is shown in Figure 1.

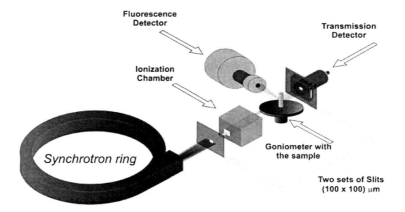

Figure 1 The experimental arrangement for an X-ray fluorescence microtomography measurement.

The sample was placed on a high precision goniometer and translation stages that allow rotating as well as translating it perpendicularly to the beam. The fluorescence photons were collected with an energy dispersive HPGe detector (CANBERRA Industries inc.) placed at 90 degrees to the incident beam, while transmitted photons were detected with a fast NaI (Tl) scintillation counter (CYBERSTAR-Oxford anfysik) placed behind the sample on the beam direction. This detector geometry allows reducing the elastic and Compton X-ray scattering from the sample due to the high linear polarization of the incoming beam in the plane of the storage ring, thus improving the signal to background ratio for the detection of trace elements [9].

A paper filter was first analyzed and used as a test sample. One piece of the filter was impregnated with a solution containing 500 ppm of zinc while the other remained untouched. Both pieces of paper filter were rolled up until it reached a cylindrical shape. The paper filter thickness was approximately 130 µm.

As a second part of the experiments, X-ray fluorescence microtomographies were performed on some tissues of rat lung as well as human lung. The rat tissues had 1 to 2 mm thickness and were stored in formaldehyde 10 percent and the human lung tissues had 1 to 2 mm too; however they were frozen dried before being analyzed.

The quality of the reconstruction is a compromise between the measuring time required for an acceptable counting statistic of the X-ray fluorescence peaks and the step size necessary to linearly move and rotate the samples. In one projection, samples were positioned in steps of 100μm (actual beam size) perpendicularly to the beam direction covering the whole transversal section of the sample proof. Each single value in a projection is obtained by measuring the fluorescent radiation emitted by all pixels along the beam [6]. The object is then rotated, and another projection is measured. Projections are obtained in steps of 3 degrees until the object has completed 180 degrees. The selected measuring time was 2s for each scanned point.

4 Results and Discussion

All the tomographies had been reconstructed using a filtered back projection algorithm and despite the self-absorption inside the samples, we have obtained good results. The X-ray fluorescence microtomography images have only qualitative character because of this self-absorption.

The X-ray transmission and X-ray fluorescence microtomography of the test sample is showed at the Figure 2. As already expected, while the transmission microtomography shows all layers of the filter, the fluorescence microtomography shows only those regions where the element of interest (Zn) was localized. These tomographies show the viability of fluorescence microtomography and confirm that this technique can be used to complement other techniques for sample characterization.

Figure 3 shows the tomographic images of a rat lung tissue sample. The transmission tomography has a poor quality owing to its low absorption coefficient.

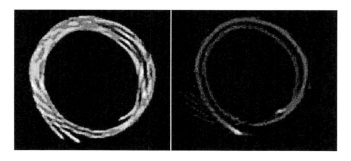

Figure 2 Tomography images of paper filter (left: X-ray transmission and right: X-ray fluorescence.

Figure 3 Tomography images of a rat lung tissue sample (a) x-ray transmission, (b) XRFCT Fe, (c) XRFCT Cu, (d) XRFCT Zn.

On the other hand, the x-ray fluorescence images show good constrast as can be observed in the elemental distribution of Zn, Cu and Fe.

Figure 4 and 5 show the reconstructed elemental maps of Fe, Cu and Zn in fetus and adult lung tissue sample respectively. The flaw that it can be visualized in the transmission mapping of adult lung sample is due to its low absorption coefficient. The fetus lung tissue sample has bigger attenuation coefficient than adult lung tissue.

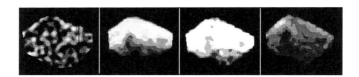

Figure 4 Tomography images of fetus human lung tissue sample (a) X-ray transmission, (b) XRFCT Fe, (c) XRFCT Cu, (d) XRFCT Zn.

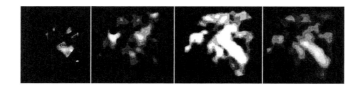

Figure 5 Tomography images of adult human lung tissue sample (a) X-ray transmission, (b) XRFCT Fe, (c) XRFCT Cu, (d) XRFCT Zn.

A special program was developed in our group to reconstruct the x-ray fluorescence tomography images putting all elements together. The results can be visualized in Figure 6, in which zinc, iron and cooper are represented in red, green and blue respectively. The images also show all possible combinations for these elements in adult human lung tissue sample. It can be seen that some regions have greater concentration of an element due to the distinction of the colors. In the combination iron plus zinc for the sample of human adult lung tissue sample can

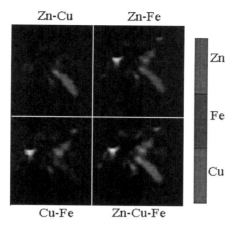

Figure 6 X-Ray Fluorescence Tomography images of adult human lung sample together.

be observed that a region with great zinc concentration exists marked with the red color.

The graphics of normalized Zn, Cu and Fe fluorescence counting in 10th projections can be visualized in the Figure 7, 8 and 9, respectively. It can be observer that the amount of these elements, despite the absorption, are different for different samples as adult lung and fetus lung tissues samples. The fetus lung sample has bigger amount of these elements than the adult lung tissue sample. The ratio of the quantity of these elements in fetus cells and in adult tissue is perhaps bigger than we can see in the graphics because the attenuation coefficient of fetus cells is greater than adult tissues, for the fluorescence energy.

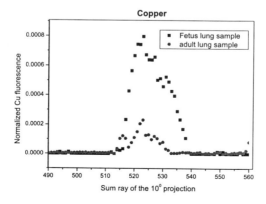

Figure 7 The graphic of normalized Cu fluorescence counting in 10th projection.

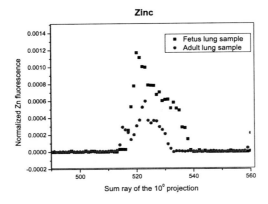

Figure 8 The graphic of normalized Zn fluorescence counting in 10th projection.

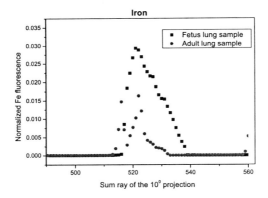

Figure 9 The graphic of normalized Fe fluorescence counting in 10th projection.

5 Conclusion

We have shown that it was possible to visualize the distribution of high atomic number elements on both, artificial and tissues samples. It was possible to compare the quantity of Zn, Cu and Fe for the lung human tissue sample and verify that these elements have a higher concentration on fetus tissues than adult tissues but it is necessary to analyze more samples to get a better statistical value. The spatial resolution of the system can be optimized in function of the application. It is necessary to correct the absorption to get the concentration of these elements in the sample and to correct the fluorescence mapping. The experimental set up at XRF-LNLS have showed to be very promising and this effort on implement the x-ray fluorescence microtomography was justified by the high quality of the images obtained.

Acknowledgements

This work was partially supported by Conselho Nacional de Desenvolvimento Científico e Tecnológico (CNPq), Fundação de Amparo a Pesquisa do Estado do Rio de Janeiro (FAPERJ) and performed at Laboratório Nacional de Luz Síncrotron (LNLS).

Bibliography

[1] D. Braz, L.M.G. da Motta, R.T.Lopes, Applied Radiation and Isotopes,the Netherlands, v. 1, n. 50, pp. 661-671, (1999).

[2] R. T. Lopes, et al, Nuclear Instruments and Methods in Physics Research, 505, pp. 604-607, (2003).

[3] R.C.Barroso, et al, Nuclear Instruments and Methods in Physics Research A, 471, pp.75-79, (1998).

[4] R. Cesareo, S. Mascarenhas, Nuclear Instruments and Methods in Physics Research A, 277, pp. 669-672, (1989).

[5] A. Brunetti , B. Golosio, Computer Physics Communications, 141, pp.412-425, (2001).

[6] J.P. Hogan , R.A. Gonsalves, A.S.Krieger, IEEE Transactions on Nuclear Science, NS v. 38, pp.1721-1727, (1991).

[7] R. Cesareo, et al, Nuclear Instruments and Methods in Physics Research A, 525, pp. 336-341, (2004).

[8] B. Golosio, et al, Journal of Applied Physics, 94, pp.145-156, (2003).

[9] M. Naghedolfeizi, et al, Journal of Nuclear Materials, 312, pp.146-155, (2003).

Anais da XXVIII Reunião de Trabalho sobre Física Nuclear no Brasil
SP, Brasil, 2005
Artigo originalmente publicado no Brazilian Journal of Physics, Vol 36 – n⁰ 4B,
Special Issue: XXVIII Workshop on Nuclear Physics in Brazil, pp.1397–1409 (2006).

Rise and Fall of Pentaquarks in the QCD Sum Rules Approach

R. D. Matheus, F. S. Navarra and M. Nielsen

Instituto de Física, Universidade de São Paulo
C.P. 66318, 05315-970 São Paulo, SP, Brazil

Abstract. We review the existing mass and decay width determinations of pentaquarks with QCD Sum Rules (QCDSR). We give special attention to the intermediate assumptions and choices which we are obliged to do in this approach. As an example, we present the full calculation of the pentaquark mass with Borel sum rules and also with Finite Energy Sum Rules (FESR). We also work out the calculation of the Θ decay width. We take the opportunity to comment our publications on this subject and include new and unpublished material.

1 Introduction

From june 2003 to august 2005 there was an intense theoretical activity devoted to understand the structure of a new family of baryons: the pentaquarks. The interest for this subject was triggered by the announcement made by the LEPS collaboration [1] reporting the observation of the $\Theta^+(1540)$. Since then, a flurry of theoretical [2] and experimental [3] papers brought the pentaquarks to the main stage of hadron physics.

The subsequent experimental searches, carried out in dozens of different machines with different energies, different beam and targets and different detection methods turned out to be inconclusive, half of the experiments observing the $\Theta^+(1540)$ and the other half not observing it. This puzzling situation began to change when, in the beginning of 2005 some groups which had previously found the pentaquark state could not see it anymore in a second round of much more careful experiments [4, 5]. Very fastly the community started not to believe in the existence of the pentaquark. In august 2005, a statement made by Ted Barnes at the HADRON05 meeting, in Rio de Janeiro, reflected this new consensus [6]. He said: "the pentaquark is dead !". Much experimental work on the subject is still needed, not only to confirm the negative result, but also to understand what was wrong before. In spite of some isolated claims that the state really exists, today it seems very likely that the final conclusion will give support to the "death sentence" pronounced by Barnes. As for explaining the previous positive results, some steps along this direction have been taken by Meier, Dzierba and Szczepaniak, who tried to demonstrate that the signal observed by the CLAS collaboration at JLAB is a consequence, among other things, of kinematical reflections [7].

On the theoretical side the situation became equally confusing. The method able to give the best answers, lattice QCD, was used by several groups to look for the pentaquark [8, 9, 10]. These "theoretical searches", in the same way as the initial experimental ones, turned out to be inconclusive, half of the groups finding a signal and half not finding it. Quark models were also extensively used in this context but they may give answer to different questions. Due to their very nature, they have to assume that a particle exists and then they may predict, for example, its spatial and color configuration.

Between lattice QCD and quark models, in an intermediate position in the theoretical scenario, we have QCD sum rules (QCDSR) [11, 12, 13], which, in principle are pure theory. However, due to the option of keeping the calculation analytic and avoiding the "brute force" numerical work; due to the option of trading a simulation of the QCD vacuum by the phenomenological parameters known as condensates, the results start to depend on certain assumptions (like, for example, the factorization hypothesis for higher dimension condensates) which are made during the calculations. The first works on $\Theta^+(1540)$ with QCDSR [14, 15, 16, 17] addressed the mass of the state and could all obtain a reasonable value for the mass. Later, a more careful analysis [18, 19, 20] revealed some problems with the previous calculations. In the mean time other pentaquarks were observed: the Ξ^{--} and the Θ_c. These were also studied with QCDSR [18, 21]. This time, with more rigorous criteria it was more difficult to reproduce the experimental data, i.e., the masses of the states (especially the Ξ^{--}). Finally, attempts to describe the extremely narrow decay width pushed the method to the limit and the conclusion was that it is very difficult to understand this decay in QCDSR [22].

If, in the near future, the non-existence of pentaquarks is confirmed, the community might address to the QCDSR practitioners the following justified and embarrassing question: "how could you so nicely calculate the correct mass of something that does not exist?"

In this work we present a critical review of these QCDSR calculations commenting their strong and weak points. In reviewing these topics we present new material, which was never published before.

The text is organized as follows. In the next section we briefly mention some interesting facts and main ideas about pentaquarks. In section 3 we present the main formulas of QCDSR, show one complete calculation of a pentaquark mass with all the inputs and assumptions. In section 4 we enumerate the requirements that must be satisfied for a QCDSR calculation to be reliable and check whether this is case for the results discussed in the preceding section. In section 5 we introduce the finite energy sum rules (FESR) and present the results for m_Θ and m_Ξ obtained with this method. In section 6 we move to the pentaquark decay width. Finally, in section 7 we present our conclusions.

2 The Pentaquark Structure

Probably the most interesting physical question related to pentaquarks is: how are these five quarks organized? The exotic baryon with the K^+n quantum numbers, the $\Theta^+(1540)$, first observed in [1], could not be a three quark state and its minimal quark content had to be $uudd\bar{s}$. These quarks could be in one of the following configurations: a) uncorrelated quarks inside a bag [23]; b) a $K-N$ molecule bound by a van-der Waals force [24]; c) a "$K-N$" bound state in which uud and $u\bar{s}$ are not separately in color singlet states [14]; d) a diquark-triquark $(ud)-(ud\bar{s})$ bound state [25] and e) a diquark-diquark-antiquark state [26]. This last possibility has been by far the most explored, also in the QCD sum rules framework.

According to Jaffe and Wilczek (JW), each diquark should have spin zero and should be in the $\bar{\mathbf{3}}$ representation of SU(3), in color and flavor. The two diquark would then combine in a P-wave orbital angular momentum to form a $\mathbf{3}$ state in color, spin $S = 0$, and $\bar{\mathbf{6}}$ in flavor. The resulting state would then be combined with the antiquark to form a flavor antidecuplet $\overline{\mathbf{10}}_f$ and octet $\mathbf{8}_f$, with spin $S = 1/2$. The Θ^+ would be at the top of the antidecuplet $\overline{\mathbf{10}}_f$ and would have an isospin $I = 0$. JW have also interpreted the lightest particle in the octet $\mathbf{8}_f$, $[ud]^2\bar{d}$, as the Roper resonance, since it has the same quantum numbers of the nucleon. The Roper resonance would be thus identical to the Θ^+ except for the substitution of the strange antiquark by a down antiquark. This would explain why the mass difference between $\Theta^+(1540)$ and $N(1440)$ is so close to the strange quark mass.

Understanding the organization of matter at the quark level is of extreme importance. This can be done with the help of lattice QCD and, to some extent, with QCDSR. In the following section we describe the calculation of a pentaquark mass with QCDSR.

3 Correlation Function, Currents and Masses

The purpose of this section is mainly to show the QCDSR machinery at work, giving emphasis to the aspects which may be potential sources of uncertainties.

In the QCDSR approach [12, 13], the short range perturbative QCD is extended by an operator product expansion (OPE) of the correlators, which results in a series in powers of the squared momentum with Wilson coefficients. The convergence at low momentum is improved by using a Borel transform. The expansion involves universal quark and gluon condensates. The quark-based calculation of a given correlator is equated to the same correlator, calculated using hadronic degrees of freedom via a dispersion relation, providing sum rules from which a hadronic quantity can be estimated. The QCDSR calculation of hadronic masses centers around the two-point correlation function given by

$$\Pi(q) \equiv i \int d^4x \, e^{iq \cdot x} < 0|T\eta(x)\bar{\eta}(0)|0 > \tag{1}$$

where $\eta(x)$ is an interpolating field (a current) with the quantum numbers of the hadron we want to study.

In the next subsections we discuss the evaluation of (1) for the cases of the Ξ^{--} and of the Θ^+.

The basic ingredients in the particle mass determinations from QCD spectral sum rules [12, 11] as well as from lattice QCD calculations are the interpolating currents used to describe the particle states. Contrary to the ordinary mesons, where the form of the current is unique, there are different choices of the pentaquark currents in the literature. We shall list below some possible operators describing the isoscalar $I = 0$ and $J = 1/2$ channel which would correspond to the experimental candidate $\Theta(1540)$.

3.1 The $\Theta(1540)$ Currents

Defining the pseudoscalar (ps) and scalar (s) diquark interpolating fields as:

$$\begin{aligned} Q_{ab}^{ps}(x) &= \left[u_a^T(x)Cd_b(x)\right] , \\ Q_{ab}^{s}(x) &= \left[u_a^T(x)C\gamma_5 d_b(x)\right] , \end{aligned} \tag{2}$$

where a, b, c are color indices and C denotes the charge conjugation matrix, the lowest dimension current built by two diquarks and one anti-quark describing the Θ as a $I = 0$, $J^P = 1/2^+$ S-wave resonance is [16]:

$$\eta_I^\Theta = \epsilon^{abc}\epsilon^{def}\epsilon^{cfg}Q_{ab}^{ps}Q_{de}^{s}C\bar{s}_g^T , \tag{3}$$

and the one with one diquark and three quarks is [14]:

$$\eta_{II}^\Theta = \frac{1}{\sqrt{2}}\epsilon^{abc}Q_{ab}^{s}\{u_e\bar{s}_e i\gamma_5 d_c - (u \leftrightarrow d)\} . \tag{4}$$

This later choice can be interesting if the instanton repulsive force arguments [27] against the existence of a pseudoscalar diquark bound state apply. Alternatively, a description of the Θ as a $I = 0$, $J^P = 1/2^+$ P-wave resonance has been proposed by [26] and used by [17] in the sum rule analysis:

$$\eta_{III}^\Theta = \left(\epsilon^{abd}\delta^{ce} + \epsilon^{abc}\delta^{de}\right)[Q_{ab}^s(D^\mu Q_{cd}^s) - (D^\mu Q_{ab}^s)Q_{cd}^s]\gamma_5\gamma_\mu C\bar{s}_e^T. \quad (5)$$

This current can be generalized by considering its mixing with the following one having the same dimension and quantum numbers:

$$\eta_{IV}^\Theta = \epsilon^{abc}\epsilon^{def}\epsilon^{cfg}Q_{ab}^{ps}Q_{de}^s\gamma_\mu(D^\mu C\bar{s}_g^T). \quad (6)$$

The evaluation of (1) with (6) revealed that [19], to leading order in α_s and in the chiral limit $m_q \to 0$, the contribution to the correlator vanishes. This result justifies a posteriori the *unique choice* of operator for the P-wave state used in [17].

3.2 The $\Xi(1862)$ Currents

Following the diquark-diquark-antiquark scheme, we can write two independent interpolating fields with the quantum numbers of Ξ^{--} ($I = 3/2$, $J^P = 1/2^-$):

$$\eta_1(x) = \frac{1}{\sqrt{2}}\epsilon_{abc}(d_a^T(x)C\gamma_5 s_b(x))[d_c^T(x)C\gamma_5 s_e(x) + s_e^T(x)C\gamma_5 d_c(x)]C\bar{u}_e^T(x)$$
$$\eta_2(x) = \frac{1}{\sqrt{2}}\epsilon_{abc}(d_a^T(x)Cs_b(x))[d_c^T(x)Cs_e(x) + s_e^T(x)Cd_c(x)]C\bar{u}_e^T(x) \quad (7)$$

where a, b, c and e are color indices and $C = -C^T$ is the charge conjugation operator.

As in the nucleon case, where one also has two independent currents with the nucleon quantum numbers [28, 29], the most general current for Ξ^{--} is a linear combination of the currents given above:

$$\eta_I^\Xi(x) = [t\eta_1(x) + \eta_2(x)], \quad (8)$$

with t being an arbitrary parameter. In the case of the nucleon, the interpolating field with $t = -1$ is known as Ioffe's current [28]. With this choice for t, this current maximizes the overlap with the nucleon as compared with the excited states, and minimizes the contribution of higher dimension condensates. In the present case it is not clear a priori which is the best choice for t.

We also employ the following interpolating field operator for the pentaquark Ξ^{--} [16, 10]:

$$\eta_{II}^\Xi(x) = \epsilon^{abc}\epsilon^{def}\epsilon^{cfg}s_a^T(x)Cd_b(x) \times s_d^T(x)C\gamma_5 d_e(x)C\bar{u}_g^T(x) \quad (9)$$

It is easy to confirm that this operator produces a baryon with $J = 1/2$, $I = 3/2$ and strangeness -2. The parts, $S^c(x) = \epsilon^{abc}s_a^T(x)C\gamma_5 d_b(x)$ and $P^c(x) =$

$\epsilon^{abc}s_a^T(x)Cd_b(x)$, give the scalar S (0^+) and the pseudoscalar P (0^-) sd diquarks, respectively. They both belong to the anti-triplet representation of the color SU(3). The scalar diquark corresponds to the $I = 1/2$ sd diquark with zero angular momentum. It is known that a gluon exchange force as well as the instanton mediated force commonly used in the quark model spectroscopy give significant attraction between the quarks in this channel.

3.3 The Ξ Mass

Inserting Eq. (8) into the integrand of Eq. (1) we obtain

$$< 0|T\eta(x)\bar{\eta}(0)|0 > = t^2\Pi_{11}(x) + t\Pi_{12}(x) + t\Pi_{21}(x) + \Pi_{22}(x) \tag{10}$$

Calling $\Gamma_1 = \gamma_5$ and $\Gamma_2 = 1$ we get

$$\begin{aligned}
\Pi_{ij}(x) &= < 0|T\eta_i(x)\bar{\eta}_j(0)|0 > \\
&= \epsilon_{abc}\epsilon_{a'b'c'}CS_{e'e}{}^T(-x)C \left\{ - \text{Tr} \left[\Gamma_i S_{bb'}^s(x)\Gamma_j CS_{aa'}^T(x)C\right] \right. \\
&\quad \times \left. \text{Tr} \left[\Gamma_i S_{ee'}^s(x)\Gamma_j CS_{cc'}^T(x)C\right] + \cdots \right.
\end{aligned} \tag{11}$$

where $S_{ab}(x)$ and $S_{ab}^s(x)$ are the light and strange quark propagators respectively. The above expression was included just to show the ingredients of the calculation. The full version of (11) can be found in [18].

In order to evaluate the correlation function $\Pi(q)$ at the quark level, we first need to determine the quark propagator in the presence of quark and gluon condensates. Keeping track of the terms linear in the quark mass and taking into account quark and gluon condensates, we get [30]

$$\begin{aligned}
S_{ab}(x) &= \langle 0|T[q_a(x)\bar{q}_b(0)]|0\rangle \\
&= \frac{i\delta_{ab}}{2\pi^2 x^4}\slashed{x} - \frac{m_q\delta_{ab}}{4\pi^2 x^2} - \frac{i}{32\pi^2 x^2}t_{ab}^A g_s G_{\mu\nu}^A(\slashed{x}\sigma^{\mu\nu} + \sigma^{\mu\nu}\slashed{x}) - \frac{\delta_{ab}}{12}\langle\bar{q}q\rangle \\
&\quad - \frac{m_q}{32\pi^2}t_{ab}^A g_s G_{\mu\nu}^A \sigma^{\mu\nu}\ln(-x^2) + \frac{i\delta_{ab}}{48}m_q\langle\bar{q}q\rangle\slashed{x} - \frac{x^2\delta_{ab}}{2^6 \times 3}\langle g_s\bar{q}\sigma\cdot\mathcal{G}q\rangle \\
&\quad + \frac{ix^2\delta_{ab}}{2^7 \times 3^2}m_q\langle g_s\bar{q}\sigma\cdot\mathcal{G}q\rangle\slashed{x} - \frac{x^4\delta_{ab}}{2^{10} \times 3^3}\langle\bar{q}q\rangle\langle g_s^2 G^2\rangle \quad (12)
\end{aligned}$$

where we have used the factorization approximation for the multi-quark condensates, and we have used the fixed-point gauge [30].

Inserting (12) into (11) we obtain a set of diagrams which contribute to the OPE side of the correlation function.

Lorentz covariance, parity and time reversal imply that the two-point correlation function in Eq. (1) has the form

$$\Pi(q) = \Pi_1(q^2) + \Pi_q(q^2)\slashed{q} \; . \tag{13}$$

A sum rule for each scalar invariant function Π_1 and Π_q, can be obtained. In [18] the chirality even structure $\Pi_q(q^2)$ was used to obtain the final results.

The phenomenological side is described, as usual, as a sum of pole and continuum, the latter being approximated by the OPE spectral density. In order to suppress the condensates of higher dimension and at the same time reduce the influence of higher resonances we perform a standard Borel transform [12]:

$$\Pi(M^2) \equiv \lim_{n,Q^2 \to \infty} \frac{1}{n!} (Q^2)^{n+1} \left(-\frac{d}{dQ^2} \right)^n \Pi(Q^2) \qquad (14)$$

$(Q^2 = -q^2)$ with the squared Borel mass scale $M^2 = Q^2/n$ kept fixed in the limit.

For current II we repeat the steps mentioned above, substituting (9) into (1), making use of the expansion (12), picking up the terms multiplying the structure $\not{q}$ and finally performing a Borel transform (14).

After Borel transforming each side of $\Pi_q(Q^2)$ and transferring the continuum contribution to the OPE side we obtain the following sum rule at order m_s:

$$\lambda_{\mp}^2 e^{-\frac{m_{\mp}^2}{M^2}} = \int_0^{s_0} e^{-\frac{s}{M^2}} \rho_i^q(s) ds. \qquad (15)$$

where $i(=\text{I, II})$ refers to the current employed and the spectral densities, up to order 6 are given by:

$$\rho_I^q(s) = c_1 \frac{s^5}{5!5!2^{12}7\pi^8} + c_3 \frac{s^3}{5!2^{10}\pi^6} m_s <\bar{s}s>$$
$$- c_4 \frac{s^3}{5!2^8\pi^6} m_s <\bar{q}q> + c_2 \frac{s^3}{5!3!2^{13}\pi^6} <\frac{\alpha_s}{\pi}G^2>$$
$$+ 7c_4 \frac{s^2}{2^{16}\pi^6} m_s <\bar{q}g_s\mathbf{ff.Gq}> + c_2 \frac{s^2}{3^2 2^{11}\pi^4} (<\bar{s}s>^2 + <\bar{q}q>^2)$$
$$+ c_4 \frac{s^2}{3!2^8\pi^4} <\bar{s}s><\bar{q}q> - c_5 \frac{s^2}{4!3!2^{11}\pi^6} m_s <\bar{s}g_s\mathbf{ff.Gs}>$$
$$- 3c_4 \frac{s^2}{4!3!2^{12}\pi^6} m_s <\bar{q}g_s\mathbf{ff.Gq}> \times \left(6\ln(\frac{s}{\Lambda_{QCD}^2}) - \frac{43}{2} \right), \qquad (16)$$

with $c_1 = 5t^2 + 2t + 5$, $c_2 = (1-t)^2$, $c_3 = (t+1)^2$, $c_4 = t^2 - 1$, $c_5 = t^2 + 22t + 1$, $\Lambda_{QCD} = 110$ MeV and

$$\rho_{II}^q(s) = \frac{s^5}{5!5!2^{10}7\pi^8} + \frac{s^3}{5!3!2^7\pi^6} m_s <\bar{s}s>$$
$$+ \frac{s^3}{5!3!2^{10}\pi^6} <\frac{\alpha_s}{\pi}G^2> + \frac{s^2}{4!3!2^9\pi^6} m_s <\bar{s}g_s\mathbf{ff.Gs}> . \qquad (17)$$

To extract the Ξ^{--} mass, m_Ξ, we take the derivative of Eq. (15) with respect to M^{-2} and divide it by Eq. (15). Repeating the same steps leading to (15), (16) and (17) for the chirality odd structure $\Pi_1(q^2)$ we arrive at

$$\lambda_\Xi^2\, m_\Xi\, e^{-\frac{m_\Xi^2}{M^2}} = \int_0^{s_0} e^{-\frac{s}{M^2}} \rho_i^1(s)\,ds. \tag{18}$$

where

$$\rho_I^1(s) = -c_1 \frac{s^4}{5!4!2^{10}\pi^6} < \bar{q}q > \; + c_1 \frac{s^3}{4!3!2^{12}\pi^6} < \bar{q}g_s\mathbf{ff.Gq} > \tag{19}$$

and

$$\rho_{II}^1(s) = -\frac{s^4}{5!4!2^7\pi^6} < \bar{q}q > \; + \frac{s^3}{4!3!2^9\pi^6} < \bar{q}g_s\mathbf{ff.Gq} > \tag{20}$$

In the numerical analysis of the sum rules, the values used for the condensates are: $\langle\bar{q}q\rangle = -(0.23)^3$ GeV^3, $\langle\bar{s}s\rangle = 0.8\langle\bar{q}q\rangle$, $< \bar{s}g_s\mathbf{ff.Gs} >= m_0^2\langle\bar{s}s\rangle$ with $m_0^2 = 0.8$ GeV^2 and $\langle g_s^2 G^2\rangle = 0.5$ GeV^4. The gluon condensate has a large error of about a factor 2, but its influence on the analysis is relatively small. We define the continuum threshold as:

$$s_0 = (1.86 + \Delta)^2 \;\; \text{GeV}^2 \tag{21}$$

In the complete theory, the mass extracted from the sum rule should be independent of the Borel mass M^2. However, in a truncated treatment there will always be some dependence left. Therefore, one has to work in a region where the approximations made are acceptable and where the result depends only moderately on the Borel variables.

A comparison between results obtained with different currents is more meaningful when they describe the same physical state, i.e., those with the same quantum numbers. Concerning spin, all currents considered in our work have the same spin ($= 1/2$). Concerning the parity, the situation is more complicated. In QCD sum rules, when we construct the current, it has a definite parity. Current (8) has parity $P = -1$ and current (9) has parity $P = +1$. However, currents can couple to physical states of different parities. As well discussed in [21], in order to know the parity of the state in QCDSR, we have to analyze the chiral-odd sum rule. If the r.h.s of this sum rule (containing the spectral density coming from QCD) is positive, then the parity of the corresponding physical state is the same as the parity of the current. If it is negative the parity is the opposite of the parity of the current. Performing this analysis we might determine in both cases the parity of the state. However it turns out that for both currents, in the chiral-odd sum rule the OPE does not have good convergence, i.e., terms containing higher order operators are not suppressed with respect to the lowest order ones. This sum rule is thus ill defined and nothing can be said about the parity of the state. The comparative study of the currents is still valid because we can compare other properties of these currents. A possible outcome of this study might be that one current has defects, which are so severe that we are forced to abandon it. If this turns out to be case, the determination of the parity of the associated states becomes irrelevant.

4 Reliability of a QCD Sum Rule Calculation

Having presented the main formulas in the last section, the next step is be to introduce numerical values for the masses and condensates, choose reasonable values for the free (or partially constrained) parameters, which are the continuum threshold, the Borel mass at which the mass sum rule is evaluated and, in the case of current (8), the value of t. As a result we obtain values for m_Ξ. In doing these calculations we must remember that *it is not enough to obtain a pentaquark mass consistent with the experimental number*. There is a list of requirements that must be fulfilled:

i the physical observables, such as masses and coupling constants, must be approximately independent of the Borel mass (this is the so called Borel stability).

ii the right hand side (RHS) of the sum rules (15) and (18) must be positive, since the left hand side (LHS) is manifestly positive.

iii the operator product expansion (OPE) must be convergent, i.e., the terms appearing in (16) and (17) must decrease with the increasing order of the operator.

iv the pole contribution must be dominant, i.e., the integral in (15) and in (18) must be at least 50 % of the integral over the complete domain of invariant masses $(s_0 \rightarrow \infty)$.

v the threshold parameter s_0 must be compatible with the energy corresponding to the first excitation of an usual baryon.

4.1 Current (8)

After an extensive search for best values of parameters and optimal Borel window, we have realized that it is extremely difficult to satisfy simultanously the conditions i)-v) given above. In particular, when we have a very good OPE convergence, the strength of the pole is very weak and vice versa. We have to look for a compromise.

In order to illustrate these results, we consider $m_s = 0.10$ GeV and $\Delta = 0.44$ GeV, and contruct, Figures 1, 2 and 3, showing the Borel mass dependence of m_Ξ, of the OPE terms (in absolute value) and of the percentage of the pole contribution respectively. For these choices the value of the current-state overlap is:

$$\lambda_I^q \simeq 5.4 \times 10^{-9} \ \text{GeV}^{13} \tag{22}$$

Restricting ourselves to the parameter combinations which satisfy the requirements i) - v) and taking the average we obtain:

$$m_\Xi = 1.85 \pm 0.05 \ \text{GeV} \tag{23}$$

We have also used a current composed by scalar diquarks only, i.e., η_1. The motivation for studying this current is to verify if it gives a smaller mass for the pentaquark than those obtained with other currents. According to the instanton description of diquark dynamics, this should be the case. We observe that, for same choices for m_s and Δ, the masses found with scalar diquark currents are only slightly smaller than the others. This means either that instanton dynamics was not captured by our choice of currents and diagrams or that the interaction between the pseudoscalar diquarks (included in the mixed currents) is more attractive than expected.

4.2 Current (9)

Using the same numerical inputs quoted in the last subsection we evaluate now the sum rules obtained with current(9). The same comments made in the previous subsection apply here. Choosing $m_s = 0.10$ GeV and $\Delta = 0.24$ GeV, we present in Figures 4, 5 and 6 the Borel mass dependence of m_Ξ, of the OPE terms and of the percentage of the pole respectively. As it can be seen, with current(9) we tend to overestimate m_Ξ, unless very low threshold parameters or quark masses are used. Besides, the pole contribution is always smaller than 40 %. On the other hand, the Borel stability seen in Fig. 4 is remarkable. This suggests that we could choose a lower value for the Borel mass, thereby increasing the pole contribution without significantly changing m_Ξ. The average over the best parameter choices leads to:

$$m_\Xi = 1.88 \pm 0.04 \ \text{GeV} \tag{24}$$

Finally, for these parameters the current-state overlap is:

$$\lambda_{II}^q \simeq 1.3 \times 10^{-9} \ \text{GeV}^{13} \tag{25}$$

Comparing (22) and (25) we observe that the coupling of current I to the the Ξ state is four times larger than the coupling of current II to this state. This speaks in favor of current I.

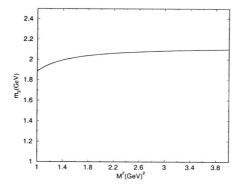

Figure 1 Ξ mass with current I. $m_s = 0.10\,\text{GeV}$, $t = 1$ and $\Delta = 0.44\,\text{GeV}$.

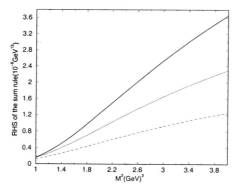

Figure 2 Leading terms of the R.H.S of (15) with current I. $m_s = 0.10\,\text{GeV}$, $t = 1$ and $\Delta = 0.44\,\text{GeV}$. Solid line: perturbative term; dotted line: operators of dimension 4; dashed line: operators of dimension 6.

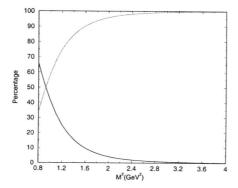

Figure 3 Relative strength of the pole (solid line) and the continuum (dotted line) as a function of the Borel mass squared with current I. $m_s = 0.10\,\text{GeV}$, $t = 1$ and $\Delta = 0.44\,\text{GeV}$.

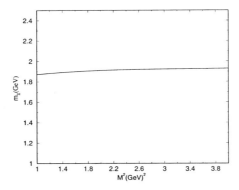

Figure 4 Ξ mass with current II. $m_s = 0.10\,\text{GeV}$ and $\Delta = 0.24\,\text{GeV}$.

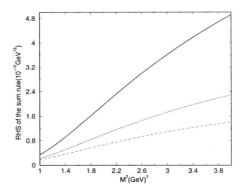

Figure 5 Leading terms of the R.H.S of (15) with current II. $m_s = 0.10\,\text{GeV}$ and $\Delta = 0.24\,\text{GeV}$. Solid line: perturbative term; dotted line: operators of dimension 4; dashed line: operators of dimension 6.

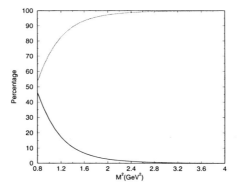

Figure 6 Relative strength of the pole (solid line) and the continuum (dotted line) as a function of the Borel mass squared with current II. $m_s = 0.10\,\text{GeV}$, $t = 1$ and $\Delta = 0.24\,\text{GeV}$.

5 Finite Energy Sum Rules

As we have seen, the description of the phenomenological side of the sum rules requires the definition of the spectral density ρ, which is written as a sum of pole and continuum contribution. The energy gap separating the ground state from the first excited state (Δ) or, equivalently, the squared mass of the first excitation, s_0, when not previously known from experiment, must be guessed. This can be considered as a weak point in the Borel sum rules. A way to reduce this arbitrariness is to work in the large Borel mass limit and try to completely determine s_0. This variant of QCDSR is called Finite Energy Sum Rules (FESR) [32, 33, 11].

We start from the general form taken by the QCDSR for any two point correlator:

$$\lambda^2 e^{-\frac{m^2}{M^2}} = \int_0^{s_0} e^{-\frac{s}{M^2}} \rho(s)\,ds. \tag{26}$$

Taking the limit $\frac{1}{M^2} \to 0$ we get:

$$\sum_n (-m)^{2n} \lambda^2 \left(\frac{1}{M^2}\right)^n = \sum_n \int_0^{s_0} (-s)^n \left(\frac{1}{M^2}\right)^n \rho(s)\,ds. \tag{27}$$

Equating the coefficients of the polynomial in $\frac{1}{M^2}$ we get simply:

$$m^{2n}\lambda^2 = \int_0^{s_0} s^n \rho(s)\,ds, \quad n = 0, 1, 2... \tag{28}$$

The mass can be easily obtained by dividing two of such equations with subsequent values of n:

$$m^2 = \frac{\int_0^{s_0} s^{n+1}\rho(s)\,ds}{\int_0^{s_0} s^n \rho(s)\,ds}, \quad n = 0, 1, 2... \tag{29}$$

In contrast to the method discussed in the previous sections, the FESR have the advantage of giving correlations between the mass (and also λ) and the QCD continuum threshold s_0, avoiding inconsistencies in the values of these parameters. Ideally, we can find a stability in the function $m(s_0)$ thus having a good criterion for fixing both s_0 and m.

On the other side, making $\frac{1}{M^2} \to 0$ takes the sum rule to a region (high M^2) where the pole contribution is almost zero (which can be seen quite clearly in figure 7).

5.1 The Θ^+ (1540) mass

We proceed by applying the FESR to the currents in Eqs. (3), (4) and (5). This has been done fully for η_I^Θ and η_{III}^Θ in [19]. So we are going to omit the analytical expressions for these and concentrate in the FESR for η_{II}^Θ, which are new. The FESR results for all three currents are presented here for the sake of comparison.

Performing the FESR analysis for the $\Pi_q(q^2)$ structure, we notice that, at the approximation where the OPE is known (dimension $\leqslant 6$), we do not have stability in s_0 for any of the currents. Therefore, we used the $\Pi_1(q^2)$ structure for the FESR.

To get the FESR for current (4) we start with the OPE expansion given in [14] and make $\frac{1}{M^2} \to 0$. We get:

$$\int_0^{s_0} \rho_{II}^i(s)ds = \frac{m_s s_0^6}{1415577600\pi^8} - \frac{\langle \bar{s}s \rangle s_0^5}{29491200\pi^6}$$
$$- \frac{7\langle \bar{q}q \rangle s_0^5}{14745600\pi^6} + \frac{\langle \bar{s}g\sigma Gs \rangle s_0^4}{4718592\pi^6} + \frac{7\langle \bar{q}g\sigma Gq \rangle s_0^4}{2359296\pi^6} \quad (30)$$

$$\int_0^{s_0} s\rho_{II}^i(s)ds = \frac{m_s s_0^7}{1651507200\pi^8} - \frac{\langle \bar{s}s \rangle s_0^6}{35389440\pi^6}$$
$$- \frac{7\langle \bar{q}q \rangle s_0^6}{17694720\pi^6} + \frac{\langle \bar{s}g\sigma Gs \rangle s_0^5}{5898240\pi^6} + \frac{7\langle \bar{q}g\sigma Gq \rangle s_0^5}{2949120\pi^6} \quad (31)$$

The mass is obtained dividing eq. (31) by eq.(30). The results are shown on figure 7 together with the results for currents (3) and (5) [19].

We can see in figure 7 that η_I^Θ has the best agreement with the experimental candidate mass. The result from η_{II}^Θ is a little high but still compatible with the experiments. The mass obtained with η_{III}^Θ is so high that we are led to think that this current couples to an angular excitation [19]. This interpretation is reinforced by the derivatives in eq.(5). We should also note that η_I^Θ stabilizes for very low values of s_0, especially when calculated in dimension 9, case in which Δs_0 is very close to zero.

Summing up the results we have (the errors have been estimated for η_I^Θ and η_{III}^Θ in [19], here we assume an error of the same magnitude for η_{II}^Θ):

$$m_I^\Theta = (1.51 \pm 0.11)\text{GeV} \quad (32)$$
$$m_{II}^\Theta = (1.63 \pm 0.16)\text{GeV} \quad (33)$$
$$m_{III}^\Theta = (1.99 \pm 0.19)\text{GeV} \quad (34)$$

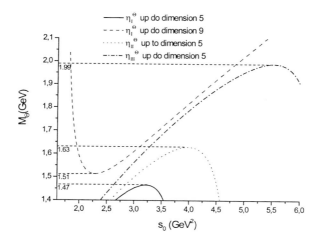

Figure 7 m_Θ obtained with the FESR for various currents up to dimension 5 condensates and up to dimension 9 in the case of η_I^Θ. The values at the stability regions are indicated on the left.

5.2 The Ξ^{--} (1862) mass

In this subsection we repeat the steps above using the currents given by Eq.(8) and Eq.(9). As before, we have stability only in the structure $\Pi_1(q^2)$. For η_I^Ξ we obtain:

$$\int_0^{s_0} \rho_I^i(s)\,ds = -\frac{c_1 \langle \bar{q}q \rangle s_0^5}{14745600\pi^6} + \frac{c_1 \langle \bar{q}g\sigma Gq \rangle s_0^4}{2359296\pi^6}. \tag{35}$$

$$\int_0^{s_0} s\rho_I^i(s)\,ds = -\frac{c_1 \langle \bar{q}q \rangle s_0^6}{17694720\pi^6} + \frac{c_1 \langle \bar{q}g\sigma Gq \rangle s_0^5}{2949120\pi^6}, \tag{36}$$

where $c_1 = 5t^2 + 2t + 5$. For η_{II}^Ξ:

$$\int_0^{s_0} \rho_{II}^i(s)\,ds = -\frac{\langle \bar{q}q \rangle s_0^5}{1843200\pi^6} + \frac{\langle \bar{q}g\sigma Gq \rangle s_0^4}{294912\pi^6} \tag{37}$$

$$\int_0^{s_0} s\rho_{II}^i(s)\,ds = -\frac{\langle \bar{q}q \rangle s_0^6}{2211840\pi^6} + \frac{\langle \bar{q}g\sigma Gq \rangle s_0^5}{368640\pi^6} \tag{38}$$

Once again the masses are obtained dividing (36) by (35) and (38) by (37). The results are show in figure 8.

The figure shows clearly that both currents give the same results in this aproximation of the OPE, and both are bellow the candidate mass. It should be noted that this result may change a lot for η_I^Ξ if we add higher dimension condensates. In this aproximation both contributions had the same polynomial in t (c_1) thus eliminating the dependence in t. This coincidence will hardly be repeated in higher dimensions.

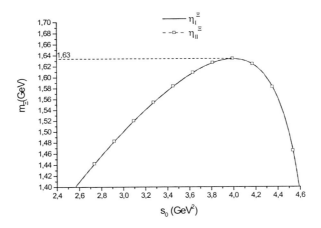

Figure 8 m_Ξ obtained with the FESR up to dimension 5 condensates. The values at the stability regions are indicated on the left.

To summarize: we have found that due to the slow convergence of the OPE and to the relative importance of the QCD continuum contribution into the spectral function, the minimal duality ansatz "one resonance + QCD continuum" is not sufficient for finding a *sum rule window* where the results are optimal. These features penalize *all* existing sum rule results in the literature, which then become unreliable despite the fact that the mass predictions reproduce quite well the expected number.

6 Θ Decay

We turn now to the pentaquark decay $\Theta \longrightarrow nK^+$. One of the most puzzling characteristics of the pentaquark is its extremely small width (much) below 10 MeV which poses a serious challenge to all theoretical models. Many explanations for this narrow width have been advanced [3]. In this section we review the calculation of the Θ decay with QCDSR.

As we have seen in the preceding sections, a common problem of all QCDSR pentaquark mass determinations is the large continuum contribution which has its origin in the high dimension of the interpolating currents and results in a strong dependence on the continuum threshold. Another problem is the irregular behavior of the operator product expansion (OPE), which is dominated by higher dimension operators and not by the perturbative term as it should be.

Here we review the sum rule determination [22] for the decay width based on a three-point function for the decay $\Theta \rightarrow nK^+$. In this way it is possible to extract the coupling $g_{\Theta nK}$, which is directly related to the pentaquark width. The pentaquark is treated as a diquark-diquark-antiquark with one scalar and one pseudoscalar diquark in a relative S-wave.

In [19, 31] it has been argued that such a small decay width can only be explained if the pentaquark is a genuine 5-quark state, i.e., it contains no color singlet meson-baryon contributions and thus color exchange is necessary for the decay. The analysis presented both in [19] and in [31] is only qualitative. The narrowness of the pentaquark width can then be attributed to the non-trivial color structure of the pentaquark which requires the exchange of, at least, one gluon. In [22] we have done a quantitative test of the conjecture advanced in [19] and [31].

6.1 The three-point functions

The investigation of the pentaquark decay width requires a three-point function which we define as

$$\Gamma(p,p') = \int d^4x \, d^4y \, e^{-iqy} \, e^{ip\cdot x} \, \Gamma(x,y) \,,$$

$$\Gamma(x,y) = \langle 0 | T\{\eta_N(x) j_K(y) \bar{\eta}_\Theta(0)\} | 0 \rangle \,,$$

(39)

where η_N, j_K and η_Θ are the interpolating fields associated with neutron, kaon and Θ, respectively [22].

We next consider the expression (39) in terms of hadronic degrees of freedom and write the phenomenological side of the sum rule. Treating the kaon as a pseudoscalar particle, the interaction between the three hadrons is described by the following Lagrangian density:

$$\mathcal{L} = i g_{\Theta nK} \bar{\Theta} \gamma_5 K n \quad \text{for } P = +$$

$$\mathcal{L} = i g_{\Theta nK} \bar{\Theta} K n \quad \text{for } P = -$$

(40)

Writing the correlation function (39) in momentum space and inserting complete sets of hadronic states we obtain an expression which depends on the following matrix elements:

$$-i V(p,p') = < n(p',s') | \Theta(p,s) K(q) > \,,$$

$$\langle 0 | \eta_N | n(p',s') \rangle = \lambda_N u^{s'}(p') \,,$$

$$\langle K(q) | j_K | 0 \rangle = \lambda_K \,,$$

$$\langle \Theta(p,s) | \bar{\eta}_\Theta | 0 \rangle = \lambda_\Theta \bar{u}^s(p) \quad \text{for } P = +$$

$$\langle \Theta(p,s) | \bar{\eta}_\Theta | 0 \rangle = -\lambda_\Theta \bar{u}^s(p) \gamma_5 \quad \text{for } P = -$$

(41)

Using the simple Feynman rules derived from (40) we can rewrite $V(p,p')$ as

$$V(p,p') = -g_{\Theta nK} \bar{u}^{s'}(p') \gamma_5 u^s(p) \quad P = +$$

$$V(p,p') = -g_{\Theta nK} \bar{u}^{s'}(p') u^s(p) \quad P = -$$

(42)

The coupling constants λ_N and λ_Θ can be determined from the QCD sum rules of the corresponding two-point functions. λ_K is related to the kaon decay constant through

$$\lambda_K = \frac{f_K m_K^2}{m_u + m_s}.\tag{43}$$

Combining the expressions above we arrive at

$$\Gamma_{phen} = \frac{-g_{\Theta n K} \lambda_\Theta \lambda_N \lambda_K}{(p'^2 - m_N^2)(q^2 - m_{K^+}^2)(p^2 - m_\Theta^2)} \times \Gamma_E + continuum \tag{44}$$

with

$$\Gamma_E = \sigma^{\mu\nu}\gamma_5 q_\mu p'_\nu - i m_N \, \slashed{q}\gamma_5 + i(m_N \mp m_\Theta) \, \slashed{p}'\gamma_5$$
$$+ i\gamma_5(p'^2 \mp m_\Theta m_N - qp') \tag{45}$$

We have worked with the $\sigma^{\mu\nu}\gamma_5 q_\mu p'_\nu$ structure because, as it was shown in [34], this structure gives results which are less sensitive to the coupling scheme on the phenomenological side, i.e., to the choice of a pseudoscalar or pseudovector coupling between the kaon and the baryons.

Coming back to (39) we write the interpolating fields in terms of quark degrees of freedom as

$$j_K(y) = \bar{s}(y) i\gamma_5 u(y),$$
$$\eta_N(x) = \epsilon^{abc}(d_a^T(x) C\gamma_\mu d_b(x))\gamma_5\gamma^\mu u_c(x),$$
$$\bar{\eta}_\Theta(0) = -\epsilon^{abc}\epsilon^{def}\epsilon^{cfg} s_g^T(0) C \times [\bar{d}_e(0)\gamma_5 C\bar{u}_d^T(0)][\bar{d}_b(0)C\bar{u}_a^T(0)]. \tag{46}$$

The pentaquark current above (proposed in [16]) contains a pseudoscalar and a scalar diquark. With these diquarks the two point function might receive a significant contribution from instantons. In [35] we have studied a situation in which these instanton contributions affected the two-point function but gave a negligible contribution to the three-point function. Moreover, in [36] we have observed that instantons give a negligible contribution to heavy baryon weak decays. Motivated by these results, in this first calculation we have neglected instantons.

Inserting the currents into (39), the resulting expression involves the quark propagator in the presence of quark and gluon condensates (12). Using it we arrive at a final complicated expression for the correlator, which is represented schematically by the sum of the diagrams of Fig. 9.

Let us consider the phenomenological side (44) and, following [37], rewrite it generically as:

$$\Gamma(q^2, p^2, p'^2) = \Gamma_{pp} + \Gamma_{pc1} + \Gamma_{pc2} + \Gamma_{cc} \tag{47}$$

where $\Gamma_{pp}(q^2, p^2, p'^2)$ stands for the pole-pole part and reads

$$\Gamma_{pp} = \frac{-g_{\Theta nK}\lambda_\Theta \lambda_N \lambda_K}{(p^2 - m_\Theta^2)(p'^2 - m_N^2)(q^2 - m_K^2)} \tag{48}$$

The continuum-continuum term Γ_{cc} can be obtained as usual, with the assumption of quark-hadron duality [22].

The pole-continuum transition terms are contained in Γ_{pc1} and Γ_{pc2}. They can be explicitly written as a double dispersion integral:

$$\begin{aligned}
\Gamma_{pc1} &= \int_{m_{K^*}^2}^\infty \frac{b_1(u, p^2)\, du}{(m_N^2 - p'^2)(u - q^2)}, \\
\Gamma_{pc2} &= \int_{m_{N^*}^2}^\infty \frac{b_2(s, p^2)\, ds}{(m_K^2 - q^2)(s - p'^2)}.
\end{aligned} \tag{49}$$

Since there is no theoretical tool to calculate the unknown functions $b_1(u, p^2)$ and $b_2(s, p^2)$ explicitly, one has to employ a parametrization for these terms. We will use two different parametrizations: one with a continuous function for the Θ and one where the pole term is singled out.

We assume here that the functions b_1 and b_2 have the following form:

$$\begin{aligned}
b_1(u, p^2) &= \widetilde{b_1}(u) \int_{m_\Theta^2}^\infty d\omega \frac{b_1(\omega)}{\omega - p^2} \\
b_2(s, p^2) &= \widetilde{b_2}(s) \int_{m_\Theta^2}^\infty d\omega \frac{b_2(\omega)}{\omega - p^2}
\end{aligned} \tag{50}$$

with continuous functions $b_{1,2}(w)$, starting from m_Θ^2. This is our **parametrization A**. The functions $\widetilde{b_1}(u)$ and $\widetilde{b_2}(s)$ describe the excitation spectra of the kaon and the nucleon, respectively. After Borel transform, the pole-continuum term contains one unknown constant factor which can be determined from the sum rules.

In order to investigate the role played by the Θ continuum, we explicitly force the phenomenological side to contain only the pole part of the Θ, both in the pole-pole term and in the pole-continuum terms. This can formally be done by choosing $b_1(\omega) = b_2(\omega) = \delta(\omega - m_\Theta^2)$ in (50) and the functions then read:

$$\begin{aligned}
b_1(u, p^2) &= \frac{\widetilde{b_1}(u)}{m_\Theta^2 - p^2}, \\
b_2(s, p^2) &= \frac{\widetilde{b_2}(s)}{m_\Theta^2 - p^2}.
\end{aligned} \tag{51}$$

This is our **parametrization B**. In this case we have the Θ in the ground state. Again, in the final expressions this gives additional constants which can be calculated.

6.2 Decay sum rules

The sum rule may be written identifying the phenomenological and theoretical descriptions of the correlation function. As mentioned above, we work with the $\sigma^{\mu\nu}\gamma_5 q_\mu p'_\nu$ structure. In the case of the three-point function considered here, there are two independent momenta and we may perform either a single or a double Borel transform. We first consider the choice:

$$(I) \qquad q^2 = 0 \qquad p^2 = p'^2 \tag{52}$$

and perform a single Borel transform: $p^2 = -P^2$ and $P^2 \to M^2$. In this case we take $m_K^2 \simeq 0$ and single out the $1/q^2$-terms. The second choice is:

$$(II) \qquad q^2 \neq 0 \qquad p^2 = p'^2 . \tag{53}$$

Here we perform two Borel transforms: $p^2 = -P^2$ and $P^2 \to M^2$ and also $q^2 = -Q^2$ and $Q^2 \to M'^2$. We have also considered the choice $q^2 = p^2 = p'^2 = -P^2$, performing one single Borel transform ($P^2 \to M^2$). However, we were not able to find a stable sum rule. Introducing the notation $G = -g_{\Theta n K}\lambda_\Theta\lambda_N\lambda_K$ and using (I) and (II) we obtain the following sum rules:

Method I:

$$\Gamma_{pp}(M^2) + \Gamma_{pc2}(M^2) = \int_0^{s_0} ds\, \rho_{th}(s)\, e^{-s/M^2} \tag{54}$$

with

$$\Gamma_{pp}(M^2) = G\frac{e^{-m_\Theta^2/M^2} - e^{-m_N^2/M^2}}{m_\Theta^2 - m_N^2} \tag{55}$$

and for the pole-continuum part we obtain

$$\begin{aligned}
\Gamma_{pc2}(M^2) &= A\, e^{-m_{N*}^2/M^2} \quad \text{param. A} \\
\Gamma_{pc2}(M^2) &= A\, e^{-m_\Theta^2/M^2} \quad \text{param. B}
\end{aligned} \tag{56}$$

In both parametrizations the term Γ_{pc1} is exponentially suppressed and, as discussed in [37], has been neglected. A is an unknown constant and can be determined from the sum rules.

Method II

$$\Gamma_{pp}(M^2, M'^2) + \Gamma_{pc2}(M^2, M'^2) = \tag{57}$$

$$\int_0^{u_0} du \int_0^{s_0} ds\, \rho_{th}(s, u)\, e^{-s/M^2}\, e^{-u/M'^2} \tag{58}$$

with

$$\Gamma_{pp} = G\, e^{-m_K^2/M'^2}\, \frac{e^{-m_\Theta^2/M^2} - e^{-m_N^2/M^2}}{m_\Theta^2 - m_N^2} \tag{59}$$

and with

$$\Gamma_{pc2} = A\, e^{-m_K^2/M'^2}\, e^{-m_{N*}^2/M^2} \tag{60}$$

for parametrization A and

$$\Gamma_{pc2} = A\, e^{-m_K^2/M'^2}\, e^{-m_\Theta^2/M^2} \tag{61}$$

for parametrization B. Also in this case Γ_{pc1} is exponentially suppressed. In the above expressions ρ_{th} is the double discontinuity computed directly from the theoretical (OPE) description of the correlation function (see [22] for details and also [38]) and s_0 is the continuum threshold of the nucleon defined as $s_0 = (m_N + \Delta_N)^2$.

6.3 Decay width

The hadronic masses are $m_N = 938$ MeV, $m_{N*} = 1440$ MeV, $m_K = 493$ MeV and $m_\Theta = 1540$ MeV. For each of the sum rules above (Eqs. (54) and (57)) we can take the derivative with respect to $1/M^2$ and in this way obtain a second sum rule. In each case we have thus a system of two equations and two unknowns (G and A) which can then be easily solved.

The couplings constants λ_N and λ_Θ are taken from the corresponding two-point functions:

$$\begin{aligned}
\lambda_N &= (2.4 \pm 0.2) \times 10^{-2}\,\text{GeV}^3 \\
\lambda_\Theta &= (2.4 \pm 0.3) \times 10^{-5}\,\text{GeV}^6
\end{aligned} \tag{62}$$

The coupling λ_K is obtained from (43) with $f_K = 160$ MeV, $m_s = 100$ MeV and $m_u = 5$ MeV:

$$\lambda_K = 0.37\,\text{GeV}^2\,. \tag{63}$$

In Fig. 9, among these OPE diagrams there are two distinct subsets. In the first two lines of the figure there is no gluon line connecting the "petals" and therefore no color exchange. A diagram of this type we call color-disconnected. In the second subset of diagrams, in the third line of the figure, we have color exchange. If there is no color exchange, the final state containing two color singlets was already present in the initial state, before the decay, as noticed in [20]. In this case the pentaquark had a component similar to a $K - n$ molecule. In the second case the pentaquark was a genuine 5-quark state with a non-trivial color structure. We may call this type of diagram a color-connected (CC) one. In our analysis we write sum rules for both cases: all diagrams and only color-connected. The former case is standard in QCDSR calculations and therefore we omit details and present only the results. The latter case implies that the pentaquark is a genuine 5-quark state and the evaluation of $g_{\Theta nK}$ is thus based only on the CC diagrams. We work in the Borel window given by $1\,\text{GeV}^2 \leqslant M^2(M'^2) \leqslant 1.5\,\text{GeV}^2$. Since the strange mass is small, the dominating diagram is Fig. 9b of dimension three with one quark condensate. In the range considered, the dimension 5 condensates are substantially suppressed compared to this term.

We have found out that the contribution from the pole-continuum part is of a similar size as the pole part. For lower values of M^2 around $1\,\text{GeV}^2$, the pole contribution dominates, however, for larger values of M^2 the importance of the pole-continuum contribution grows and eventually becomes larger than the pole part. This is an additional reason to restrict the analysis to small values for the Borel parameters.

We have evaluated the sum rules for the coupling constant computed with all diagrams of Fig. 9 and we have found that they are very stable. We give the values of the coupling extracted at $M^2 = 1.5\,\text{GeV}^2$ and $M'^2 = 1\,\text{GeV}^2$ in Table I. We present our results for the coupling constant $g_{\Theta nK}$ obtained with the color connected diagrams only. In Fig. 10 we show the coupling, given by the solution of the sum rule I A (54), as a function of the Borel mass squared M^2. Different lines show different values of the continuum threshold Δ_N. As it can be seen, $g_{\Theta nK}$ is remarkably stable with respect to variations both in M^2 and in Δ_N. In Fig. 11 we show the coupling obtained with the sum rule II A (57). We find again fairly stable results which are very weakly dependent on the continuum threshold. In Fig. 12 we show the results of the sum rule I B. In Fig. 13 we present the result of the sum rule II B. The meaning of the different lines is the same as in the previous figures. The results are similar to the cases before.

In Table I we present a summary of our results for $g_{\Theta nK}$ giving emphasis to the difference between the results obtained with all diagrams and with only the color-connected ones. For the continuum thresholds we have employed $\Delta_N = \Delta_K = 0.5\,\text{GeV}$.

Table 1 $g_{\Theta nK}$ for various cases

case	$\|g_{\Theta nK}\|$ (CC)	$\|g_{\Theta nK}\|$ (all diagrams)
I A	0.71	2.59
II A	0.82	3.59
I B	0.84	3.24
II B	0.96	4.48

For our final value of $g_{\Theta nK}$ we take an average of the sum rules I A - II B. It is interesting to observe that the influence of the continuum threshold is relatively small, especially when compared to the corresponding two-point functions.

Considering the uncertainties in the continuum thresholds, in the coupling constants $\lambda_{K,N,\Theta}$ and in the quark condensate we get an uncertainty of about $50\,\%$.

Our final result then reads:

$$|g_{\Theta nK}|(\text{all diagrams}) = 3.48 \pm 1.8,$$
$$|g_{\Theta nK}|(CC) = 0.83 \pm 0.42. \tag{64}$$

Including all diagrams, the prediction for Γ_Θ is then 13 MeV (652 MeV) for a positive (negative) parity pentaquark. In the CC case we get a width of 0.75 MeV (37 MeV) for a positive (negative) parity pentaquark. The measured upper limit of the width is around 5-10 MeV both in the Kn channel (considered here) and in the Kp channel.

To summarize: we have presented a QCD sum rule study of the decay of the Θ^+ pentaquark using a diquark-diquark-antiquark scheme with one scalar and one pseudoscalar diquark. Based on the evaluation of the relevant three-point function, we have computed the coupling constant $g_{\Theta nK}$. In the operator product expansion we have included all diagrams up to dimension 5. In this particular type of sum rule a complication arises from the pole-continuum transitions which are not exponentially suppressed after Borel transformation and must be explicitly included. The analysis was made for two different pole-continuum parametrizations and in two different evaluation schemes. The results are consistent with each other. In addition, we have tested the ideas presented in [31, 19] by including only diagrams with color exchange. Our final results are given in eq. (64). We conclude that for a positive parity pentaquark a width much smaller than 10 MeV would indicate a pentaquark which contains no color-singlet meson-baryon contribution. For a negative parity pentaquark, even under the assumption that it is a genuine 5-quark state, we could not explain the narrow width of the Θ.

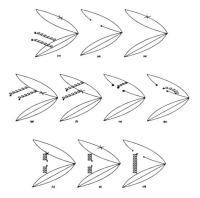

Figure 9 The main diagrams which contribute to the theoretical side of the sum rule in the relevant structure. a) - g) are the color-disconnected diagrams, whereas h) - j) are the color-connected diagrams. The cross indicates the insertion of the strange mass.

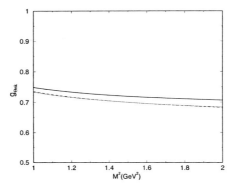

Figure 10 $|g_{\Theta nK}|$ in case I A with three different continuum threshold parameters. Solid line: $\Delta_N = 0.5$ GeV, dotted line: $\Delta_N = 0.4$ GeV, dash-dotted line: $\Delta_N = 0.6$ GeV.

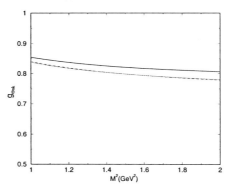

Figure 11 $|g_{\Theta nK}|$ in case II A. Solid line: $\Delta_N = 0.5$ GeV. Dotted line: $\Delta_N = 0.4$ GeV. Dashed line: $\Delta_N = 0.6$ GeV. $M'^2 = 1$ GeV^2.

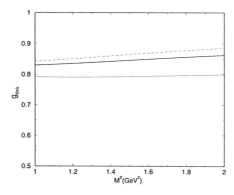

Figure 12 $|g_{\Theta nK}|$ in case I B with three different continuum threshold parameters. Solid line: $\Delta_N = 0.5$ GeV, dotted line: $\Delta_N = 0.4$ GeV, dash-dotted line: $\Delta_N = 0.6$ GeV.

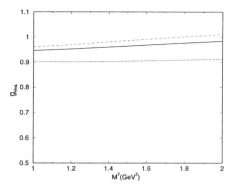

Figure 13 $|g_{\Theta nK}|$ in case II B. Solid line: $\Delta_N = 0.5$ GeV. Dotted line: $\Delta_N = 0.4$ GeV. Dashed line: $\Delta_N = 0.6$ GeV. $M'^2 = 1$ GeV^2.

7 Conclusion

In this review we have discussed only our works on pentaquarks in a somewhat critical perspective. Most of the other calculations, most of them quoted here, have the same successes and difficulties as ours. Looking back and taking distance, we might say that the work done over the last two years has undergone continuous improvements in quality. At the very beginning, in the heat of the discovery hours, some works were done in rush and with a certain negligence in various aspects. For example, in the very first paper [14] reproducing the Θ mass, no analysis of the OPE convergence was presented and neither an estimate of the continuum contribution was performed. The second round of calculations went much deeper in the details of QCDSR procedures. However it was not just a matter of "doing better" what we already knew how to do. The method had to face new challenges. For example: in the pentaquark study, for the first time, we were dealing with a system that could be composed by independent subsystems, like two non-interacting hadrons. In [20] this configuration was dubbed "two-hadron reducible" component. It has been a subject of debate how to disentangle and subtract this component from the final results. Also, the more quarks we have, the less unique is the definition of the interpolating current. Increasing the number of lines introduces new technical complications for the evaluation of the OPE. The efforts of the community to overcome all these problems were very productive. All in all, we can say that pentaquarks have done more for QCDSR than these have done for pentaquarks.

To conclude we come back to the question raised in the introduction."How could we calculate the correct mass of something that does not exist?" In the light of the discussion presented in the last sections, a sober answer would be: although we started reproducing unfounded experimental results, it was just a matter of time until we would reach a situation where, reproducing these data would be so artificial as it was to use the notion of "aether" in the years of the birth of special relativity. At some point we would be obliged to push and twist the method so far, that some more audacious groups would be brave enough to go against the "experimental evidence" and put doubts on the experiments. This attitude was already taken by some phenomenologists [7, 39], by some experimentalists [40] and by lattice theorists.

The final feeling is that all this work was fruitful and there is nothing to be regretted.

Acknowledgements

We are grateful to FAPESP and to CNPq for financial support. Discussions with Markus Eidemüller, Rômulo Rodrigues da Silva and Stephan Narison are gratefully acknowledged.

Bibliography

[1] T. Nakano *et al.*, LEPS Coll., Phys. Rev. Lett. **91** (2003) 012002.

[2] F. Close, Nature **435** (2005) 287; M. Ostrick, Prog. Part. Nucl. Phys. **55** (2005) 337; D. Barna, Acta Phys. Hung. A **22** (2005) 187; M. Danilov, hep-ex/0509012. K. Hicks, AIP Conf.Proc. 756 (2005) 195.

[3] For recent reviews see: M. Karliner, Int. J. Mod. Phys. A **20** (2005) 199; Shi-Lin Zhu, Int. J. Mod. Phys. A **20** (2005) 1548; M. Oka, J. Sugiyama and T. Doi, Nucl. Phys. A **755** (2005) 391; A. Hosaka, hep-ph/0506138; S. H. Lee, H. Kim and Y. s. Oh, J. Korean Phys. Soc. **46** (2005) 774.

[4] M. Battaglieri, R. De Vita and V. Kubarovsky [CLAS Collaboration], AIP Conf. Proc. **806** (2006) 48.

[5] M. Battaglieri, R. De Vita, V. Kubarovsky and P. Stoler [CLAS Collaboration], JLAB-PHY-06-09 *Contributed to 25th International Symposium on Physics in Collision (PIC 05), Prague, Czech Republic, 6-9 Jul 2005*

[6] T. Barnes, hep-ph/0510365.

[7] A. R. Dzierba, C. A. Meyer and A. P. Szczepaniak, J. Phys. Conf. Ser. **9** (2005) 192 hep-ex/0412077.

[8] F. Csikor, Z. Fodor, S. D. Katz and T. G. Kovacs, Int. J. Mod. Phys. A **20** (2005) 4562 and references therein.

[9] F. Csikor *et al.*, JHEP 0311 (2003) 70; T.W. Shiu, T.H. Hsieh, hep-ph/0403020; N. Matur *et al.*, hep-ph/0406196.

[10] S. Sasaki, hep-lat/0310014.

[11] For a review and references to original works, see e.g., S. Narison, *QCD as a theory of hadrons, Cambridge Monogr. Part. Phys. Nucl. Phys. Cosmol.* **17** (2002) 1; *QCD spectral sum rules , World Sci. Lect. Notes Phys.* **26** (1989) 1; *Acta Phys. Pol.* **26** (1995) 687; *Riv. Nuov. Cim.* **10N2** (1987) 1; *Phys. Rep.* **84** (1982) 1.

[12] M.A. Shifman, A.I. and Vainshtein and V.I. Zakharov, Nucl. Phys. **B147** (1979) 385.

[13] L.J. Reinders, H. Rubinstein and S. Yazaki, Phys. Rep. **127** (1985) 1.

[14] S.-L. Zhu, Phys. Rev. Lett. **91** (2003) 232002.

[15] R.D. Matheus, F.S. Navarra, M. Nielsen, R. Rodrigues da Silva and S.H. Lee, Phys.Lett. **B578** (2004) 323.

[16] J. Sugiyama, T. Doi, M. Oka, Phys. Lett. **B581** (2004) 167.

[17] M. Eidemüller, Phys. Lett. **B597** (2004) 314.

[18] R.D. Matheus, F.S. Navarra, M. Nielsen, R. Rodrigues da Silva, Phys. Lett. **B602** (2004) 185.

[19] R.D. Matheus, S. Narison, hep-ph/0412063.

[20] Y. Kondo, O. Morimatsu, T. Nishikawa, Phys. Lett. **B611** (2005) 93; Y. Kwon, A. Hosaka and S. H. Lee, hep-ph/0505040.

[21] Hungchong Kim, Su Houng Lee, *Phys. Lett.* **B595** (2004) 293.

[22] M. Eidemüller, F. S. Navarra, M. Nielsen and R. Rodrigues da Silva, Phys. Rev. D **72** (2005) 034003.

[23] D. Strottman, Phys. Rev. **D20** (1979) 748.

[24] This configuration was considered in the case of a $uudc\bar{c}$ state in: S.J. Brodsky, I. Schmidt, G.F. de Teramond, Phys. Rev. Lett. **64** (1990) 1011.

[25] M. Karliner and H. Lipkin, hep-ph/0307243.

[26] R. Jaffe and F. Wilczek, Phys. Rev. Lett. **91** (2003) 232003.

[27] E. Shuryak and I. Zahed, Phys. Lett. **B589** (2004) 21

[28] B. L. Ioffe, Nucl. Phys. **B188**, 317 (1981); **B191**, 591(E) (1981).

[29] Y. Chung, H. G. Dosch, M. Kremer, D. Schall, Phys. Lett. **B102** (1981) 175; Nucl. Phys. **B197** (1982) 55.

[30] K.-C. Yang *et al.*, Phys. Rev. **D47** (1993) 3001.

[31] B. L. Ioffe and A. G. Oganesian, JETP Lett. **80** (2004) 386; A. G. Oganesian, hep-ph/0410335.

[32] R.A. Bertlmann, C.A. Dominguez, G. Launer and E. de Rafael, *Nucl. Phys.* **B 250** (1985) 61; R.A. Bertlmann, G. Launer and E. de Rafael, *Z. Phys.* **C 39** (1988) 231.

[33] K. Chetyrkin, N.V. Krasnikov and A.N. Tavkhelidze, *Phys. Lett.* **B 76** (1978) 83; N.V. Krasnikov, A.A. Pivovarov and A.N. Tavkhelidze, *Z. Phys.* **C 19** (1983) 301.

[34] H. Kim, S.H. Lee and M. Oka, Phys. Lett. **B453** (1999) 199; Phys. Rev. **D60** (1999) 034007; M.E. Bracco, F.S. Navarra, M. Nielsen, Phys. Lett. **B454** (1999) 346.

[35] H. G. Dosch, E. M. Ferreira, F. S. Navarra and M. Nielsen, Phys. Rev. D **65** (2002) 114002.

[36] R. S. Marques de Carvalho, F. S. Navarra, M. Nielsen, E. Ferreira and H. G. Dosch, Phys. Rev. D **60** (1999) 034009.

[37] B. Ioffe and A. Smilga, Nucl. Phys. **B232** (1984) 109.

[38] M. E. Bracco, M. Chiapparini, A. Lozea, F. S. Navarra and M. Nielsen, Phys. Lett. B **521** (2001) 1.

[39] J. Haidenbauer and G. Krein, Phys. Rev. C **68** (2003) 052201.

[40] H. G. Fischer and S. Wenig, Eur. Phys. J. C **37** (2004) 133.

Anais da XXVIII Reunião de Trabalho sobre Física Nuclear no Brasil
SP, Brasil, 2005
Artigo originalmente publicado no Brazilian Journal of Physics, Vol 36 – n° 4B,
Special Issue: XXVIII Workshop on Nuclear Physics in Brazil, pp.1345–1348 (2006).

Accumulation and Long-Term Behavior of Radiocaesium in Tropical Plants

C. Carvalho, B. Mosquera, R. M. Anjos*, N. Sanches, J. Bastos, K. Macario, R. Veiga

Instituto de Física, Universidade Federal Fluminense,
Av. Litorânea s/n, Gragoatá, Niterói, RJ, Brazil, CEP 24210-340

Abstract. The accumulation and distribution of ^{40}K and ^{137}Cs in tropical plant species were studied through measurements of gamma-ray spectra from mango, avocado, guava, pomegranate, chili pepper, papaya and manioc trees. Our goal was to infer their differences in the uptake and translocation of ions to the aboveground plant parts and to establish the suitability of using radiocaesium as a tracer for the plant uptake of nutrients such as K^+.

Keywords. ^{137}Cs and ^{40}K distributions; tropical trees.

1 Introduction

In recent years, there has been a growing interest in the evaluation of nutrient fluxes and radioactive contaminants in forest and agricultural ecosystems [1-6]. For being chemically similar to available nutrients in soil (such as potassium), radiocaesium is absorbed by plants, entering the food chain and therefore representing a significant pathway of human radiation exposure. Thus, in order to contribute to the understanding of the relative behavior of K^+ and Cs^+ ions, the purpose of this work was to examine the concentration levels of ^{137}Cs and ^{40}K in several tropical plants. The plants were analyzed within the context of a radiological or nuclear accident, investigating the ^{137}Cs contamination by root uptake from contaminated soil. Additionally, two of the plants were transferred to uncontaminated soil (i.e. soil free from the specific ^{137}Cs contamination caused by the radiological accident) in such way that the main source of the contamination of new leaves and fruits is the fraction of the available radiocaesium in the body of the

---

* meigikos@if.uff.br Phone: 55-21-26295770, Fax: 55-21-26295887

plant. Biological half-lives of ^{137}Cs for this second situation, and analysis of ^{137}Cs and ^{40}K distributions throughout these tropical plants are also shown in this paper.

2 Material and Methods

Samples of soils, roots, trunks, twigs, leaves and edible parts were collected from mango (*Mangifera indica*), avocado (*Persea americana*), guava (*Psidium guajava*), pomegranate (*Punica granatum*), chili pepper (*Capsicum frutescens*), papaya (*Carica papaya*) and manioc (*Manihot esculenta*) trees. These plants used to be cultivated together at one of the sites where the Goiânia radiological accident occurred [7]. Additionally, chili pepper and pomegranate trees were transplanted to another site with uncontaminated soil. Samples of green and red pepper, and pomegranate were collected over one and three years, respectively. At the end of these periods, samples of root, trunk, twig, leaves and fruit of these plants were also analyzed.

Sample preparation and analysis were carried out at the Laboratory of Radioecology (LARA) of the Physics Institute of the Universidade Federal Fluminense. The amount of ^{137}Cs and ^{40}K in the samples was determined by standard gamma-ray spectroscopy using HPGe and NaI(Tl) detectors. Figure 1 shows a typical spectrum from an analyzed sample. Technical details of sample preparation and analysis can be obtained in references [1-5].

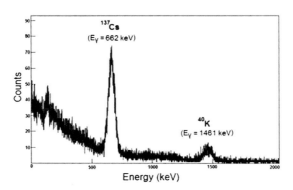

Figure 1 Typical spectrum of a analyzed sample obtained with the NaI(Tl) detector during a counting period of 5.8 x 10^4 seconds.

3 Results

Available information in the literature show that potassium can be readily absorbed by plant roots from soil and can be translocated to the aboveground plant parts, such as twigs, leaves, flowers and fruits. Similarly to potassium, caesium has a high mobility within a plant, exhibiting the highest values of concentration in the growing parts of the tree: fruits, leaves, twigs, barks and the outer growth layers [1-6]. In this way, the results presented in Table 1 for the mean values of ^{137}Cs and ^{40}K concentrations in some compartments of tropical plants contaminated by caesium suggests that tissues with low K concentrations (on a dry weight basis) have also low Cs concentration. On the other hand, the results shown in this table indicate that only for wood trees contaminated by caesium, both ^{137}Cs and ^{40}K have simultaneously higher concentrations in the younger parts than in the older parts. This result implies that the Cs distribution within plants of some species (such as manioc) does not behave exactly as that of K and, therefore, in such cases, ^{137}Cs could not be a good tracer of plant nutrients. For Papaya trees, although they do not follow the same pattern of wood trees, Cs and K have a similar behavior. Additionally, in what concerns the decontamination process, when some tropical trees were transplanted to a site with uncontaminated soil, there is an inversion of ^{137}Cs levels in the different organs of the plant. The youngest parts present specific ^{137}Cs activity lower than other older parts, but the potassium behavior is not changed since their ions remain being absorbed by roots from soil. This caesium behavior suggests that there is an amount of radiocaesium in the body of the plant that remains retained in the main trunk and that just ends up by the physical decay of ^{137}Cs. For the chili pepper tree the enhancement observed for the potassium concentrations was due to the different concentrations in the transplantation soil, increasing the availability of potassium to the plant.

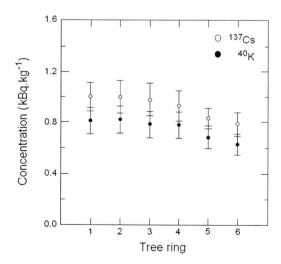

Figure 2 Radial distribution of mean values of ^{137}Cs and ^{40}K concentrations within the tree rings in main trunk of a guava tree.

Table 1 Mean values of ^{137}Cs and ^{40}K concentrations for the younger and older parts of the tropical plants contaminated by caesium and for decontaminated plants. Values in parentheses represent the standard deviation from the mean.

Tree	Sample	Mean values of concentrations $(kBq.kg^{-1})$	
		^{137}Cs	^{40}K
Contaminated plants by cesium			
Guava	Roots	0.8(0.1)	0.4(0.1)
	Main trunk	0.9(0.2)	0.4(0.1)
	Leaves	2.1(0.3)	0.7(0.1)
	Fruits	2.0(0.4)	0.9(0.1)
Mango	Main trunk	1.0(0.1)	0.15(0.02)
	Leaves	1.9(0.6)	0.32(0.05)
	Fruits	3.2(0.4)	0.27(0.04)
Avocado	Main trunk	1.8(0.4)	0.23(0.04)
	Leaves	2.3(0.2)	0.4(0.1)
Chili pepper	Roots	2.1(0.2)	0.24(0.04)
	Main trunk	1.1(0.1)	0.21(0.03)
	Leaves	6.1(0.5)	0.93(0.01)
	Fruits	2.5(0.3)	0.48(0.13)
Papaya	Roots	1.0(0.4)	2.0(0.5)
	Main trunk	0.6(0.2)	1.4(0.5)
	Leaves	0.4(0.1)	1.6(0.2)
	Fruits	0.5(0.3)	1.7(0.5)
Manioc	Roots	0.12(0.04)	0.22(0.02)
	Main trunk	0.21(0.06)	0.14(0.22)
	Leaves	0.36(0.09)	0.15(0.03)
Decontaminated Plants			
Chili pepper	Roots	0.13(0.01)	0.45(0.05)
	Main trunk	0.28(0.02)	0.50(0.06)
	Twigs	0.08(0.01)	1.2(0.2)
	Leaves	0.04(0.01)	2.2(0.4)
	Fruits	0.01(0.01)	1.2(0.2)
Pomegranate	Roots	1.2(0.1)	0.13(0.01)
	Main trunk	0.8(0.1)	0.09(0.01)
	Twigs	0.19(0.04)	0.19(0.02)
	Leaves	0.34(0.06)	0.54(0.06)
	Fruits	0.23(0.02)	0.47(0.06)

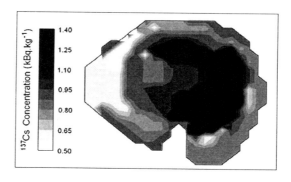

Figure 3 Bi-dimensional diagram of the radial distributions of ^{137}Cs concentration mean values within the tree rings in main trunk of the guava tree (same of Figure 2).

Figure 2 shows conventional one-dimensional radial distributions of both elements within the tree rings in the main trunk of a guava tree. From this figure, one notices that allegedly both ^{137}Cs and ^{40}K are concentrated inside the trunk, and they have similar behavior. However, the bi-dimensional diagrams allow observing interesting details that cannot be noticed in one-dimensional distributions. For instance, the bi-dimensional diagram of ^{137}Cs radial distribution in the same main trunk presented in Figure 3, shows that radiocaesium can be mobile, migrating within the rings, and tending to the side most exposed to sunlight. Although such behavior had already been observed for other plant species [1, 2, 5], its origin is still not clear. In order to improve the understanding of such issue more experimental investigations would be of great importance. For potassium, on the other hand, the performed analysis did not show the same pattern, since distributions of ^{40}K remained concentrated inside the trunk. In this way, the results suggest that although the two elements may have physicochemical similarities, they do not always behave alike. So, even though Cs and K show similar behavior along wood plants, there are subtle differences between them that should be taken into account if radiocaesium is to be used to trace the plant uptake of nutrients such as K$^+$. The biological half-life of ^{137}Cs can be evaluated from Figure 4a for pomegranate and Figure 4b for both red and green chili peppers, because the main source of the contamination of new leaves and fruits was the fraction of the available radiocaesium in the body of the plant. It is expected for ^{137}Cs concentrations to decrease faster than its physical half-life because of the loss of ^{137}Cs by leaves or fruits, plant growth, transfer back to the soil, etc. This loss can be estimated using a compartment model for long-term radiocaesium contamination of fruit trees proposed by Antonopoulos-Domis et al. [8]. According to this model, the loss of ^{137}Cs concentration in fruits or leaves can be described by the sum of two

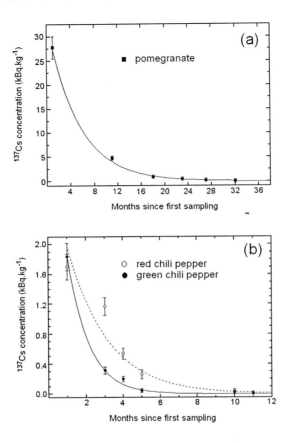

Figure 4 Time dependence of ^{137}Cs concentrations: a) in pomegranate fruits, for which the solid line represents a single exponential function fit, providing a λ_T value of 0.19 ± 0.01 months^{-1}. The fruits samplings consisted on five replicates of one fruit each. The error bars represent the standard deviation of the mean; b) in green and red chili peppers, for which the solid and dotted lines represent single exponential function fits, providing λ_T values of 0.89 ± 0.06 months^{-1} and 0.47 ± 0.02 months^{-1} for green and red chili peppers, respectively. About 100 units of green and red chili peppers were collected from each sampling, resulting in 3 replicates from each of them. The error bars represent the standard deviation of the mean.

exponentials:

$$C^{Cs}(t) = A.e^{-\lambda_T . t} + B.e^{-\lambda_u . t} \quad (1)$$

where, A, B, $\lambda_T > 0$ and $\lambda_u > 0$ are constants. The first term of Equation (1) corresponds to the translocation of cesium from the tree reservoir, and the second is due to root uptake. Given that this experiment was performed with the aim of investigating how a fraction of the available radiocaesium in the body of the plant can be redistributed to other permanent organs, the second term can be neglected.

From λ_T there were obtained the following biological half-life values due to ^{137}Cs translocation from the tree reservoir (BHL$_T$): 0.30, 0.12 and 0.07 years for pomegranate and red and green peppers, respectively. It is interesting to note that the (BHL$_T$) of pomegranate is about 100 times lower than the physical decay constant of ^{137}Cs ($T_{1/2} = 30.2$ years). In the case of chili peppers this ratio is 250 times lower. Knowing the rate at which such contaminant decline within various ecosystems components can be important for evaluating the length and severity of potential risks to resident species.

4 Conclusions

Our results have shown that ^{137}Cs and ^{40}K have similar distributions among some tree compartments. However, establishing the suitability of radioactive isotopes to trace plant uptake of nutrients is still an unsolved problem, and detailed studies for the development of this technique are still required.

The study of biological half-life due to ^{137}Cs translocation from the tree reservoir indicates that its values are of the order of months. Such information is important for the reclaiming of agricultural ecosystems after nuclear fallout. Knowing how tropical trees decontaminate after a period of growth in a ^{137}Cs contaminated area could be of interest in terms of risk assessment, and in terms of understanding the process involved in translocation of small ions in tropical or subtropical species.

Acknowledgements

The authors would like to thank the Brazilian funding agencies FAPERJ, CNPq and CAPES for their financial support.

Bibliography

[1] B. Mosquera, R. Veiga, L. Estellita, L. Mangia, D. Uzeda, C. Carvalho, T. Soares, B. Violini, R. M. Anjos. *^{137}Cs Distribution in guava trees.* Brazilian Journal of Physics 34 (2004) 841-844.

[2] B. Mosquera, C. Carvalho, R. Veiga, L. Mangia, R. M. Anjos. *^{137}Cs Distribution in tropical fruit trees after soil contamination.* Environmental and Experimental Botany 55 (2006) 273-281.

[3] C. Carvalho, B. Mosquera, R. Veiga, R. M. Anjos. *Radioecological Investigations in Brazilian Tropical Plants*. Brazilian Journal of Physics 35 (2005) 808-810.

[4] C. Carvalho, R. M. Anjos, B. Mosquera, K. Macario, R. Veiga. *Radiocesium contamination behavior and its effect on potassium absorption in tropical or subtropical plants*. Journal Environmental Radioactivity 86 (2006) 241-250.

[5] R. M. Anjos, B. Mosquera, C. Carvalho, N. Sanches, J. Bastos, R. Veiga, K. Macario. *Accumulation and long-term decline of radiocaesium contamination in tropical fruit trees*. Nuclear Instruments and Methods in Physics Research (2006) in press.

[6] Y-G. Zhu and E. Smolders. *Plant uptake of radiocaesium: a review of mechanisms, regulation and application*. Journal of Experimental Botany 51 (2000) 1635-1645.

[7] R. M. Anjos, N. K. Umisedo, A. Facure, E. M. Yoshimura, P. R. S. Gomes, E. Okuno. *Goiânia: 12 Years after the ^{137}Cs Radiological Accident*. Radiation Protection Dosimetry 101 (2002) 201-204.

[8] M. Antonopoulos-Domis, A. Clouvas and A. Gagianas. *Compartment model for long-term contamination prediction in deciduous fruit trees after a nuclear accident*. Health Physics 58 (1990) 737-741.

Anais da XXVIII Reunião de Trabalho sobre Física Nuclear no Brasil
SP, Brasil, 2005

Implications of the $\rho(770)^0$ Mass Shift Measured at RHIC

P. Fachini, R. Longacre, Z. Xu and H. Zhang

Brookhaven National Laboratory, Upton, NY, 11973, USA

Abstract. $\rho(770)^0$ mass shifts of about -40 MeV/c^2 and -70 MeV/c^2 were measured in p+p and peripheral Au + Au collisions at RHIC, respectively. The possible explanations for a mass shift are phase space, medium modification, Bose-Einstein correlations, and interference. We will show that the phase space does not account for the $\rho(770)^0$ mass shift measured at RHIC, CERN, and LEP. In addition, we will discuss the phase space in p + p collisions and conclude that there are significant scattering interactions in p + p reactions.

PACS. 2 5.75.Dw,13.85.Hd

1 Introduction

In-medium modification of the ρ meson due to the effects of increasing temperature and density has been proposed as a possible signal of a phase transition of nuclear matter to a deconfined plasma of quarks and gluons, which is expected to be accompanied by the restoration of chiral symmetry [1].

The ρ^0 meson measured in the dilepton channel probes all stages of the system formed in relativistic heavy-ion collisions because the dileptons have negligible final state interactions with the hadronic environment. Heavy-ion experiments at CERN indicate an enhanced dilepton production cross section in the invariant mass range of 200-600 MeV/c^2 [2]. The hadronic decay measurement at RHIC [3], $\rho(770)^0 \to \pi^+\pi^-$, was the first of its kind in heavy-ion collisions. Since the ρ^0 lifetime of $c\tau$ = 1.3 fm is small with respect to the lifetime of the system formed in Au + Au collisions, the ρ^0 meson is expected to decay, regenerate, and rescatter all the way through kinetic freeze-out. Therefore, the measured ρ^0 mass at RHIC should reflect conditions at the late stages of the collisions. [4, 5]. However, its has been shown that the ρ^0 width goes to zero with the ρ^0 mass dropping near the chiral phase transition, which is called the the vector manifestation and it can occur near T_c in the hot and dense matter [6]. In this scenario, the ρ^0 lifetime becomes very large and the ρ^0 produced at the early stages can also be measured in the hadronic channel.

135

2 Discussion ρ Mass Average From the Particle Data Group (PDG)

We will first discuss the ρ mass average from the PDG [7]. The ρ^0 mass average 775.8 ± 0.5 MeV/c^2 from e^+e^- was obtained from either $e^+e^- \to \pi^+\pi^-$ or $e^+e^- \to \pi^+\pi^-\pi^0$. This means that the ρ^0 mass average was obtained from *leptonic exclusive interactions*. Similarly, the $\rho^\pm$ mass average 775.5 ± 0.5 MeV/c^2 was also was obtained from *leptonic exclusive interactions*.

The ρ averages reported by the PDG from reactions other than leptonic interactions are systematic lower than the value above by ~ 10 MeV/c^2 [7]. These hadronic interactions are both inclusive and exclusive. In the case of inclusive productions, the phase space was take into account when the ρ mass was measured (e.g. [8]).

These observations lead us to conclude that the ρ mass changes whether it is measured in a hadronic or leptonic interaction, or whether it is measured in an exclusive or inclusive production. Since a leptonic reaction and exclusive measurement of the ρ^0 mass lead to a negligible modification of any kind of the ρ^0 mass, the average 775.8 ± 0.5 MeV/c^2 [7] from e^+e^- should correspond to the ρ^0 mass in the vacuum.

3 ρ^0 Mass Shifts at RHIC, CERN, and LEP

The ρ^0 was measured in the hadronic decay channel $\rho^0 \to \pi^+\pi^-$ at RHIC, CERN, and LEP. At RHIC, the STAR collaboration measured the ρ^0 at $\sqrt{s_{NN}} = 200$ GeV and observed mass shifts of the position of the ρ^0 peak of about -40 MeV/c^2 and -70 MeV/c^2 in minimum bias p + p and peripheral Au + Au collisions, respectively [3]. The invariant mass distributions from [3] is shown in Fig. 1. The STAR collaboration took into account the phase space (PS) [4, 9, 10, 11, 12, 13, 14, 15, 16] and the ρ^0 distribution was fit to the BW × PS function [3], where BW is the relativistic Breit-Wigner function [3]. The ρ^0 mass obtained from the BW × PS fit is depicted in Fig. 2, where it is clear that the phase space did not account for the measured mass shifts of the position of the ρ^0 peak.

The K*(892) was also measured at RHIC by the STAR collaboration and a mass shift of ~ 10 MeV/c^2 was observed for $p_T < 1$ GeV/c^2 [17]. Previous experiments have not observed a mass shift for the K*(892), for a review see [7]. However, these experiments have not performed the detailed study of the K* mass that the STAR collaboration achieved [17]. Actually, these previous experiments reported the K* mass *integrated* in p_T, x_F, or x_p [7].

At CERN, NA27 measured the ρ^0 in minimum bias p+p at $\sqrt{s} = 27.5$ GeV and reported a mass of 762.6 ± 2.6 MeV/c^2 [8]. The invariant $\pi^+\pi^-$ mass distribution after subtraction of the mixed-event reference distribution was fit to the BW×PS function [8] and is shown in Fig. 3. The vertical dashed line represent the average

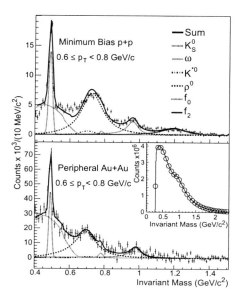

Figure 1 The raw $\pi^+\pi^-$ invariant mass distributions after subtraction of the like-sign reference distribution for minimum bias p + p (top) and peripheral Au + Au (bottom) interactions measured by STAR. For details see [3]

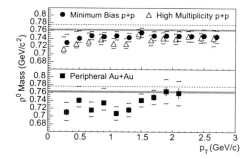

Figure 2 The ρ^0 mass as a function of p_T for minimum bias p + p (filled circles), high multiplicity p + p (open triangles), and peripheral Au + Au (filled squares) collisions measured by STAR. The error bars indicate the systematic uncertainty. Statistical errors are negligible. The ρ^0 mass was obtained by fitting the data to the BW×PS functional form described in [3]. The dashed lines represent the average of the ρ^0 mass measured in e^+e^- [7]. The shaded areas indicate the ρ^0 mass measured in p + p collisions [8]. The open triangles have been shifted downward on the abscissa by 50 MeV/c for clarity.

of the ρ^0 mass 775.8 ± 0.5 MeV/c² measured in e^+e^- [7]. The vertical solid line is the ρ^0 mass 762.6 ± 2.6 MeV/c² reported by NA27 [8]. As shown by Fig. 3, the position of the ρ^0 peak is shifted by $\sim$ 40 MeV/c^2 compared to the ρ^0 mass in the vacuum 775.8 ± 0.5 MeV/c^2 [7]. As mentioned, NA27 obtained the ρ^0 mass by fitting the invariant $\pi^+\pi^-$ mass distribution to the BW×PS function, and they reported a mass of 762.6 ± 2.6 MeV/c^2, which is $\sim$ 10 MeV/c^2 lower than the ρ^0 mass in the vacuum. Ideally, the PS factor should have accounted for the shift on the ρ^0 peak, and the mass obtained from the fit should have agreed with the ρ^0 mass in the vacuum. However, this was not the case, since the phase space did not account for the mass shift on the position of the ρ^0 peak.

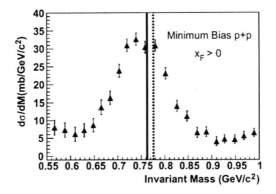

Figure 3 The invariant $\pi^+\pi^-$ mass distribution after subtraction of the mixed-event reference distribution for minimum bias p + p collisions measured by NA27. For details see [8]. The vertical dashed line represent the average of the ρ^0 mass 775.8 ± 0.5 MeV/c^2 measured in e$^+$e$^-$ [7]. The vertical solid line is the ρ^0 mass 762.6 ± 2.6 MeV/c^2 reported by NA27 [8].

At LEP, OPAL, ALEPH, and DELPHI measured the ρ^0 in inclusive e$^+$e$^-$ reactions at $\sqrt{s}$ = 90 GeV [18, 19, 20, 21]. Even though OPAL never reported the value of the ρ^0 mass, OPAL reported a shift on the position of ρ^0 peak by $\sim$ 70 MeV/c^2 at low x$_p$ and no shift at high x$_p$ (x$_p$ $\sim$ 1) [18, 19]. OPAL also reported a shift in the position of the $\rho^\pm$ peak from -10 to -30 MeV/c^2, which was consistent with the ρ^0 measurement [18, 19]. ALEPH reported the same shift on the position of ρ^0 peak observed by OPAL [20]. DELPHI fit the invariant $\pi^+\pi^-$ mass distribution after background subtraction to the BW×PS functionin a particular x$_p$ region (0.1 < x$_p$ < 0.4) and reported a ρ^0 mass of 757 ± 2 MeV/c^2 [21], which is five standard deviations below than the ρ^0 mass in the vacuum (775.8 ± 0.5 MeV/c^2). Bose-Einstein correlations were used to describe the shift on the position of ρ^0 peak. However; high (even unphysical) Λ-parameter (Λ $\sim$ 2.5) were needed [18, 19, 20]

4 Phase Space in Au+Au Collisions

In Au + Au collisions, the phase space accounts for hadrons scattering and forming resonances. In the case of the ρ^0, $\pi^+\pi^- \to \rho^0 \to \pi^+\pi^-$. The effect of the phase space on the ρ^0 mass distribution can be studied using transport model calculations, such as UrQMD [14]. Figure 4 depicts the ρ^0 distribution (solid line) from 10^5 central (b < 3fm) Au + Au UrQMD events at $\sqrt{s_{NN}} = 200$ GeV for $0.2 \leqslant p_T < 0.4$ at midrapidity ($|y| \leqslant 0.5$). The vertical solid line is the input value of 769 MeV/c^2 for the ρ^0 mass in UrQMD. A shift of ~ 30 MeV on the position of the ρ^0 peak is observed. The dashed line in Fig. 4 is the fit to the BW×PS function [3] and the mass of 765 MeV/c^2 obtained from the fit is equivalent to the input value of 769 MeV/c^2.

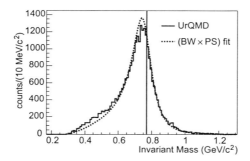

Figure 4 ρ^0 distribution (solid line) from 10^5 central (b < 3fm) Au + Au UrQMD events at $\sqrt{s_{NN}} = 200$ GeV for $0.2 \leqslant p_T < 0.4$ at $|y| \leqslant 0.5$. The vertical solid line is the input value of 769 MeV/c^2 for the ρ^0 mass in UrQMD. The dashed line is the fit to the BW×PS function [3].

5 Phase Space in p+p Collisions

In multiparticle production processes, single-inclusive (e.g. p+p), invariant particle spectra are typically exponential in p_T. The reason for the exponential behavior is the phase space of the final multiparticle state [22]. The slope parameter in p + p collisions are independent of the particle mass [23]. At RHIC, the ρ^0 spectra in minimum bias and high multiplicity p + p are exponential in p_T up to 1.1 GeV/c^2 with slope parameters of ~ 180 MeV. The ρ^0 spectrum in minimum bias p + p [3] is depicted in Fig. 5. For reference, the slope parameter of π^- is ~ 160 MeV [24]. Note that the slope parameter in p + p is independent of the particle mass and also independent of the multiplicity.

We should discuss the effect of the phase space on the ρ^0 mass distribution. At RHIC, the ρ^0 mass measured in p + p is multiplicity dependent (the mass shift in high multiplicity is higher than in minimum bias p + p collisions) [3], which is opposite to the slope parameter that is multiplicity independent. According to quantum mechanics, a resonance has the form of a non-relativistic Breit-Wigner distribution [25]. Furthermore, following the rules of quantum field theory, an unstable particle has the form of a relativistic Breit-Wigner function. As a consequence, in single-inclusive multiparticle production processes, the phase space of the final multiparticle state *cannot* influence the resonance mass distribution.

Since the phase space of the final multiparticle state *cannot* explain the distortion of the ρ^0 line shape, one can conclude that the phase space in p + p collisions also accounts for hadrons scattering and forming resonances. In the case of the ρ^0, $\pi^+\pi^- \to \rho^0 \to \pi^+\pi^-$. This can be pictured in the string fragmentation particle production scenario, where the string breaks several times, two pions are formed, they scatter, and form a ρ^0. Such interactions are significant and modify the ρ^0 line shape in e^+e^- and p + p from a relativistic p-wave Breit-Wigner function.

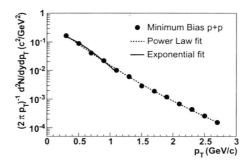

Figure 5 The p_T distribution at $|y| < 0.5$ for minimum bias p + p collisions [3]. The black circles are the data, the dashed line is the power-law fit and the solid line is the exponential fit. See [3] for the fit functions. The errors shown are statistical only and smaller than the symbols.

6 Conclusions

We showed that a shift on the position of the ρ^0 peak has been measured before and that the phase space does not account for the $\rho(770)^0$ mass shift measured at RHIC, CERN, and LEP. In addition, we discussed the phase space in p+p collisions and concluded that there are significant scattering interactions in p + p reactions. These interactions modify the ρ^0 line shape in p + p and e^+e^- interactions from a relativistic p-wave Breit-Wigner function.

Acknowledgements

This work was supported in part by the HENP Divisions of the Office of Science of the U.S. DOE.

Bibliography

[1] R. Rapp and J. Wambach, Adv. Nucl. Phys. **25**, 1 (2000).

[2] G. Agakishiev *et al.*, Phys. Rev. Lett. **75**, 1272 (1995); B. Lenkeit *et al.*, Nucl. Phys. A **661**, 23 (1999).

[3] J. Adams *et al.*, Phys. Rev. Lett. **92** 092301 (2004).

[4] E.V. Shuryak and G.E. Brown, Nucl. Phys. A **717**, 322 (2003).

[5] R. Rapp, Nucl.Phys. A **725**, 254 (2003).

[6] M. Harada and K. Yamawaki, Phys. Rept. 381, 1 (2003).

[7] S. Eidelman *et al.*, Phys. Lett. B **592**, 1 (2004).

[8] M. Aguilar-Benitez *et al.*, Z. Phys. C **50**, 405 (1991).

[9] H.W. Barz *et al.*, Phys. Lett. B **265**, 219 (1991).

[10] P. Braun-Munzinger (private communication).

[11] P.F. Kolb and M. Prakash, nucl-th/0301007.

[12] R. Rapp, Nucl. Phys. A **725**, 254 (2003).

[13] W. Broniowski *et al.*, Phys. Rev. C **68**, 034911 (2003).

[14] M. Bleicher and H. Stöcker, J. Phys. G **30**, S111 (2004).

[15] S. Pratt and W. Bauer, Phys. Rev. C **68**, 064905 (2003).

[16] P. Granet *et al.*, Nucl. Phys. B **140**, 389 (1978).

[17] J. Adams *et al.*, Phys. Rev. C **71** 064902 (2005).

[18] P.D. Acton *et al.*, Z. Phys. C **56**, 521 (1992).

[19] G.D. Lafferty, Z. Phys. C **60**, 659 (1993); (private communication).

[20] D. Buskulic *et al.*, Z. Phys. C **69**, 379 (1996).

[21] K. Ackerstaff *et al.*, Eur. Phys. J. C **5**, 411 (1998).

[22] R. Hagedorn, Relativistic Kinematics, W.A. Benjamin, 1963; E. Byckling and K. Kajantie, Particle Kinematics, Wiley, 1973.

[23] I.G. Bearden *et al.*, Phys. Rev. Lett. **78** 2080 (1997).

[24] J. Adams *et al.*, Phys. Rev. Lett. **92** 112301 (2004).

[25] J.J. Sakurai, Modern Quantum Mechanics, Addison Wesley, 1985.

Anais da XXVIII Reunião de Trabalho sobre Física Nuclear no Brasil
SP, Brasil, 2005

Real-Time Radiographic Imaging With Thermal Neutrons

Marcelo J. Gonçalves[1], Ricardo Tadeu Lopes[1,*], Maria Ines Silvani[2],
Gevaldo L. de Almeida[2], Rosanne C. A. A. Furieri[2]

[1] UFRJ - COPPE

[2] Instituto de Engenharia Nuclear - CNEN
C.P. 68550, Ilha do Fundão,
21945-970 Rio de Janeiro, RJ, Brazil
e-mail: msouza@ien.gov.br

Abstract. This paper describes the accomplishments achieved by an work addressing the development of 2D and 3D Image Acquisition Systems capable to perform *quasi* real-time non-destructive assays. It is comprised of a thermal neutron source provided by the Argonauta Reactor at the *Instituto de Engenharia Nuclear /CNEN*, Brazil, capable to furnish a beam of 10^5 n $\cdot$ cm^{-2} $\cdot$ s^{-1}, a converter-scintillating screen, and CCD-based video camera optically coupled to the screen through a proper device. A non-refrigerated video-camera - the refrigeration could reduce the electronic noise improving thus the image quality - has been used. None kind of light-intensifier was employed as well to enhance the system sensitivity. In spite of this simple configuration, digital images of acceptable quality have been obtained with this video-camera coupled to a microcomputer carrying out the signal processing. The performance of the system has been assessed through its *Modulation Transfer Function - MTF*, a proper curve to express the changes in resolution with the spatial frequency. In order to determine this curve, unique collimators designed to simulate different spatial frequencies have been manufactured. Besides that, images of some objects have been acquired with the system being developed, as well as with a system using conventional radiographic films, allowing thus a qualitative comparison between them.

1 Introduction

Thermal neutrons have a great potential capability to perform non-destructive assays thanks to their high penetrability. This is specially true for heavy materials,

---

* ricardo@lin.ufrj.br

which attenuate intensely X-rays, becoming hence opaque to them, which precludes an inspection of their inner structure.

Another remarkable characteristic of the thermal neutrons is that the absorption cross-section does not increase regularly with atomic number, as X-rays do. Such a behavior makes sometimes possible the differentiation between neighbor elements, a task ordinarily difficult or even impossible for X-rays.

The utilization of a CCD-based video-camera in a image acquisition system presents several relevant advantages over the conventional radiographic film as follows. First, the exposure time is dramatically shortened, reaching up to 1% of that required by conventional film radiography. Such a short exposure time allows a *quasi* real time image acquisition of moving objects. Second, unlike systems employing radiographic films, the images are obtained already in the digital form, facilitating thus its treatment and storage.

This work presents some preliminary results obtained with the system being developed comprised of a CCD-based video camera - commercially available for conventional tasks such as surveillance and security - and a converter-scintillating screen as shown in Fig.1. These results are then compared with equivalent ones obtained with systems using radiographic films.

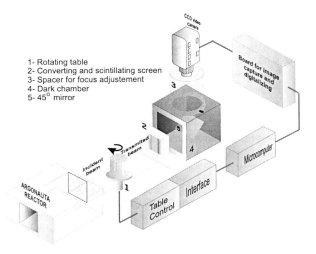

Figure 1 Brief scheme of the Image Acquisition System under development.

2 Neutron Conversion and Detection

The thermal neutron beam used in this work was produced by the Argonauta Research Reactor installed at the *Instituto de Engenharia Nuclear/CNEN* - Brazil. Its main properties are presented in Table 1.

The converter-scintillating screen acts as a transducer detecting the neutrons and emitting a visible bluish light with a peak at 450nm. It is constituted by a thin 15x10 cm^2 plastic plate with a silver-activated Li^{6}F(ZnS) layer deposit on its surface, being commercially known as NE426.

Table 1 Properties of the Neutron Beam.

Thermal Flux	$4.46 \times 10^5\,\mathrm{n} \cdot \mathrm{cm}^{-2} \cdot \mathrm{s}^{-1}$
Epithermal Flux	$6.00 \times 10^3\,\mathrm{n} \cdot \mathrm{cm}^{-2} \cdot \mathrm{s}^{-1}$
Neutron/Gamma Ratio	$3.00 \times 10^6\,\mathrm{n} \cdot \mathrm{cm}^{-2} \cdot \mathrm{mR}^{-1}$
L/D Ratio	63.25
Cadmium Ratio	25
Beam Divergence	$1°16'$

Its luminescence decay time is short enough to allow the imaging of moving objects by using a suitable video camera. In this work a Panasonic CCD video camera model AW-E650 has been employed. It requires a minimum illumination threshold of 5×10^{-5} lx and allows a maximum integration time of 2 seconds. The optical coupling between the screen and the camera is performed by a dark chamber provided with a first surface mirror deflecting the light from the screen to the vertical direction where the camera axis is positioned. This arrangement prevents the direct incidence of neutrons to the camera which could damage it.

3 Image Quality

The *Modulation Transfer Function - MTF* [2] is an adequate tool to evaluate the performance of an imaging system. It shows how the capability of a system to resolve neighbor features is impaired by their spatial frequency. Relevant and useful information concerning the system can be inferred from this function, such as spatial resolution and contrast.

The MTF can be obtained by using special, and very expensive (when available at all), multi-slit collimators, or alternatively by a Fast Fourier Transform - FFT of the *Line Spread Function - LSF*. This function expresses the response of an imaging system to an ideal line source: a blurred strip instead of a single line. Methods to get the LSF can be found elsewhere [2,3].

Besides the resolution, the contrast plays an important role in the overall image quality. Among other factors, it depends mainly on the *modulation* - defined as the *Amplitude-to-Average Value* ratio - resulting from the overlap between neighbor LSFs. As one can infer from Fig.2, high spatial frequencies associated with low detector resolutions (broad LSFs) lead to low modulations and consequently to a poor image contrast.

An image is considered of acceptable quality when the modulation reaches at least 10 %. [2].

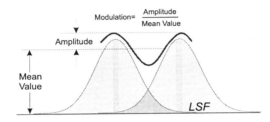

Figure 2 Overlap of 2 neighbor *Line Spread Functions*. High spatial frequencies and poorer detector resolutions produce low modulations degrading the image contrast.

4 Methodology to Determine the MTF

In order to determine the MTF using the slit collimator approach, instead of the FFT technique, it is necessary measure the modulation for several points of the spatial frequency domain. Since a single collimator of this kind, i.e., a piece containing constant-width slits at different gaps, was not available, several individual *three-slits* collimators have been manufactured. A photographic top view of one of these devices is shown in Fig.3. Their slit width ranged from 0.0125 to 0.560 mm, covering thus the spatial frequency 1.8 - 80 mm^{-1}. The modulation has been determined by mapping the optical density of neutrongraphic images of the collimator. A plot of the modulation versus spatial frequency, as defined by the slit widths, yields the aimed MTF. Further details can be found elsewhere [4].

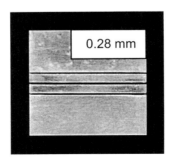

Figure 3 Photographic top view of a *three-slits* collimator with a 0.28 mm slit.

Table 2 Resolution and required acquisition times for the developed system and a conventional one employing radiographic films.

SYSTEM	Video Camera	Radiographic Film
Resolution(mm)	0.26	0.015
Acquisition Time	2 s	40 min + Development Time

5 Quantitative Results

The MTF for the developed system presented in Fig.4 exhibits a 10 % cutoff-value at the spatial frequency of 3.9 mm^{-1}. Therefore the system is capable to distinguish features up to 0.26 mm apart from each other.

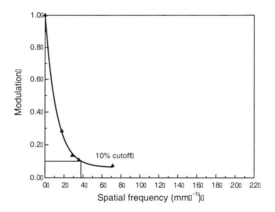

Figure 4 MTF for the developed system.

When compared with a conventional *radiographic film* imaging system, the system being developed using a converter-scintillating screen & video camera exhibits a much higher cutoff-value as shown in Table 2. Indeed, while the first system could resolve features up to 0.015 mm apart [4], the second one only would succeed for a gap about 20 times larger. Nevertheless, the exposure time required by the conventional system surpasses that required by the system using a converter-scintillating screen by a factor 1,200, excluding the time required for chemical processing.

6 Qualitative Results

A high resolution is solely one of the desirable properties that determine the overall performance of an imaging system. In some circumstances, other characteristics may become more important as they could impact directly on efforts and costs. Although the ultimate goal of this work is the development of an imaging system capable to deal with moving objects, some images of static objects have been taken as well, addressing a proper system characterization. Within this frame, Fig.5 shows images of an automobile fuel injector, taken with the system being developed (a) and with a conventional one employing radiographic films (b). A conventional photography of this device is shown as well for comparison.

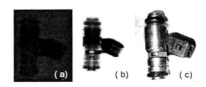

Figure 5 Neutrongraphic images of an automobile fuel injector taken with the developed system (a) and with a conventional one employing radiographic films (b). A conventional photography of the device is also shown as reference (c).

The conventional system produces an image of superior quality. Yet, the developed system exhibits an image that can be regarded as acceptable. Indeed, it reveals the same basic inner structure of the device as its companion.

Fig. 6 shows images of an automotive oil pressure probe taken with both systems for comparison. As in the previous example, the higher image quality produced by the conventional system is evident.

A very interesting feature of the imaging system employing a converter-scintillating screen optically coupled to a video camera is the possibility to study the kinetics of systems as those involving moving objects or materials. Fig.7 for instance, shows some frames picked up from an actual quasi real-time movie of an ordinary lighter (placed upside down during the take), submitted to an intentional gas leakage. Thanks to the hydrogen contents of the hydrocarbon fuel, it is possible to get a high image contrast defining the fluid level.

The faster movement of a mechanical chronometer has been filmed with the developed system as well. For this purpose, a thin cadmium strip has been glued up to its major pointer to increase its neutron attenuation. Three chosen frames corresponding to elapsed times 0, 17 and 20 seconds are shown in Fig. 8. The pointer appears somewhat blurred due to the long integration time (2 seconds) required to obtain a reasonable exposure. The utilization of a camera with a higher sensitivity, or the employing of equipment to enhance the signal-to-noise ratio, such as light intensifiers or cooling devices would certainly improve the image quality.

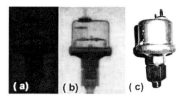

Figure 6 Neutrongraphic images of an automotive oil-pressure probe obtained with the developed system (a), and with the conventional radiographic film system (b). A conventional photo of the device is also shown as reference (c).

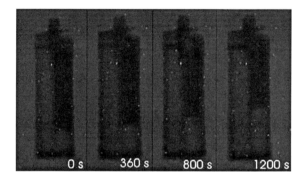

Figure 7 Some frames picked up from the movie of an ordinary lighter - placed upside down during the take - with a gas leakage.

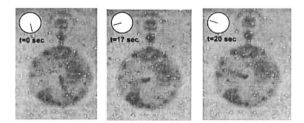

Figure 8 Frames picked up from the movie of a mechanical chronometer. The time elapsed between the left and the right frame was 20 seconds.

7 Conclusions

A low-cost *quasi* real-time imaging system comprised of a converter-scintillating screen optically coupled to a general use video-camera has been assembled, tested and evaluated on both qualitative and quantitative aspects. The quality of the images produced by this system cannot compete with those given by a conventional imaging system employing radiographic films. However, it is capable to deal with moving objects, a task obviously impossible for the conventional system. It employs a reusable detector that furnishes final images already in the digital form without need of chemical processing in a dark room for its development. In addition, it requires an exposure time several orders of magnitude shorter than a conventional radiographic film system.

Bibliography

[1] M. j. Gonçalves, R. T. Lopes, M. i. Silvani, G. L. Almeida and R. C. A. A. Furieri, and R. T. Lopes, Brazilian J. of Physics, 35,3B,763-766(2005).

[2] ASTM E1441-95 and 1570-95a, Non-Destructive Testing, Radiation Methods, Computed Tomography. Guide for Imaging and Practice for Examination, ISSO/TC 135/SC, N118 USA (1996).

[3] G. L. Almeida, M. I. Silvani and R. T. Lopes, ENAN 2002, Rio de Janeiro-Brasil, 2002, E10-142.

[4] M. j. Gonçalves, R. T. Lopes, M. i. Silvani, G. L. Almeida and R. C. A. A. Furieri, and R. T. Lopes, ENAN 2005, Santos-Brasil, 2005.

Anais da XXVIII Reunião de Trabalho sobre Física Nuclear no Brasil
SP, Brasil, 2005

A QTDA Model for Double Beta Decay

A. R. Samana[1,2], F. Krmpotić[2,3,4]

[1] Centro Brasileiro de Pesquisas Físicas, CEP 22290-180, Rio de Janeiro, Brazil

[2] Departamento de Física Matemática, Instituto de Física da Universidade de São Paulo, Caixa Postal 66318, 05315-970 São Paulo, SP, Brazil

[3] Instituto de Física La Plata, CONICET, 1900 La Plata, Argentina

[4] Facultad de Ciencias Astronómicas y Geofísicas, Universidad Nacional de La Plata, 1900 La Plata, Argentina

Abstract. The Quasiparticle Tamm-Dancoff Approximation (QTDA) is applied to describe the nuclear double beta decay. The QTDA allows for the explicit evaluation of energy distributions of the double-charge-exchange transition strengths and of their sum rules, and can be straightforwardly applied to single- and double-closed shell nuclei. Several serious inconveniences found in the Quasiparticle Random Phase Approximation (QRPA) are not present in the QTDA. As example, the decay $^{48}\mathrm{Ca} \rightarrow {}^{48}\mathrm{Ti}$ is discussed within the particle-hole limit of QTDA .

PACS. 21.60.-n, 21.10.-k

Keywords. Energy spectra; double beta decay

1 Introduction

The experimental evidence of neutrino oscillations has opened a new exciting era in neutrino physics. Until this moment we do not know two crucial features of the neutrino physics, essentially: (i) the absolute mass scale, and (ii) whether the neutrino is a Majorana or a Dirac particle. This fact has motivated an attractive next generation of experiments [1] for many different isotopes, including $^{48}\mathrm{Ca}$, $^{76}\mathrm{Ge}$, $^{100}\mathrm{Mo}$, $^{116}\mathrm{Cd}$, $^{130}\mathrm{Te}$, $^{136}\mathrm{Xe}$, $^{150}\mathrm{Nd}$, and $^{160}\mathrm{Gd}$.

There are in nature about 50 nuclear systems in which the single β-decay is energetically forbidden, and therefore the $\beta\beta$-decay turns out to be the only possible mode of disintegration $(\mathrm{N,Z}) \xrightarrow{\beta\beta^-} (\mathrm{N\text{-}2,Z+2})$. The nuclear pairing force causes such "anomaly", by making the odd-odd isobar, within the isobaric triplet $(\mathrm{N, Z})$, $(\mathrm{N}-1, \mathrm{Z}+1)$, $(\mathrm{N}-2, \mathrm{Z}+2)$, to have a higher mass than the even-even neighbors. The modes of these disintegrations are: (a) the two-neutrino double beta $(2\nu\beta\beta)$

decay, that can occur by two successive β decays, passing through the intermediate virtual states of the $(N-1, Z+1)$ nucleus, and (b) the neutrinoless $\beta\beta$ ($0\nu\beta\beta$) decay. These two modes of disintegration are related between them through the nuclear structure effects. The $0\nu\beta\beta$-decay can provide the information about the absolute mass scale of neutrino and if the neutrino is a Majorana or a Dirac particle. Their half-lives cast in the form:

$$T_{2\nu}^{-1} = \mathcal{G}_{2\nu}\mathcal{M}_{2\nu}^2, \qquad T_{0\nu}^{-1} = \mathcal{G}_{0\nu}\mathcal{M}_{0\nu}^2 \langle m_\nu \rangle^2, \tag{1}$$

where $\mathcal{G}$'s are geometrical phase space factors, and $\mathcal{M}$'s are nuclear matrix elements which present many similar features, to the fact that we shall not understand the $0\nu\beta\beta$-decay unless we understand the $2\nu\beta\beta$-decay.

Independently of the nuclear model which is used, and when only the allowed transitions are considered, the $2\nu\beta\beta$ matrix element for the $|0_f^+\rangle$ final state reads

$$\mathcal{M}_{2\nu}(f) = \sum_{\lambda=0,1} (-)^\lambda \sum_\alpha \left[\frac{\langle 0_f^+ || \mathcal{O}_\lambda^{\beta^-} || \lambda_\alpha^+ \rangle \langle \lambda_\alpha^+ || \mathcal{O}_\lambda^{\beta^-} || 0^+ \rangle}{\mathcal{D}_{\lambda_\alpha^+, f}} \right],$$
$$\equiv \mathcal{M}_{2\nu}^F(f) + \mathcal{M}_{2\nu}^{GT}(f), \tag{2}$$

where the summation goes over all intermediate virtual states $|\lambda_\alpha^+\rangle$, and

$$\mathcal{O}_\lambda^{\beta^-} = (2\lambda+1)^{-1/2} \sum_{pn} \langle p || O_\lambda || n \rangle \left(c_p^\dagger c_{\tilde{n}} \right)_\lambda,$$

$$\text{with} \qquad \begin{cases} O_0 = 1 & \text{, for F} \\ O_1 = \sigma & \text{, for GT} \end{cases}, \tag{3}$$

are the Fermi (F) and Gamow-Teller (GT) operators for β^--decay, and $c^\dagger$ (c) are the particle creation (annihilation) operators. The corresponding β^+-decay operators are $\mathcal{O}_\lambda^{\beta^+} = \left(\mathcal{O}_\lambda^{\beta^-} \right)^\dagger$, and

$$\mathcal{D}_{\lambda_\alpha^+, f} = E_{\lambda_\alpha^+} - \frac{E_0 + E_{0_f^+}}{2} = E_{\lambda_\alpha^+} - E_0 - \frac{E_{0_f^+} - E_0}{2}, \tag{4}$$

is the energy denominator. E_0 and $E_{0_f^+}$ are, respectively, the energy of the initial state $|0^+\rangle$ and of the final states $|0_f^+\rangle$.

The $\beta\beta$ decays occur in medium-mass nuclei that are often far from closed shells, and as a consequence most of the recent attempts to evaluate $\mathcal{M}_{2\nu}$ and $\mathcal{M}_{0\nu}$ are:

- The QRPA makes a large fraction of nucleons to take part in a large single-particle space, but within a modest configuration space, so it is by

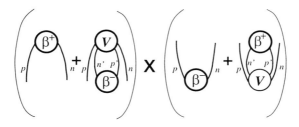

Figure 1 Graphical representation of the numerator in (2) within the QRPA. The first and second terms match, respectively, to $\langle 0_1^+ || \mathcal{O}_\lambda^{\beta^-} || \lambda_\alpha^+ \rangle$ and $\langle \lambda_\alpha^+ || \mathcal{O}_\lambda^{\beta^-} \rangle 0^+ \rangle$. The residual interaction brings about the ground-state correlations in the vertex V.

far simpler computationally than the Shell Model (SM). Moreover, it is clear as well that the QRPA equation of motion corresponds exactly to the full many-body Schrödinger equation if and only if the ground state in the equation of motion is the true ground state and a phonon creation operator exhausts the whole Hilbert space, e.g. all excitations: 1p-1h, 2p-2h, 3p-3h, and so on. Till now only 1p-1h excitations are used, and the ground state is approximated by the BCS vacuum. The graphical representation of the numerator in (2) within the QRPA is shown in Fig. (1).

- The SM deals with a small fraction of the nucleons in a limited single-particle space, but allows them to correlate in arbitrary ways within a large configuration space (1p-1h, 2p-2h, 3p-3h, and so on). In conventional shell model calculations, a complete set of many-body bases is used and this diagonalize the Hamiltonian matrix. The applicability of such a method has been limited to relatively light nuclei or those near the closed shell, because of the explosive increase of the basis dimension and occasionally the non-conservation of the sum rule in charge-exchange reactions. Owing to recent developments in computational facilities and numerical calculation techniques, most of the pf-shell nuclei are now in the scope of exact $0\hbar\omega$ calculations. The current frontier of such direct calculations is in the middle of the pf-shell ($A \approx 60$), where the maximum M-scheme dimension reaches to two billion [12].

Summarizing, the kinds of correlations that these methods include are not the same. Various and different QRPA have been solved and mounted, but as it is shown in [2], that none of the amendments of the QRPA, proposed so far to rescue this nuclear model, were able to change qualitatively the behavior of the amplitude $\mathcal{M}_{2\nu}$ (the model collapses in the physical region of the coupling constant for the pp channel) unless we agree to assume the violation of the Ikeda's Sum Rule, which could be dangerous as we have no control on how this affects the definite value of $\mathcal{M}_{2\nu}$.

2 Quasiparticle Tamm-Dancoff Approximation

We present a simple nuclear model for evaluating the $\beta\beta$ decay rates, based on the well-known Quasiparticle Tamm-Dancoff Approximation QTDA [4], where the main difference in comparison to the QRPA comes from how one describes the final $(N-2, Z+2)$ nucleus. As in the QRPA [3], we express the total Hamiltonian as

$$H = H_p + H_n + H_{pn} + H_{pp} + H_{nn} \equiv H_0 + H_{res}, \tag{5}$$

where H_p and H_n are, respectively, the effective proton and neutron single-quasiparticle Hamiltonians, and H_{pn}, H_{pp}, and H_{nn} are the matching effective two-quasiparticle interaction Hamiltonians for the valence quasiparticles. We assume that:

(i) the initial state is the BCS vacuum in the (N, Z) nucleus $|0^+\rangle = |BCS\rangle$,

(ii) the intermediate states are two quasiparticle excitations on this vacuum:

$$|\lambda_\alpha^+\rangle = \sum_{pn} X_{pn;\lambda_\alpha^+} |pn; \lambda^+\rangle,$$

with

$$|pn; \lambda^+\rangle = [a_p^\dagger a_n^\dagger]_{\lambda^+} |BCS\rangle,$$

(iii) the final states are four quasiparticle excitations on this vacuum:

$$|0_f^+\rangle = \sum_{p_1p_2n_1n_2J} Y_{p_1p_2n_1n_2J;0_f^+} |p_1p_2, n_1n_2; J\rangle,$$

with

$$|p_1p_2, n_1n_2; J\rangle = N(p_1p_2)N(n_1n_2)\{[a_{p_1}^\dagger a_{p_2}^\dagger]_J [a_{n_1}^\dagger a_{n_2}^\dagger]_J\}_0 |BCS\rangle,$$

and $N(ab) = (1 + \delta_{ab})^{-1/2}$. Here $a^\dagger$ (a) is the quasiparticle creation (annihilation) operator relative to the BCS vacuum. One can find the matrix elements of H_{pn}, H_{pp} and H_{nn} in [4]. The explicit results for the one-body matrix elements that appear in (2), are:

$$\langle \lambda_\alpha^+ || \mathcal{O}_\lambda^{\beta^-} || 0^+ \rangle = \sum_{pn} \Lambda_+^0 (pn; \lambda) X_{pn;\lambda_\alpha^+}, \tag{6}$$

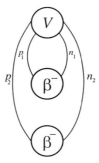

Figure 2 Graphical representation of $\mathcal{M}_{2\nu}$ in the QTDA. The first and second vertices match, respectively, with the matrix elements (2) and (3), and the third vertex represents the residual interaction in the final state.

and

$$\langle 0_f^+ \| O_\lambda^{\beta^-} \| \lambda_\alpha^+ \rangle = \\ -\sum_{pn} X_{pn;\lambda_\alpha^+} \sum_{p_1 p_2 n_1 n_2 J} Y_{p_1 p_2 n_1 n_2 J; 0_f^+} N(p_1 p_2) N(n_1 n_2) \bar{P}(p_1 p_2 J) \\ \times \bar{P}(n_1 n_2 J) \sqrt{2J+1}(-)^{n_1+p_2+J+\lambda} \\ \times \begin{Bmatrix} p_1 & n_1 & \lambda \\ p & n & J \end{Bmatrix} \Lambda_+^0(p_1 n_1; \lambda) \delta_{p_2 p} \delta_{n_2 n}, \quad (7)$$

where $\bar{P}(p_1 p_2 J) = 1 - (-)^{p_1+p_2+J} P(p_1 \leftrightarrow p_2)$ is the permutation operator. The energies in the denominator (2) are

$$E_{\lambda_\alpha^+} = E_0 + \omega_{\lambda_\alpha^+} + \lambda_p - \lambda_n, \\ E_{0_f^+} = E_0 + \omega_{0_f^+} + 2\lambda_p - 2\lambda_n, \quad (8) \\ \mathcal{D}_{\lambda_\alpha^+,f} = \omega_{\lambda_\alpha^+} - \frac{\omega_{0_f^+}}{2}.$$

where $\omega_{\lambda_\alpha^+}$ and $\omega_{0_f^+}$ are the eigenvalues of the Hamiltonian (5) for the intermediate states $|\lambda_\alpha^+\rangle$ and of the final states $|0_f^+\rangle$, respectively, and λ_p and λ_n are the chemical potentials.

For the residual interaction we adopted the delta force,

$$V = -4\pi(\nu_s P_s + \nu_t P_t)\delta(r)$$

which has been used extensively in the literature [5]. The ν_s and ν_t are the interaction strength, P_s and P_t are the projection operators for the singlet and triplet channel respectively. The Fig. (2) shows the graphical representation of $\mathcal{M}_{2\nu}$ in the QTDA.

3 Results and Conclusions

Our propose here is to explain how the QTDA quenching-mechanism works and for the first calculation we take the particle-hole limit when $v_p \to 0, v_n \to 1$ (the $v_{p(n)}^2$, merged from the BCS equations, give the occupation probability for the quasiparticle in the p (n) level).

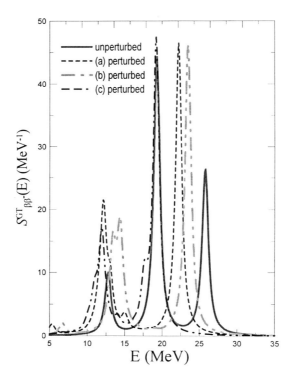

Figure 3 Graphical representation of the energy dependence of the double GT strengths $S_{GT}^{\beta\beta}$, measured from the ^{48}Ca ground state. The coupling constant in the perturbed calculations values were $v_s = 35$, $v_t = 65$ case (a), and $v_s = 40$, $v_t = 60$ case (b), and $v_s = 41$, $v_t = 64$ case (c) for a new set of SPE from [6]. More details in the text.

One delicate issue is the choice of the effective single-particle energies (SPE). We will assume them to be $\epsilon_{j_n}^0$ from the experimental SPE for ^{40}Ca, taken from the Fig. 2 in Ref. [10], namely: $\epsilon_{f_{7/2}}^0 = 0$, $\epsilon_{f_{5/2}}^0 = 6.5$, $\epsilon_{p_{3/2}}^0 = 2.1$, and $\epsilon_{p_{1/2}}^0 = 4.1$, in units of MeV, and they were corrected using the self-energies that account for 8 neutrons in the $f_{7/2}$ shell, and using the Coulomb displacement energy in ^{48}Ca for the protons. The values of coupling constant in the perturbed calculations were

$v_s = 35$, $v_t = 65$ in the case (a), and $v_s = 40$, $v_t = 60$ in the case (b). The graphical representation of the energy dependence of the double GT strengths $S_{GT}^{\beta\beta}$, measured from the ^{48}Ca ground state are shown in Fig. (3). This figure shows the GT distributions for the cases (a) and (b) for the particle-hole limit of the QTDA, and for a new case (c) obtained with the coupling constants $v_s = 41$, $v_t = 64$ for a new set of SPE, in the same model space that the ones cases (a) and (b), namely: $\epsilon_{f_{7/2}} = -9.62$, $\epsilon_{f_{5/2}} = -5.80$, $\epsilon_{p_{3/2}} = -6.93$, $\epsilon_{p_{1/2}} = -4.58$ for protons, and $\epsilon_{f_{7/2}} = -9.94$ for neutron, in units of MeV. This new set, previously used in the QRPA [6], is important because it will be the trial set to implement the full QTDA in a future work.

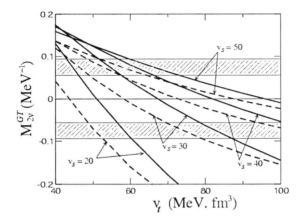

Figure 4 Graphical representation of the $\beta\beta$ amplitudes $\mathcal{M}_{2\nu}^{GT}$ in the QTDA as a function of the coupling constants v_s and v_t. The solid lines were obtained with the SPE mentioned in the text, and the dashed line with the SPE from [6]. The lined boxes in blue represent the experimental value of $|\mathcal{M}_{2\nu}|$ [8]

Let us comment briefly some results for the $\beta\beta$-strength. For the unperturbed case, the DIAS (Double Isobaric Analog State), at energy $2\Delta_c = 12.78$ MeV, have only 12% of the total GT intensity. The major part of the GT part of the double unperturbated GT strength concentrates with 55% and 33% rates in the levels that lie, respectively, at energies $2\Delta_C + \epsilon_{f_{5/2}}^0$, and $2\Delta_C + 2\epsilon_{f_{7/2}}^0$, which as seen from the Figure (3). For the lowest final 0^+ state the unperturbed denominators (4) are null, and both $\mathcal{M}_{2\nu}^F$ and $\mathcal{M}_{2\nu}^{GT}$, to become ∞. The situation changes when the residual interaction is switched on. First, as we show in Figure (4) (the solid lines were drawn with the SPE above mentioned, whereas the dashed lines were drawn with the SPE used in the QRPA [6]), there are a great variety of physical sound values for v_s and v_t that allow the model to account for the phenomenological nuclear matrix elements (2) and (3). In another words, same as the QRPA, the QTDA restore the Wigner SU(4) symmetry, quenching in this way the nuclear matrix elements. This is the mechanism we have been searching for.

For the perturbated case, we note that the residual interaction acts differently on the double F and GT transition strengths. In the first place, all intensity continues concentrated at the energy $2\Delta_C$, independently of the values of v_s and v_t, but now there is only one DIAS, in the second case the double GT resonance (DGTR), placed at an energy that is only slightly lower than $2\Delta_C + \epsilon^0_{f_{5/2}}$, carries a large part of the strength. That is $S^{\beta\beta^-}_{GT}$ (DGTR): 72.9% in the case (a) and 72.6% in the case (b). The first observed single particle levels in ^{48}Sc, within these approximation are presented in Fig. (5). The particle-hole limit of QTDA for (a),(b) and (c) set of parameters reproduced reasonably well the experimental data.

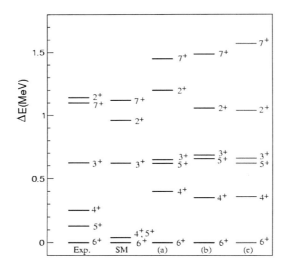

Figure 5 Comparison between the experimental and theoretical energy levels of ^{48}Sc for the cases (a) $v_s = 35$, $v_t = 65$, (b) $v_s = 40$, $v_t = 60$, and (c) $v_s = 41$, $v_t = 64$ for other set of SPE. We also display the shell-model results from Ref. [11].

The results for GT, F and total $2\nu\beta\beta$ matrix elements in units of $(MeV)^{-1}$, are presented in the Table (2) where we use the kinematical factor $\mathcal{G}_{2\nu}$ of Ref.[6], and the axial-vector coupling constant $g_A = 1$. The theoretical values obtained for the $\mathcal{M}_{2\nu}$ are so closed to the experimental one. We consider the complete pf particle-hole space. Of course, this configuration space is limited in comparison with the SM ones within the pf single-particle space [7], but this is only an exploratory study to explain the way the QTDA quenching-mechanism works and the SM is able to yield a more realistic description of the $\beta\beta$-decay [8] and of the related charge-exchange reactions [9] than the present model.

Table 1 Results for GT, F and total $2\nu\beta\beta$ matrix elements in units of $(\mathrm{MeV})^{-1}$.

case	$\mathcal{M}_{2\nu}^{GT}$	$\mathcal{M}_{2\nu}^{F}$	$\mathcal{M}_{2\nu}$	$|\mathcal{M}_{2\nu}|_{\exp}$ [8]
(a)	0.043	0.008	0.035	
(b)	0.080	0.011	0.069	$0.075^{+0.015}_{-0.019}$
(c)	0.047	0.014	0.061	

In summary, we have demonstrated that the simple version of the QTDA proposed here is able to account for the suppression of $\mathcal{M}_{2\nu}$, which was the major merit of the QRPA. Several serious inconveniences found in the QRPA, such as: i) the extreme sensitivity of the $2\nu\beta\beta$ decay amplitudes $\mathcal{M}_{2\nu}$ on the residual interaction in the particle-particle channel, ii) the ambiguity in treating the intermediate states, and iii) the need for performing a second charge-conserving QRPA to describe the $\beta\beta$-decays to the excited final states, are not present in the QTDA. The QTDA never collapses, and therefore the amplitude $\mathcal{M}_{2\nu}$ does not have a pole anymore, as it happens in the QRPA. The energy distributions of the double-charge-exchange transition strengths and their sum rules are explicitly evaluated by the QTDA, let us to use this approximation to single- and double-closed shell nuclei, as seen in the example shown for the decay ^{48}Ca $\rightarrow {}^{48}$Ti.

We feel that the new model comprises all essential nuclear structure ingredients that are needed for describing the $\beta\beta$-decay processes. Of course, whether this is totally true or only partly true, the QTDA has still to be tested. The QTDA is actually been implemented in one extended space in ^{48}Ca and ^{76}Ge.

Acknowledgements

F.K. acknowledges the support of the CONICET, Argentina. A.R.S. appreciates support received from Conselho Nacional de Ciência e Tecnologia (CNPq) and Fundação Carlos Chagas Filho de Amparo à Pesquisa do Estado do Rio de Janeiro (FAPERJ).

Bibliography

[1] J.N. Bahcall, H. Murayama, and C. Pena-Garay, Phys. Rev. D **70** (2004), 033012.

[2] F. Krmpotić, Fizica **B** 14 (2005) 2, 139.

[3] J.A. Halbleib and R.A. Sorensen, Nucl. Phys. **A98**(1967), 524.

[4] M.K. Pal, Y.K. Gambgir, and Ram Raj, Phys. Rev. **155** (1966), 1144.

[5] J. Hirsch and F. Krmpotić, Phys. Rev. C **41**,304 (1990) and references therein.

[6] F. Krmpotić and S. Shelly Sharma, Nucl. Phys. **A572** (1994), 329.

[7] L. Zhao, B.A. Brown, and W.A. Richter, Phys. Rev. C **42** (1990), 1120.

[8] V. B. Brudanin, N. I. Rukhadze, Ch. Briançon, V. G. Egorov, V. E. Kovalenko, A. Kovalik, A. V. Salamatin, I. Imagetekl, V. V. Tsoupko-Sitnikov, Ts. Vylov and P. Imageermák, Phys. Lett. **B495** (2000), 63.

[9] S. Rakers, C. Bäumer, A. M. van den Berg, B. Davids, D. Frekers, D. De Frenne, Y. Fujita, E.-W. Grewe, P. Haefner, M. N. Harakeh, M. Hunyadi, E. Jacobs, H. Johansson, B. C. Junk, A. Korff, A. Negret, L. Popescu, H. Simon, and H. J. Wörtche, Phys. Rev. C **70** (2004), 054302.

[10] B.A. Brown, and W.A. Richter, Phys. Rev. C **58** (1998), 2099.

[11] E. Caurier, A. P. Zuker, A. Poves, and G. Martínez-Pinedo, Phys. Rev. C **50** (1994), 225.

[12] M. Honma, T. Otsuka, T. Mizusaki, Y. Utsuno and N. Shimizu, Advanced in Calculational Methods, (2001) 517; http://t16web.lanl.gov/nd2004/abstracts-pdf/517.pdf.

Anais da XXVIII Reunião de Trabalho sobre Física Nuclear no Brasil
SP, Brasil, 2005
Artigo originalmente publicado no Brazilian Journal of Physics, Vol 36 – nº 4B,
Special Issue: XXVIII Workshop on Nuclear Physics in Brazil, pp.1349–1353 (2006).

Quasiparticle-Rotor Model Description of Carbon Isotopes

T. Tarutina[1], A. R. Samana[2], F. Krmpotić[1,3,4], M. S. Hussein[1]

[1] Departamento de Física Matemática, Instituto de Física da Universidade de São Paulo, Caixa Postal 66318, CEP 05315-970, São Paulo, SP, Brazil

[2] Centro Brasileiro de Pesquisas Físicas, CEP 22290-180, Rio de Janeiro, Brazil

[3] Instituto de Física La Plata, CONICET, 1900 La Plata, Argentina

[4] Facultad de Ciencias Astronómicas y Geofísicas, Universidad Nacional de La Plata, 1900 La Plata, Argentina

Abstract. In this work we perform quasiparticle-rotor coupling model calculations within the usual BCS and the projected BCS for the carbon isotopes ^{15}C, ^{17}C and ^{19}C using ^{13}C as the building block. Owing to the pairing correlation, we find that ^{13}C as well as the cores of the other isotopes, namely ^{14}C, ^{16}C and ^{18}C acquire strong and varied deformations. The deformation parameter is large and negative for ^{12}C, very small (or zero) for ^{14}C and large and positive for ^{16}C and ^{18}C. This finding casts a doubt about the purity of the supposed simple one-neutron halo nature of ^{19}C.

PACS. 21.60.-n, 21.10.-k

Keywords. ^{13}C, ^{15}C, ^{17}C, ^{19}C nuclei; Energy spectra; particle projected BCS model; core excitation; quasiparticle-rotor model

1 Introduction

Many experimental and theoretical investigations for over two decades were concentrated upon study of light nuclei near neutron drip line (see references in recent reviews [1, 2]). Pairing correlations play an important role in the structure of these nuclei, because of the proximity of the Fermi surface to the single-particle continuum. This gives rise to many interesting phenomena and creates a challenge for conventional nuclear structure models. The experimental evidence for the N = 8 shell melting [3] and the appearance of the $1s_{1/2}$ intruder state in ^{11}Be are famous indications of the complicated structure of light nuclei. In Refs.[4, 5] it was shown that the increase in pairing correlations and the shallow single-particle potentials for nuclei close to the driplines may result in a more uniformly spaced spectrum of single particle states.

To mention a few recent works on this subject, a series of interesting articles appeared in the literature which study the pair correlation in spherical and deformed nuclei near the drip line using a simplified HFB model in coordinate representation with the correct asymptotic boundary conditions. In Ref.[6] the effects of continuum coupling have been studied, it was shown that for small binding energies, the occupation probability decreases considerably for neutrons with low orbital momentum. In Ref.[7] the weakly bound neutrons in $s_{1/2}$ state have been studied. The effective pair gap was found to be much reduced compared with that of neutrons with larger orbital momentum. In the presence of pair correlations, the large rms radius was obtained for neutrons close to the Fermi level, thus favouring the halo formation.

Nowadays it is generally accepted that nucleus tends to form a halo when the valence particles are loosely bound and the relative angular momentum is small. In this case, the last nucleons and the core are to a large extent separable and therefore these nuclei can be approximated as an inert core plus the halo formed by the valence particles. This supports the use of the cluster models for such systems. But for real nuclei, the admixtures between the core and valence particles have to be taken into account. One of the simple models that include core degrees of freedom is a *core + particle* cluster model for one-neutron halo systems including core excitation via deformation assuming a rotational model for the core structure [8, 9]. Although this model allowed successful description of the $1s_{1/2}$ intruder state in ^{11}Be, it has a serious drawback as it does not treat Pauli Principle correctly.

In this work we incorporate the residual interaction between nucleons moving in an overall deformed potential and perform a Bogoljubov-Valatin canonical transformations from particles to quasiparticles, thus modifying the band-head energies and the non-diagonal particle core matrix elements. The resulting model is known as the quasiparticle-rotor coupling model (QRCM). The inclusion of pairing effects allows a correct treatment of Pauli Principle. This model is then applied to the series of odd-mass carbon isotopes.

Heavy carbon isotopes have recently been studied extensively, and ^{17}C and especially ^{19}C are suggested to be candidates to possess neutron halo. To calculate the quasiparticle energies for ^{13}C and heavy carbon isotopes the BCS calculation with particle projection procedure was used (PBCS). It is known that in BCS formalism the number of particles is not conserved, which becomes a serious drawback for the case of light nuclei. It was shown in Refs.[12, 15], that the projection procedure is very important in such nuclei. The Fig. 1 shows the BCS and PBCS calculations confronted with experimental data, using the results obtained in [12], where we evaluate the ground states energies for the remaining odd-mass

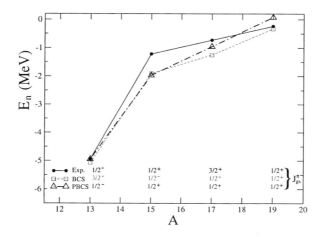

Figure 1 Comparison between the experimental and the calculated ground state energies in BCS and PBCS in odd mass carbon isotopes. The experimental data were taken from Ref. [10] for ^{13}C, ^{15}C, and Ref. [11] for ^{17}C and ^{19}C.

carbon isotopes, by only modifying the numbers of neutrons: $N = 8, 10$ and 12 for ^{15}C, ^{17}C and ^{19}C, respectively. One immediately sees that the pairing interactions account for the main nuclear structure features in these nuclei.

2 The QRCM Model

Most applications of the particle-rotor coupling model (PRCM) use the simplest picture of the last odd nucleon moving outside the even-even core, and any kind of residual interaction between nucleons is neglected (see, for example, Refs.[9, 13, 14]). The core is assumed to be an axially symmetric rigid rotor. The Hamiltonian of the core-neutron system can be written as

$$H_{p-rot} = H_{sp} + H_{rot} + H_{int}, \tag{1}$$

where $H_{sp} = T + U_0$ is single-particle (shell-model) Hamiltonian, that is the sum of central Woods-Saxon potential and standard spin orbit interaction, whose eigenvalues e_j are specified by the particle angular momentum **j**. We consider here the quadrupole deformation only, specified by the deformation parameter β. In this case, the core-neutron coupling hamiltonian

$$H_{int} = -\beta k(r) Y_{20}(\theta', 0); \quad k(r) \equiv r \frac{dU_0(r)}{dr}, \tag{2}$$

where $k(r)$ is the radial part of the interaction.

The total angular momentum is $\mathbf{I} = \mathbf{j} + \mathbf{R}$ (in this work we consider only $I = 0, 2$) and the coupling between $\mathbf{j}$ and $\mathbf{R}$ is evidenced in the weak-coupling representation of the total wave function, which reads

$$|jR; IM\rangle = \sum_{M_R, m} (jmRM_R|IM)|jm\rangle|RM_R\rangle, \tag{3}$$

where $|RM_R\rangle$ is the rotor wave function. The matrix elements of the Hamiltonian (1) are:

$$\langle j'R'; IM|H_{p-rot}|jR; IM\rangle = \left[\frac{\hbar^2 R(R+1)}{2\mathcal{J}} + e_j\right] \delta_{RR'}\delta_{jj'}$$

$$- \beta(-)^{j'+R+I}\sqrt{2R+1}(R020|R'0)\left\{\begin{matrix} I & R' & j' \\ 2 & j & R \end{matrix}\right\} \langle j'\|kY_2\|j\rangle. \tag{4}$$

The resulting eigenfunctions are of the form:

$$|I_n\rangle = \sum_{jR} C^I_{jR}|jR; I\rangle, \tag{5}$$

with energies $\mathcal{E}_{I_n}$, and where the C^I_{jR} are constants that result from the solution of the coupled channel system of equations obtained by substituting the total wave function into the Schöedinger equation and projecting out the basis wave functions.

We now incorporate the residual interaction, V_{res}, between nucleons moving in an overall deformed potential. The total Hamiltonian becomes $H_{qp-rot} = H_{p-rot} + V_{res}$. After carrying out a Bogoljubov-Valatin canonical transformation from particle to quasiparticle operators, one gets that the matrix elements of H_{qp-rot} for odd-mass systems continue being of the form (4), except for (i) the band-head energies for a particle state are modified as $e_j \rightarrow E_j$ in BCS, and $e_j \rightarrow \epsilon^K_j$ in PBCS, where E_j is the quasiparticle BCS energy, and $\epsilon^K_j = \frac{R^K_0(j) + R^K_{11}(jj)}{I^K(j)} - \frac{R^K_0}{I^K}$, where quantities R^K and I^K, for $K = Z, N$, are defined in Ref.[15]; and (ii) the non-diagonal matrix elements are renormalised by the following overlap factors

$$\mathcal{F}_{j'j} = u_{j'}u_j - v_{j'}v_j \text{ (BCS)}$$

or

$$= \frac{u_{j'}u_j I^K(j'j) - v_{j'}v_j I^{K-2}(j'j)}{[I^K(j')I^K(j)]^{1/2}} \text{ (PBCS)}, \tag{6}$$

where u_j and v_j are the occupational numbers for the state j. $I^K(j'j)$, $I^{K-2}(j'j))$ and $I^K(j)$ are the PBCS number projection integrals (see Refs.[16, 12]).

The energies to be confronted with the experimental data are

	BCS	PBCS	State
$\epsilon_{I_n}^{(+)}$	$\lambda + \epsilon_{I_n}$	$\epsilon_{I_n}^{N}$	particle
$\epsilon_{I_n}^{(-)}$	$\lambda - \epsilon_{I_n}$	$-\epsilon_{I_n}^{N-2}$	hole

where λ is the BCS chemical potential. Of course, in the resulting quasiparticle-rotor coupling model (QRCM) the state $|jm\rangle$ is now a quasiparticle state. The QRCM differs in several important aspects from the PRCM. First, the BCS (PBCS) energies $\epsilon_j^{\pm}$ are quite different from the single-particle energies e_j. Second, the factors $\mathcal{F}_{jj'}$ correctly take into account the Pauli principle. In addition, the particle-like states do not couple to the hole-like states. As a brief test, in Fig. (2) we show a comparison of the energy levels for ^{13}C and ^{11}Be in the Pure Rotor Model with those in the QRCM: Rotor + BCS and Rotor + PBCS as a function of the parameter of deformation β for the single particles taken from the Vinh-Mau's work [17] for ^{13}C and ^{11}Be. From a careful scrutinising of the energy levels in Fig. (2), one can easily convince oneself that, none of our particle-rotor coupling calculations is able to reproduce the experimentally observed spin sequence $1/2^+ - 1/2^- - 5/2^+$ in ^{11}Be, for such a value of β neither positive nor negative. The constant of pairing for ^{13}C is $v_s = 37.80$ in both BCS and PBCS, while that in ^{11}Be they are $v_s = 21.02$ in BCS and $v_s = 24.25$ in PBCS.

3 Results and Conclusions

In our procedure we use the nucleus ^{13}C as a building block of the calculation, assuming core ^{12}C deformation $\beta = -0.6$. In our calculations we adjust the single particle state energies e_j (for $1p_{3/2}$, $1p_{1/2}$, $1d_{5/2}$ and $2s_{1/2}$) so that the resulting quasiparticle energies including the core excitation give correct binding energies of four bound states in ^{13}C. The same set of single particle energies are then used in PBCS calculation to obtain quasiparticle energies of ^{15}C, ^{17}C and ^{19}C, occupation numbers for the states in these nuclei and the reduction factors of the matrix elements. Subsequently, the quasiparticle energies and reduction factors then enter the coupled channel system to include core excitation and give the binding energies of ^{15}C, ^{17}C and ^{19}C. For ^{15}C, ^{17}C and ^{19}C we adjust the deformation parameter β to obtain a correct one-neutron separation energy in each of the isotopes. Using this deformation parameter we calculate the energy of the first excited state. For simplicity, in the present calculation we use fixed values for averaged radial matrix elements between different states, that is, $\langle j'|k(r)|j\rangle = 25$, 30 and 35 MeV. These values provide a reasonable estimation of non-diagonal part of the interaction and

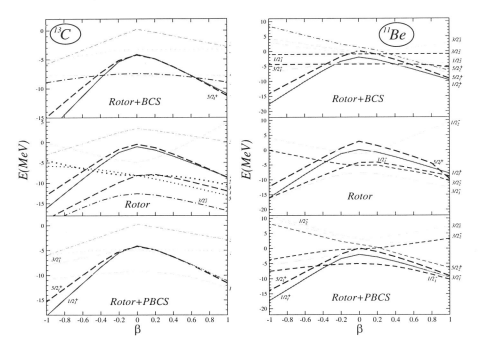

Figure 2 (Color online) Energies of the lowest states in ^{13}C and ^{11}Be as functions of deformation parameter β calculated in the Pure Rotor model (middle panels), Rotor + BCS model(upper panels) and Rotor + PBCS (lower panels).

correctly reproduce the spin-orbit splitting between initial single-particle $1p_{3/2}$ and $1p_{1/2}$ states. The energies obtained within the Rotor + PBCS model on different carbon isotopes for $\langle k(r) \rangle = 25$ MeV (all energies are in MeVs) are presented in the Table 1.

Table 1 Energies obtained within the Rotor + PBCS model on different carbon isotopes for $\langle k(r) \rangle = 25$ MeV. All energies are in MeVs.

State	^{13}C	^{15}C	^{17}C	^{19}C
$1p_{3/2}$	0.63	6.95	9.66	11.86
$1p_{1/2}$	-4.75	1.92	4.38	6.32
$1d_{5/2}$	-0.10	-0.94	-0.30	0.32
$2s_{1/2}$	-0.75	-1.25	-0.36	0.45

The results for the odd-mass carbon isotopes can be summarised as:

^{15}C: In our calculation we obtain a zero deformation for the core ^{14}C in case of $\langle k(r) \rangle = 25$ MeV and a binding energy of 1.25 MeV. The calculation with larger average matrix element 30 and 35 MeV, gives an oblate shape with deformation parameter equal to -0.23 and -0.32 respectively. (In this calculation we introduced a lower 2^+ rotational energy in ^{14}C $\epsilon_{2+} = 3$ MeV). This confirms the idea that the core nucleus ^{14}C is nearly spherical and supports the possibility of halo formation for ^{15}C. Finally, our calculation also predicts a first excited state $\frac{5}{2}^+$ with the binding energy -936 keV, -991 keV and -867 keV for $\langle k(r) \rangle$ equal to 25 MeV, 30 MeV and 35 MeV, respectively. The binding of the first excited state is overestimated in our model.

^{17}C: Small separation energy of ^{17}C 0.728 MeV suggests a possible halo structure of this isotope. There is an uncertainty about ground state spin assignment for this nucleus. In our model we obtain $\frac{1}{2}^+$ ground state for ^{17}C, although the experimental value is $\frac{3}{2}^+$ [11]. A simple explanation for this experimental result could be found in the so called $J = j - 1$ anomaly discussed by Bohr and Mottelson [18]. Core-particle calculations without pairing similar to those in Ref.[13] used deformation parameter $\beta = 0.55$.

We obtain the β equal to 0.505, 0.583 and 0.609 for average radial matrix element $\langle k(r) \rangle$ equal to 25, 30 and 35 MeV, respectively. We see that the deformation changes slowly with the value of the matrix element and is rather large. The first excited state of ^{17}C in our model is $\frac{5}{2}^+$ state. The binding energy of this state is as follows: -0.616 for $\langle k(r) \rangle = 25$ MeV, -0.527 MeV for $\langle k(r) \rangle = 30$ MeV and -0.430 for $\langle k(r) \rangle = 35$ MeV. These energies are close to the experimental separation energy of -0.433 MeV.

^{19}C: Ground state with spin-parity $\frac{1}{2}^+$ and small one-neutron separation energy favours the formation of the halo in ^{19}C. In our analysis we adjust the deformation parameter in order to obtain the adopted separation energy of 0.58 MeV. Core-particle calculations without pairing used deformation parameter $\beta = 0.5$. $\beta = 1.27, 1.51$ and 1.82. That is, we obtain very a large prolate deformation for the core ^{18}C. The first excited state we obtain to be $\frac{5}{2}^+$ state with the binding energy -0.21MeV ($\langle k(r) \rangle = 25$ MeV).

The Fig. 3 summarises the above results and compares them with the results of other models and experimental data.

Finally, in Fig. 4 we show the dependence of the deformation parameter on the number of neutrons in the core N for three values of $\langle k(r) \rangle$. It is seen that in each case, the deformation changes from negative values (for cores ^{12}C and ^{14}C) to large positive values (for cores ^{16}C and ^{18}C). The dependence is nearly linear and crosses the zero deformation in the region of the core ^{14}C.

To summarise, in this work we analyse the structure of heavy carbon isotopes ^{15}C, ^{17}C and ^{19}C in the quasiparticle-rotor model using the ^{13}C as the building block of the calculation. The quasiparticle energies and the reduction factors in the matrix elements are calculated in the projected BCS model. As it was mentioned in the

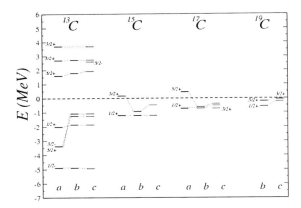

Figure 3 Comparison between the calculated (a) core-particle, (b) QRCM, and (c) experimental level scheme for odd mass carbon.

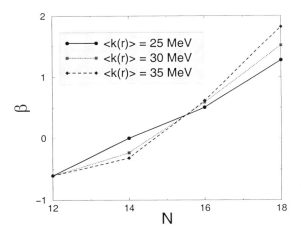

Figure 4 The adjusted value of deformation parameter β of cores ^{12}C, ^{14}C, ^{16}C and ^{18}C for three values of $\langle k(r) \rangle$.

introduction, we first explore the results of the pure BCS and PBCS models, that use the single-particle energies and pairing strengths of Ref. [12] fixed so that the experimental binding energies of ^{11}C and ^{13}C, together with the ^{13}C energy spectra, are reproduced by the calculations. With these parameters we evaluate next the low-lying states in ^{15}C, ^{17}C and ^{19}C, by augmenting correspondingly the number of neutrons only. We found that proceeding in this way both models are capable to explain fairly well the decrease of the binding energies in going from ^{13}C to ^{19}C. Second, the QRCM was implemented, and in this approximation we obtain ^{15}C, ^{17}C and ^{19}C with $\frac{1}{2}^+$ ground state and $\frac{5}{2}^+$ first excited state. For ^{15}C we obtain zero or small oblate deformation. For ^{17}C large prolate deformation is given. Finally ^{19}C has a very large prolate deformation, indicating that the nature of this nucleus is more complicated than the simple one-neutron halo picture of this nucleus would suggest.

The further analysis of this model and the details of its application to describe the low-lying spectra of light nuclei model will be presented in our next work [19].

Acknowledgements

F.K. acknowledges the support of the CONICET, Argentina. A.R.S. appreciates support received from Conselho Nacional de Ciência e Tecnologia (CNPq) and Fundação de Amparo à Pesquisa do Estado do Rio de Janeiro (FAPERJ). T.T. and M.S.H. would like to thank Fundação de Amparo à Pesquisa do Estado de São Paulo (FAPESP) and CNPq.

Bibliography

[1] B. Jonson, Phys. Rep. **389**, 1 (2004).

[2] A.S. Jensen, K.Riisager, D.V. Fedorov, E. Garrido, Rev. Mod. Phys. **76**, 215 (2004).

[3] H. Iwasaki *et al.*, Eur. Phys. J. A **13**, 55 (2000).

[4] J. Dobaczewski, I. Hamamoto, W. Nazarewicz, J.A. Sheikh, Phys. Rev. Lett. **72**, 981 (1994).

[5] J. Dobaczewski, W. Nazarewicz, T.R. Werner, J.F. Berger, C.R. Chinn, J. Dechargé, Phys. Rev. C **53**, 2809 (1996).

[6] I. Hamamoto and B.R. Mottelson, Phys. Rev.C **68**, 034312 (2003).

[7] I. Hamamoto and B.R. Mottelson, Phys. Rev.C **69**, 064302 (2004).

[8] F.M. Nunes, *Core Excitation in Few Body Systems: Application to halo nuclei* , (PhD thesis, University of Surrey, 1995, unpublished).

[9] F.M. Nunes, I.J. Thompson and R.C. Johnson, Nucl. Phys. **A596**, 171 (1996).

[10] F. Ajzenberg-Selove, Nucl. Phys. **A433**, 1 (1985) ; TUNL Nuclear Data Evaluation Project, available WWW: http:// www.tunl.duke.edu/nucldata/

[11] Z. Elekes, Z. Dombradi, R. Kanungo, H. Baba, Z. Fulop, J. Gibelin, A. Horvath, E. Ideguchi, Y. Ichikawa, N. Iwasa, H. Iwasaki, S. Kanno, S. Kawai, Y. Kondo, T. Motobayashi, M. Notani, T. Ohnishi, A. Ozawa, H. Sakurai, S. Shimoura, E. Takeshita, S. Takeuchi, I. Tanihata, Y. Togano, C. Wu, Y. Yamaguchi, Y. Yanagisawa, A. Yoshida, and K.Yoshida, Phys. Lett. **B614**, 174 (2005).

[12] F. Krmpotić, A. Mariano, A. and A. Samana, Phys. Lett. **B541**,298 (2002).

[13] D. Ridikas, M. H. Smedberg, J. S Vaagen, and M. V. Zhukov, Nucl. Phys. **A628**, 363 (1998).

[14] T. Tarutina, I.J. Thompson, J.A. Tostevin, Nucl. Phys. **A733**, 53 (2004).

[15] F. Krmpotić, A. Mariano, A. and A. Samana, Pys. Rev. C **71**, 044319 (2005).

[16] F. Krmpotić, A. Mariano, T.T.S. Kuo, and K. Nakayama, Phys. Lett. **B319**, 393 (1993).

[17] N. Vinh Mau, Nuc. Phys. **A582**, 33 (1995).

[18] A. Bohr and B.R. Mottelson, *Nuclear Structure*, vol.II, (W.A. Benjamin, Inc. 1975).

[19] A.R. Samana, T. Tarutina, F. Krmpotić and M.S. Hussein, in preparation.

Anais da XXVIII Reunião de Trabalho sobre Física Nuclear no Brasil
SP, Brasil, 2005

Analytical Sensitivity Analysis and Optimization of a System Employing a Si-PIN Detector for the Study of Environmental Samples

Melquiades F. L.[1,2,], Parreira P. S.[2,†], Appoloni C. R.[2,‡], Lopes F.[2,§]*

[1] Departamento de Física/Universidade Estadual do Centro Oeste
Rua Presidente Zacarias, 875 - Cx. Postal 3010, CEP 85015-430 - Guarapuava - PR

[2] Departamento de Física/Universidade Estadual de Londrina
Campus Universitário Cx. Postal 6001, CEP 86051-990 - Londrina - PR

Abstract. Sensitivity is an indicative of a X ray spectrometer performance. EDXRF technique was used to evaluate the variation in the analytic sensitivity of a Si-PIN detector system. For sample excitation a portable mini X ray tube (Ag target, 4 W), under different operation conditions, was used with the objective of optimize the analytical sensitivity for trace elements determination in environmental samples. The sensitivity curves were obtained through standard samples irradiation. The concentration of two certified reference materials SRM1832 and SRM1833 were determined with the purpose of method validation. Ag, Al and Pb collimators were used on the detector, and in the X ray tube exit were used Mo and Ag filters. The results indicated the presence of some contaminants (Mn and Fe in the Al collimator and Au in the Ag collimator) due to the metallic alloy from which the collimators and filters were made. This fact is a limitation for qualitative identification and for the detection limit in quantitative and semi-quantitative analysis of elements. The best results were obtained for Ag and Pb collimators operating at 25 kV and 20 µA, with Mo filter.

1 Introduction

A current tendency is the development of new analytical techniques and methods with ability to identify and quantify complex samples constituents as those related

---

* fmelquiades@unicentro.br
† parreira@uel.br
‡ appoloni@uel.br
§ bonn@uel.br

to food protections, environmental problems and archaeology and art samples, and to be able to give fast analytical response and/or to turn possible "in situ" analysis[1]. In the last case, portable systems have been used thoroughly to define sampling areas or to classify samples to be analyzed with more details in a later stage. Therefore, instead of trying to quantify the pollutants as first objective, it could be enough to certify if they are really present above or below some pre-established concentration level, as those determined by environmental protection agencies[2].

The use of the Energy Dispersive X ray Fluorescence (EDXRF), in portable systems, is one of the techniques that really allows to obtain fast results due its characteristic as an instrumental, multielementar and simultaneous technique. Due to these characteristics, these systems have increasingly been applied "in situ" for trace metal analysis in different kinds of samples [3].

The basic principle of EDXRF is to excite a sample with an incident radiation (X rays or gamma rays) and to measure directly the characteristic X-ray energies emitted by the sample in a detector.

The system ability in giving a quantitative result is measured by the sensitivity. It is one of the main characteristics to analyze the performance of X-ray spectrometers [4]. So, this work objective is to study the sensitivity variation for a portable excitation-detection system, employing a Si-PIN detector and a mini X ray tube, under different operating conditions, in order to optimize the condition for trace metals determination in environmental samples.

2 Material and Methods

The measurement system comprises a Si-PIN detector (221 eV FWHM at 5.9 keV and 25 µm Be window), coupled to a pre-amplifier, both thermoelectrically cooled, a high voltage source with amplifier, multichannel analyzer and a palmtop for data acquisition [5]. The samples excitation were done with a mini X ray tube (Ag target, 4 W) [6]. The whole system is portable and can be used in the sampling place.

Individual thin film certified standards of K, Ca, Sc, Ti, Mn, Fe, Cu, Zn, Pb and a blank deposited in polycarbonate membranes, were measured, all from MicroMatter Inc. Were also measured two certified reference material from National Institute of Standard and Technology (NIST), SRM1832 and SRM1833. The measuring time was 200 s for each standard. The obtained spectra were analyzed by WinQXAS software [7].

The X ray tube was operated at different high voltage and current conditions (25 kV – 40 µA, 25 kV – 20 µA, 25 kV – 10 µA).Continuous X-ray beam and monochromatic beam filtered with Ag (metallic, 50 µm and 10 µm thickness) and

Mo (resin, 1 mm thickness) filters were tested in the tube end. Collimators made of Al, Pb and Ag (3mm diameter aperture) were tested in the detector.

The relationship between the fluorescence intensity of characteristic Kα or Lα lines and the concentration of an element in the sample is given by the fundamental parameters equation[6], for thin samples:

$$I_i = S_i C_i \qquad (1)$$

were I_i is the characteristics X rays net intensity (cps), C_i is the concentration ($\mu g.g^{-1}$), and S_i the elementary sensitivity of the analyzed element i (cps.g^{-1}.cm^2).

3 Results

Sensitivity curves were obtained for each tested condition after irradiation with MicroMatter standards. Figure 1 exemplify one of the curves, with Mo filter.

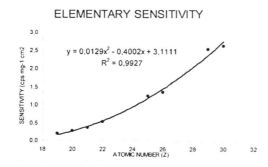

Figure 1 Elementary sensitivity plot with Mo filter, Ag collimator, 25 kV and 20 μA.

Comparative measurements of the blank standard spectra from MicroMatter, using different collimators, are shown at Figure 2. As can be seen there is a Ni contamination due to an internal rod that hold the Si PIN crystal [7], so that nickel will be an element always present in the spectra.

Blank spectra obtained with Ag and Mo filter are shown at Figure 3. 10 μm Ag filter is very thin to monochromatize the beam. 50 μm Ag filter is the one that presents the smallest background response.

It was verified that for Al collimator there are three small peaks in the spectrum at 5.9 keV, 6.4 keV and 8.04 keV energies which refers, respectively, to Mn, Fe and Cu aluminum alloy contaminants.

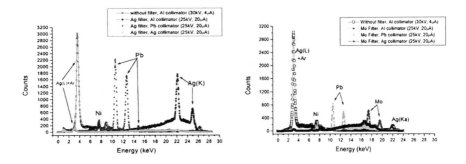

Figure 2 Blank spectra obtained with different geometry and excitation conditions.

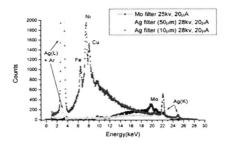

Figure 3 Blank spectra obtained with Mo and Ag filters.

Concentrations results for SRM1832 and SRM1833 standards upon different conditions are shown at Table 1. The condition where the majority of elements are within the confidence interval of the certified values were obtained for Pb or Ag collimators, using 25 kV e 20 µA. For Pb collimator the elements measured were Ti, V, Mn, Co, Cu, Zn, and for the Ag collimator the elements were Ca, V, Co, Cu, Zn and Pb.

Values for K (3,3 keV) could not be determined due to the interference of Ag-L_α (3,0 keV) and Ar-K_α (3,0 keV) lines.

Deviation for SRM certified values were from 5 to 10% and our results presented deviation from 10 to 15%.

4 Conclusion

Tests showed that the system sensitivity is good enough for roughly determination of heavy metals on membrane filters. From 10 elements detected in standard SRM, 6 were determined within the confidence interval of the certified values,

Table 1 Concentration results with the different tested conditions. In the last column, The marked values are related to the SRM1833 concentrations (C) and the others to the SRM1832.

Element	Mo filter Pb Collimator 25 kV 40 µA	Mo filter Pb Collimator 25 kV 20 µA	Mo filter Al Collimator 25 kV 40 µA	Mo filter Al Collimator 25 kV 20 µA	Mo filter Al Collimator 25 kV 10 µA	Certified Value
(Atomic Num.)	C ($\mu g.cm^{-2}$)	C ($\mu g.cm^{-2}$)	C ($\mu g.cm^{-2}$)	C ($\mu g.cm^{-2}$)	C ($\mu g.cm^{-2}$)	C ($\mu g.cm^{-2}$)
K(19)	11,02	14,50	10,51	11,14	30,48	**17,4 (1,6)**
Ca(20)	16,53	17,67	16,68	18,59	51,80	**19,99 (1,30)**
Ti(22)	11,07	11,10	9,91	10,06	12,20	**12,9 (1,8)**
V(23)	3,96	4,33	4,38	4,47	14,74	**4,5 (0,5)**
Mn(25)	4,69	5,03	5,58	5,37	31,47	**4,5 (0,5)**
Fe(26)	12,73	13,11	12,02	12,51	31,12	**14,3 (0,5)**
Co(27)	1,10	1,00	1,00	1,02	3,12	**0,99 (0,1)**
Cu(29)	3,20	2,58	2,54	2,43	7,43	**2,52 (0,16)**
Zn(30)	3,44	3,82	3,02	3,05	8,64	**3,8 (0,2)**
Pb(82)			12,79	12,65	39,98	**15,9 (0,9)**

Elemento (N. Atômico)	Mo filter Ag Collimator 25 kV 40 µA $C\ (\mu g.cm^{-2})$	Mo filter Ag Collimator 25 kV 20 µA $C\ (\mu g.cm^{-2})$	Mo filter Ag Collimator 25 kV 10 µA $C\ (\mu g.cm^{-2})$	Ag-50 filter Ag Collimator 28 kV 20 µA $C\ (\mu g.cm^{-2})$	Ag-10 Filter Ag Collimator 28 kV 20 µA $C\ (\mu g.cm^{-2})$	Certified Value $C\ (\mu g.cm^{-2})$
K(19)	6,24	9,74	22,64			**17,4 (1,6)**
Ca(20)	16,77	17,83	15,15	15,25	0,64	**19,99 (1,30)**
Ti(22)	11,40	10,98	11,90	9,38	0,36	**12,9 (1,8)**
V(23)	4,73	4,63	4,69	4,68	6,03	**4,5 (0,5)**
Mn(25)	5,17	5,31	5,38	3,65	0,19	**4,5 (0,5)**
Fe(26)	14,86	15,93	14,95	10,97	0,57	**14,3 (0,5)**
Co(27)	1,38	1,16	1,12	1,12	1,05	**0,99 (0,1)**
Cu(29)	3,01	2,69	3,02	2,03	0,50	**2,52 (0,16)**
Zn(30)	3,84	3,83	3,73	2,81	0,40	**3,8 (0,2)**
Pb(82)	15,47	14,67	14,57	8,83	4,07	**15,9 (0,9)**

using the multielementar EDXRF technique. The obtained results can be improved increasing the measurement repetitions as well as the time of excitation-detection.

Employing aluminum collimator, the determination of Mn and Fe are committed compromised because these elements are present in the aluminum alloy.

Best results were obtained for silver and lead collimators at 25 kV and 20 μA with Mo filter. The choice for the best collimator will depend on the elements and concentrations levels of interest.

Acknowledgements

To Prof. Dr. Virgílio Franco Nascimento Filho, Laboratório de Instrumentação Nuclear (CENA/USP), for his collaboration and loan of the reference materials.

Bibliography

[1] Hou X., He Y, Jones B.T., "Recent Advances in Portable X-Ray Fluorescence Spectrometry", *Appl. Spectr. Reviews*, **39(1)**, 1-25, (2004)

[2] Trullols E., Ruisánchez I., Xavier Rius F., "Validation of qualitative analytical methods" *T. Anal. Chem.*, **23(2)**, 137-145,(2004).

[3] F.L. Melquiades and C.R. Appoloni, "Application of XRF and field portable XRF for environmental analysis", *Radiat. Phys. Chem.*, **262(2)**, 533-541, (2004).

[4] M. Ferretti, "Fluorescence from the collimator in Si-PIN and Si-drift detectors:problems and solutions for the XRF analysis of archeological and historical materials". *Nucl. Instrum. Methods Phys. Research B*, **226**, (2004).

[5] Operating manual – XR100CR X ray detector system and PX2CR. Power supply, Amptek INC, 1998.

[6] Operating Manual – Miniature Bullet X ray Tube, Moxtek INC, 2003.

[7] Software WinQxas, *http://www.iaea.or.at/programmes/ripc/physics/faznic/win-qxas.htm*, visited in 06/06/2005.

[8] S.E., Manahan. *Environmental Chemistry*. 6 ed. Boca Raton: CRC PRESS, 1994, 843p.

[9] P. Kump, M. Necemer, Z. Rupnik, "Development of the quantification procedures for in-situ XRF analysis", IAEA/AL/130 REPORT, Vienna, 12-16 March, Attachment 5.10 - SLO-29042, 2001.

Anais da XXVIII Reunião de Trabalho sobre Física Nuclear no Brasil
SP, Brasil, 2005

Caracterização Química de Minerais em um Suplemento Vitamínico Comercial por Fluorescência de raios X com Equipamento Portátil

Fabio Lopes[1], Fábio L. Melquiades[1,2,*], Paulo S. Parreira[1], Carlos R. Appoloni[1]

[1] Laboratório de Física Nuclear Aplicada, Departamento de Física, Universidade Estadual de Londrina, Rodovia Celso Garcia Cid – Pr 445 - Km 380, Campus Universitário Cx. Postal 6001, 86051-990 - Londrina – PR, bonn@uel.br

[2] Departamento de Física, Universidade Estadual do Centro Oeste, Rua Presidente Zacarias, 875 - Cx. Postal 3010, 85015-430Guarapuava – PR

Resumo. Vitaminas e suplementos minerais são amplamente consumidos pela população adulta, entretanto pesquisas indicam que uma inadequada ingestão de suplementos pode ocasionar mais danos que benefícios ao corpo humano. Desta forma é importante a verificação da composição química elementar destes produtos. A técnica utilizada para quantificação foi a Fluorescência de raios X por dispersão em energia (EDXRF) com equipamento portátil. A validação da metodologia foi realizada através da análise de dois padrões de referência certificados, SRM-1832 e 1833. Três amostras de comprimidos revestidos foram selecionadas e moídas e em seguida montadas na forma de pastilha. As concentrações para os elementos encontrados, em $\mu g\, g^{-1}$, com os respectivos erros foram: Cl $(22,4 \pm 4,40)$, K $(26,0 \pm 1,15)$, Ca $(148,7 \pm 3,25)$, Mn $(2,9\pm0,54)$, Fe $(6,3\pm0,36)$, Cu$(4,0\pm0,14)$, Zn $(70,9\pm1,88)$, Se $(0,1\pm0,01)$, Br $(0,2\pm0,02)$, Sr $(0,2\pm0,01)$ e Mo $(0,1\pm0,01)$. Foi possível determinar o limite de detecção para os elementos Ca $(0,29)$, Fe $(0,09)$, Cu$(0,05)$ e Zn$(0,22)$, em $\mu g\, g^{-1}$. Dos 11 elementos encontrados no suplemento três deles (Cl, Br e Sr) não constavam na bula do produto e outros três (V, Cr, Ni) não foram encontrados.

1 Introdução

Vitaminas e suplementos minerais são amplamente consumidos pela população adulta. Especialmente entre os jovens tem se difundida amplamente a utilização de suplementos, compostos à base de carboidratos, lipídios, aminoácidos, vitaminas

---

* fmelquiades@unicentro.br

e minerais, comercializados e consumidos indiscriminadamente, sem orientação competente, sem que se conheçam seus efeitos, muitos dos quais aindª estão sob investigação [1,2]. Mais da metade da população adulta americana usa suplementos [3]. Segundo *Food and Drug Administration* – FDA [4], os suplementos não são considerados drogas, são produtos com vitaminas, minerais, plantas e substâncias derivadas de plantas, aminoácidos e concentrados, metabólitos, bem como constituintes e extratos dessas substâncias.

Um dos problemas é que não há estudos sobre o uso de suplementos, e seus efeitos são desconhecidos em adolescentes. A indústria e os distribuidores têm interesse financeiro em encorajar o uso desses produtos, e treinadores e atletas almejam o melhor desempenho, mais resistência e força física, a qualquer custo [5]. No entanto, resultados de estudos científicos sobre os seus efeitos no desempenho físico e mental, mostram que ainda não foi possível comprová-los. Testes com estudantes utilizando vitaminas e ferro, não mostraram melhora em seu desempenho [6].

Não há evidência científica de que um maior consumo de vitaminas e minerais melhore o desempenho [7], pelo contrário, pesquisas indicam que uma inadequada ingestão de suplementos pode ocasionar mais danos que benefícios ao corpo humano [8].

Um outro problema é que durante o processo de fabricação desses suplementos podem ocorrer contaminações introduzindo assim elementos que não são adequados ao consumo humano.

No Brasil, o Ministério da Saúde através das portarias 32 e 33 de 13 de janeiro de 1998, regulamenta e controla a fabricação e distribuição de Suplementos Vitamínicos e ou de Minerais, bem como adota valores para níveis de Ingestão Diária Recomendada (IDR) para as vitaminas, minerais e proteínas [9,10].

O objetivo deste trabalho é analisar polivitaminicos comercializados no mercado utilizando a técnica fluorescência de raios X por dispersão em energia (EDXRF) com equipamento portátil, por ser uma técnica multielementar, simultânea e rápida, exigindo uma preparação mínima da amostra, e desta forma quantificar os elementos que forem encontrados.

2 Materiais e Métodos

Preparação das amostras: Três amostras de comprimidos revestidos de uma mesma marca foram selecionadas e triturados em um almofariz com pistilo em porcelana. Em seguida foi utilizado 1 grama de cada comprimido e compactados manualmente na forma de pastilha que foram acomodadas em porta amostras próprias para analises por Fluorescência de Raios X (Chemplex Inc.), recobertos com filme de Mylar.

Instrumentação Nuclear: As amostras foram excitadas com raios X de 22,1 keV, proveniente de um mini tudo de raios X (FTC 100) de potência máxima 4 W [11]; com alvo de Ag e filtro de Ag de 50µm, operando com 28 kV e 20 µA. Para a aquisição dos dados foi utilizado um detector semicondutor do tipo Si-PIN com resolução em energia de 221 eV para energia de 5,9 keV, (XR-100CR) refrigerado pelo efeito Peltier [12]. O cristal do detector possui dimensões de 2,4 mm por 2,8 mm totalizando uma área efetiva de 7mm^2 e espessura de 300µm, protegido por uma janela de berílio de espessura 25µm.

Os espectros foram obtidos por um analisador multicanal acoplado a um palmtop e analisados através do programa WINQXAS, obtendo-se as áreas líquidas para os raios X característicos dos elementos presentes nas amostras. O tempo de aquisição para todas as amostras foi de 500 segundos, em triplicata.

Análise quantitativa por EDXRF: Para a quantificação utilizou-se a equação dos parâmetros fundamentais (1):

$$I_i = C_i \cdot S_i \tag{1}$$

onde I_i representa a intensidade líquida do raio X característicos (em cps), C_i representa a concentração ($\mu g\ cm^{-2}$), e S_i a sensibilidade elementar ($cps\ g^{-1}\ cm^2$) do elemento analisado.

O cálculo do limite de detecção para os filtros foi obtido de acordo com a equação definida por (Bertin, 1975)[14].

$$LD = \frac{3\sqrt{BG/t}}{S} \tag{2}$$

Tabela 1 Elementos encontrados nos padrões SRM.

	SRM 1832		SRM 1833
Z	Valor certificado ($\mu g/cm^2$)	Z	Valor certificado ($\mu g/cm^2$)
$_{13}$Al	14,94		
$_{14}$Si	36,60	$_{14}$Si	32,8
$_{20}$Ca	19,99	$_{19}$K	17,4
$_{23}$V	4,5	$_{22}$Ti	12,9
$_{25}$Mn	4,5	$_{26}$Fe	14,3
$_{27}$Co	0,99	$_{30}$Zn	3,8

onde **BG** representa a intensidade (cps) do *continuum* sob o pico do analito, t o tempo (s) de excitação/detecção e **S** a sensibilidade elementar do analito (cps g^{-1} cm^2).

A validação da metodologia foi realizada através da análise de dois padrões de referências certificados, SRM-1832 e SRM-1833, mostrados na tabela 1.

3 Resultados e Discussão

A excitação e detecção dos raios X foram realizadas em atmosfera de ar e foram quantificados 11 elementos químicos nos comprimidos na faixa do Cl ao Mo.

A curva de sensibilidade foi obtida pela irradiação de padrões monoelementares de K, Ca, Sc, Ti, Mn, Fe, Cu, Zn, em membranas de policarbonato da Micromatter [13]. A Figura 1 mostra a curva de sensibilidade com filtro e colimador de Ag.

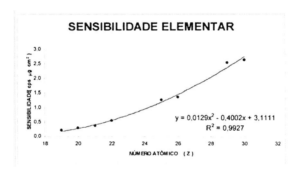

Figura 1 Curva de sensibilidade elementar obtida com padrões monoelementares.

Na Tabela 2 estão os valores estabelecidos pelo Ministério da Saúde para os níveis de IDR com relação a alguns elementos [10]. Os elementos encontrados nas amostras analisadas com seus respectivos desvios estão apresentados na Tabela 3.

Como a massa de cada cápsula de suplemento é em torno de 2 gramas, logo a ingestão de 1 cápsula por dia, conforme indica a embalagem, não extrapola o valor recomendado pela portaria nacional para nenhum dos elementos.

Os limites encontrados usando a equação (2), foram: Ca (0,29), Mn (0,08), Fe (0,09), Cu (0,05) e Zn (0,22), em $\mu g\ g^{-1}$, pois para os demais elementos não foi possível determinar o valor da sensibilidade experimental devido à falta de padrões. Os altos valores encontrados para o Ca, e Zn quando comparados com os demais elementos devido a contaminação do espectro pela linha L da Ag e pela linha K do Zn, elementos esses que são constituintes do tubo de raios X.

Tabela 2 Ingestão diária recomendada (IDR) para adultos.

Nutriente	Unidade	IDR
Cálcio	mg	800
Ferro	mg	14
Zinco	mg	15
Cobre*	mg	03
Iodo	µg	150
Selênio*	µg	70
Molibdênio*	µg	250
Cromo*	µg	200
Manganês*	mg	05

Fontes: Resolução GMC número 18/94 Mercosul e (*) RDA/NAS, 1989.

Tabela 3 Concentração dos elementos encontrados na amostra

Elemento	Concentração(μg g-1)	Desvio padrão
Cl	22,4	4,4
K	26,0	1,2
Ca	148,7	3,2
Mn	2,9	0,5
Fe	6,3	0,4
Cu	4,0	0,1
Zn	70,9	1,9
Se	0,12	0,01
Br	0,23	0,02
Sr	0,20	0,01
Mo	0,14	0,01

Os elementos Cl, Br e Sr medidos nos comprimidos não constavam da bula do suplemento polivitamínico e três outros (V, Cr e Ni) não foram determinados, embora estavam presentes na bula.

4 Conclusão

A técnica de EDXRF empregada possibilitou a rápida análise e identificação qualitativa e quantitativa dos elementos presentes nos comprimidos polivitamínicos sem a necessidade de processos químicos para preparação da amostra. Esta metodologia pode ser utilizada para analise in situ e agilizar o processo de fiscalização por parte dos órgãos reguladores.

Foi possível identificar e quantificar 11 elementos nas amostras, dentre eles o Cl, Br e Sr não constavam na bula do suplemento analisado.

Os resultados mostram que embora a concentração de metais, neste caso, esteja abaixo do limite estabelecido pela portaria do Mistério da Saúde, foram encontrados elementos que podem vir a ser prejudiciais à saúde, o que enfatiza a necessidade de controle e fiscalização sobre estes produtos.

Agradecimentos

Ao Prof. Dr. Virgílio Franco Nascimento Filho, coordenador do Laboratório de Instrumentação Nuclear (CENA/USP) por sua colaboração e empréstimo dos padrões usado nas medidas.

Referências Bibliográficas

[1] SAWADA, L.A.; COSTA, A.S.; MARQUESI, M.L.; LANCHA JR, A.H. Suplementação de aminoácidos e resistência à insulina. **Rev. Bras. Nutr. Clin.**, 1999, 14: 31-39.

[2] CARVALHO, Tales de (Coord.). Modificações dietéticas, reposição hídrica, suplementos alimentares e drogas: comprovação de ação ergogênica e potenciais riscos para a saúde. **Rev. Bras. Med. Esporte,** V.9, n.2, Mar/Abr, 2003, p. 43-56.

[3] KURTZWEIL, P. An FDA Guide to Dietary Supplements. **U.S. Food and Drug Administration.** Washington. September-October 1998. Disponível em: < http://www.cfda.gov/^dms/fdsuppch.html >. Acesso em 13 fev 2006.

[4] FDA. FOOD AND DRUG ADMINISTRATION. Supplements associated with illnesses and injuries. **U.S. Food and Drug Administration.** Washington. September-October 1998. Disponível em: < http://www.fda.gov/fdac/features/1998/dietchrt.html >. Acesso em 13 fev 2006.

[5] METZL, J.D. Strength training and nutritional supplement use in adolescents. **Curr. Opin. Pediatr.,** v.11, n.4, August 1999, p. 292-299.

[6] DOKKUM, V.W. Nutrition, doping and sport. **Rev. Nutrição - Saúde & Performance.** São Paulo, ano 4, n.17, abr/mai/jun. 2002, p. 4-7.

[7] CASTILLO, V.D. La alimentacion del deportista. **Educación Física y Deportes,** año 3, n.9. Buenos Aires, Marzo 1998. Disponível em: < http://www.efdeportes.com,/efd9/nutric9.htm >. Acesso em 13 FEV 2006.

[8] Tripp, F. The use of dietary supplements in the elderly: current issues and recommendations. **Journal of American Dietetic Association**, 10, 1997.

[9] Portaria nº 32, de 13 de janeiro de 1998, emitida pelo SVS/MS - Ministério da Saúde. Secretaria de Vigilância Sanitária

[10] Portaria nº 33, de 13 de janeiro de 1998, emitida pelo SVS/MS - Ministério da Saúde. Secretaria de Vigilância Sanitária

[11] Manual de operações – Miniature Bullet X-ray Tube, Moxtek INC, 2003.

[12] Manual de operações – XR100CR X ray detector system and PX2CR. Power supply, Amptek INC, 1998.

[13] Micromatter Corporation, Micromatter X Ray Fluorescence Calibration Standars. Deer Harbor, WA, 1991. 14 BERTIN, E.P. **Principles and practice of X-ray spectrometric analysis**. London: Plenum Press, 1975. 1079p.

Minimum Level of Component Discrimination (MLCD) for Characterization of Biphasic Mixtures (Oil / Brine or Gas) Using Gamma and X-Ray Transmission

Luiz Diego Marestoni, Filipe Sammarco, Carlos Roberto Appoloni, Jair Romeu Eichlt

Departamento de Física, Universidade Estadual de Londrina, Londrina-PR

Abstract. In this work we propose the concept of Minimum Level of Component Discrimination (MLCD) to determine the components discrimination in a biphasic mixture. A mathematical modeling followed by experimental corroboration illustrates the concept for different mixtures using transmission measurements of five lines (13.9; 17.8; 20.8; 26.35 and 59.54 keV) from a ^{241}Am source. The simulate mixtures were: oil/brine or gas. Four oil types were used, denominated A, B, Bell and Generic, one type of natural gas and water with salinity of 35.5 kg.m^{-3} of NaCl.

The single beam simulation was performed using one box with acrylic walls and another box with epoxy walls reinforced with carbon fiber. The epoxy with carbon fiber was used because it is a material that offers little photon attenuation and resists to great pressures. Based on the simulations, minimum discrimination tables were done, for each possible biphasic mixture with oil, gas and brine.

These discrimination tables are the ideal theoretical predictions to lead experimental measurement, since they supply the minimum discriminative percentage for each energy line, as well as the ideal energy to study each mixture or situation. The simulate discrimination levels were tested by performing experiments with conditions and materials similar to the ones used in the simulation, using the box with epoxy walls reinforced with carbon fiber, for the 20.8 and 59.54 keV energy lines. The obtained results were good.

For the brine/paraffin mixture (simulating oil), an experimental minimum discrimination of 10 % of brine for statistics with deviation of 5 % in I and Io was obtained, while the theoretical simulation predicted the same discrimination for statistics with deviation of 3 %.

---

1 Introduction

Brazil possesses 6.43×10^6 km^2 of sedimentary basins, with 4.83×10^6 km^2 on earth and 1.55×10^6 km^2 in continental platform. The main source for the oil formation was the phytoplankton of the Jurassic-Cretaceous period, which, when dead, was deposited on the bottom of the oceans and became part of the sediments and formed the leaflets rich in organic matter.

For the oil formation it is necessary that the basins have been formed in very specific conditions. For example, source rocks subject to temperatures of 50°C for 30000 years would not be sufficiently appropriate to generate oil, but only biogenic methane. In a similar period, rocks heated up to 190°C could generate natural gas.

Variations in the geothermic gradient, depth and burial duration can generate variations in the sedimentary basin (STRAPASSON, 2002), so different compositions for oil of different places will exist. The processes of oil extraction are evolving and becoming more and more accelerated, such as the need for more and more modern and effective equipments and processes for all the areas, from the extraction to the final product and besides, the cost has to be taken in consideration.

It is common, on the oil extraction process at off-shore, the extraction of brine and/or natural gas together with oil. The work was focused in this mixture type, because most part of the oil stations in Brazil is in the ocean.

The traditional solution to determine the amount of each component of the mixture is to separate the components and then measure the amount of each one using a single phase instrument (JOHANSEN et. al, 2000). However, this process is slow and expensive. A possible solution for this problem is the on line measurement of the mixture, through the process of gamma or X-ray densitometry.

Since there are studies in the literature to determine the substances in a mixture with the use of a radioactive source, the most part of the papers using [241]Am source (TJI.JGUM et. al., 2001), in this work the aim was to develop a mathematical modeling to determine which line of energy from the [241]Am source best discriminates a substance inside of a mixture and which is the minimum level of component discrimination for the mixture oil/brine/gas, as well as to perform measurements to test the mathematical equations.

2 Methodology

The X and gamma-ray densitometry considers the attenuation of the X and gamma-rays that cross the sample $\left(\frac{I}{I_0}\right)$. For the determination of the fraction of a substance in a biphasic mixture a single beam is enough. The Figure 1 represents the diagram for the mathematical modeling, using a biphasic mixture of two hypothetical substances s and m.

Where M represents the thickness of the m substance inside the sample, $\chi a'$ and $\chi a''$ the thickness of the air layers between source/sample and sample/source, respectively, X is the total thickness of the sample, that in this work is fixed in $X = 2.5$ cm, S is the thickness of the s substance inside the sample, $d1 + d2 = d$

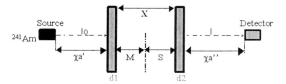

Figure 1 Schematic model used to perform the mathematical modeling.

represents the box total thickness (acrylic or epoxy), that in this work was fixed in d = 0.4 cm, Io and I are the incident and transmitted beam intensity by the sample, respectively.

From the arrangement shown in the Figure 1, of the beam fraction that was attenuated, of the attenuation coefficients μ and the densities ρ of the involved materials, the fraction of a substance S can be determined from the equation:

$$\left(\frac{I}{Io}\right) = \exp\left[-\mu_{epoxy}\rho_{epoxy}d - \mu_s\rho_s S - \mu_a\rho_a X_a + \mu_m\rho_m(S-X)\right] \quad (1)$$

The program WinXcom (GERWARD et. al., 2001) was used to calculate the cross sections and attenuation coefficients for combinations and desired mixtures. For the measurements, it was used a radioactive source of ^{241}Am, a 2×2"NaI(Tl) detector, both with collimators of 2 mm, coupled to a standard nuclear electronics. The samples, composed of variable amounts of paraffin and brine with salinity of 35.5 kg.cm^{-3}, were conditioned in an epoxy box reinforced with carbon fiber.

The measurements were performed using the 59.54 and 20.8 keV energy lines from ^{241}Am. The paraffin was used to simulate oil, because its attenuation coefficient is similar to the types of oil used. The measurements of each sample were repeated 5 times each. For the 20.8 keV energy line, 10 and 30 % of brine with paraffin were used. For the 59.54 keV energy line, the samples were prepared with 20, 70 and 90 % for the same mixtures.

The criterion for MLCD was created so it would be possible to assure that certain percentile of a substance can be discriminated in pure material. The criteria can be summarized in the following way:

- To compare the values of $\frac{I}{Io}$ of the mixture with $\frac{I}{Io}$ of the pure sample, taking as significant numerals until the position corresponding to two significant numerals of the standard deviation;

- To compare the band of values of $\frac{I}{Io}$ of the pure sample and of the considered mixture, using the interval $\left(\frac{I}{Io} \pm 1.96\sigma_{\frac{I}{Io}}\right)$;

- To be possible to consider that the mixture is discriminated in relation to the pure sample, the two intervals of values of $\frac{I}{I_0}$ need to be compared, and it must have a difference of two units in the last significant numeral.

3 Results

Since the epoxy used is a special material, it was necessary to perform experiments to obtain some physical informations. It was determined, through the gravimetric conventional method with ten repetitions, that its density was $\rho_{epoxy} = (1.4519 + 0.0057)$ g.cm^{-3}. To determine the linear attenuation coefficient, for the 59.54 keV energy line, ten epoxy plates were used, with average thickness of 2 mm each.

Initially one plate was used, then two, three and so forth until ten plates, with eight repetitions in each measurement of the attenuated beam in relation to the initial ratio $\left(\frac{N}{N_0}\right)$. Each measure lasted one minute. Interpolating the values of the thickness for the $\ln\left(\frac{N}{N_0}\right)$ value obtained for each group of plates, it was obtained that $\mu = (0.4951 + 0.0011)$ cm^{-1}, as shown in Figure 2.

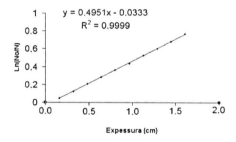

Figure 2 Plot to determine the linear attenuation coefficient of expoxy reinforced with carbon fiber.

For comparison effect and validation, using the reduced chemical formula of the epoxy, $C_{39}H_{82}O_{10}Cl_{28}$, it was determined the mass attenuation coefficient with the program WinXcom, for the 59.54 keV energy line. Once the value of mass attenuation coefficient was obtained, it was multiplied by the measured density of the epoxy, resulting in $\mu_{epoxy} = (0.4965 + 0.0009)$ cm^{-1}. As there was good agreement between the experimental and theoretical values for the linear attenuation coefficient, this theoretical procedure was then used to calculate the attenuation coefficients for the other energy lines of ^{241}Am. The results are shown in Table 1.

In the same way, it was determined the density and the linear attenuation coefficient of the paraffin, where five pieces of paraffin of varied thickness were

Table 1 Linear and mass attenuation coefficients of the epoxy.

Energy (keV)	μ_{Mass} (cm^2g^{-1})	μ_{linear} (cm^{-1})
59.54	0.3420	0.4965
26.35	2.1800	3.1650
20.80	4.2300	6.1414
17.80	6.6000	9.5823
13.90	13.5000	19.6000

Table 2 Values of the mass and linear attenuation coefficients of the paraffin.

Energy (keV)	μ_{Mass} (cm^2g^{-1})	μ_{linear} (cm^{-1})
59.54	0.2000	0.1672
26.35	0.3050	0.2550
20.80	0.4060	0.3395
17.80	0.5240	0.4382
13.90	0.8780	0.7342

used. The value obtained for the density of the paraffin was $\rho = (0.8362 + 0.0073)$ g.cm^{-3} and for the linear attenuation coefficient attenuation was $\mu = (0.1770 + 0.0009)$ cm^{-1}, also for the 59.54 keV energy line. The values of the attenuation coefficients obtained with the program WinXcom and the measured density value are shown in Table 2. As can be seen in Table 3, the value of the attenuation coefficient of the four types of oil is similar to the paraffin, therefore its value was used in the experiment.

Plots were made for the ratio $\left(\frac{I}{I_0}\right)$ versus the fraction of brine in oil, for the five lines of energy from the ^{241}Am, with several salinities and using the acrylic box. Figure 3 shows the plot for the 26.35 keV energy line. It was observed that the energy lines 26.35 and 20.80 keV were the ones that presented a better discrimination, because the lines of 13.90 and 17.80 keV were very attenuated, while the line 59.54 keV was little attenuated.

Tables 4, 5 and 6 present the data referring to the theoretical calculation of the MLCD for three different biphasic mixtures, with the five energy lines from the ^{241}Am. For the oil case, only the table with A oil will be presented, because they are all similar. For the epoxy box, only the three higher energy lines will be presented.

Table 3 Values of the attenuation mass coefficient ($cm^2 g^{-1}$) of the elements used in this work.

Material Energy		59.54 (keV))	23.35 (keV))	20.80 (keV))	17.80 (keV))	13.90 (keV))
Generic Oil		0.196	0.322	0.448	0.594	1.030
A Oil		0.199	0.306	0.409	0.53	0.893
B Oil		0.201	0.306	0.407	0.525	0.879
Bell Oil		0.178	0.295	0.412	0.550	0.963
Gas		0.212	0.310	0.402	0.509	0.830
Acrylic		0.193	0.357	0.529	0.732	1.340
Epoxy		0.342	2.180	4.230	6.600	13.500
Water		0.207	0.462	0.741	1.070	2.060
	35.5	0.212	0.532	0.884	1.300	2.530
	50	0.214	0.560	0.940	1.390	2.720
Brine	100	0.221	0.648	1.120	1.670	3.310
$kg.m^3$	150	0.227	0.729	1.290	1.940	3.850
	200	0.232	0.804	1.440	2.180	4.350
	250	0.237	0.872	1.580	2.400	4.810
	300	0.242	0.935	1.700	2.600	5.230
Air		0.187	0.424	0.681	0.984	1.890

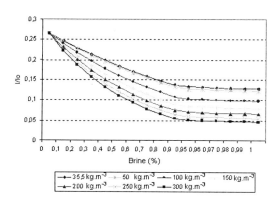

Figure 3 Discrimination of brine in A oil for 26.35keV energy line.

Table 4 Minimum Level of Component Discrimination (MLCD) of brine (35.5 kg m^{-3}) in gas using an acrylic box.

Energy (keV)	Statistics of I and I_0	MLCD (%)	I/I_0 of MLCD
	15	5	1.22 e-01
13.9	5	3	1.36 e-01
	2	1	1.52 e-01
	15	20	2.01 e-01
17.8	10	10	2.62 e-01
	3	5	2.99 e-01
	2	1	3.32 e-01
	15	30	2.62 e-01
	10	20	3.11 e-01
20.8	5	10	3.69 e-01
	2	5	4.02 e-01
	0.5	1	4.31 e-01
	15	95	2.23 e-01
	10	40	3.74 e-01
26.35	5	20	4.51 e-01
	2	10	4.95 e-01
	0.5	2	5.34 e-01
	5	90	5.40 e-01
59.54	3	50	5.79 e-01
	2	40	6.13 e-01
	0.5	10	6.61 e-01

Table 5 Minimum Level of Component Discrimination (MLCD) of gas in A oil.

Energy (keV)	Statistics of I and I_0	MLCD (%) Acrylic Box	I/I_0 of MLCD Acrylic Box	MLCD (%) Epoxy Box	I/I_0 of MLCD Epoxy Box
	15	20	9.05 e- 2	—	—
13.9	10	10	8.43 e-02	—	—
	5	5	8.07 e-02	—	—
	15	80	2.29 e-01	—	—
	10	50	2.04 e-01	—	—
	5	40	1.96 e-01	—	—
17.8	3	20	1.81 e-01	—	—
	2	10	1.73 e-01	—	—
	0.5	4	1.69 e-01	—	—
	0.1	2	1.68 e-01	—	—
	15	90	2.80 e-01	—	—
	10	80	2.71 e-01	90	3.99 e-01
	5	50	2.48 e-01	50	3.54 e-01
20.8	3	20	2.27 e-01	30	3.34 e-01
	2	10	2.21 e-01	20	3.24 e-01
	0.5	—	—	5	3.10 e-01
	0.1	1	2.15 e-01	—	—
	5	60	3.01 e-01	80	4.47 e-01
	3	50	2.94 e-01	60	4.29 e-01
26.35	2	20	2.77 e-01	50	4.20 e-01
	1	10	2.71 e-01	—	—
	0.5	—	—	10	3.86 e-01
	0.1	1	2.66 e-01	—	—
	3	90	3.69 e-01	—	—
	2	40	3.48 e-01	70	5.14 e-01
59.54	1	20	3.39 e-01	—	-
	0.5	—	—	20	4.84 e-01
	0.1	1	3.32 e-01	—	—

Table 6 Minimum Level of Component Discrimination (MLCD) of brine in A oil.

Energy (keV)	Statistics of I and I_0	MLCD (%) Acrylic Box	I/I_0 of MLCD Acrylic Box	MLCD (%) Epoxy Box	I/I_0 of MLCD Epoxy Box
13.9	15	2	7.15 e-02	—	—
	3	1	7.49 e-02	—	—
17.8	15	20	1.07 e-01	—	—
	10	10	1.33 e-01	—	—
	3	2	1.59 e-01	—	—
	0.5	1	1.63 e-01	—	—
20.8	15	20	1.61 e-01	20	2.30 e-01
	5	10	1.86 e-01	10	2.65 e-01
	3	3	2.05 e-01	—	—
	2	—	—	3	2.93 e-01
	0.5	2	2.08 e-01	1	2.63 e-01
26.35	15	40	1.98 e-01	50	2.63 e-01
	10	30	2.13 e-01	40	2.83 e-01
	5	—	—	20	3.27 e-01
	3	20	2.29 e-01	—	—
	2	10	2.47 e-01	—	—
	0.5	2	2.61 e-01	3	3.70 e-01
59.54	5	95	2.92 e-01	—	—
	3	80	2.98 e-01	90	4.20 e-01
	2	40	3.14 e-01	70	4.31 e-01
	0.5	20	3.23	20	4.60 e-01
	0.1	2	3.30 e-01	—	—

Table 7 Minimum Level of Component Discrimination (MLCD) of gas in paraffin.

Energy (keV)	Statistics of I and I_0 (%)	MLCD (%)	I/I_0 of MLCD
20.8	10	30	0.4761
	5	10	0.4875
	3	90	0.4719
29.54	2	70	0.4997
	2	20	0.5040

Table 8 Minimum Level of Component Discrimination measured for brine in paraffin.

Energy (keV)	Statistics of I and I_0 (%)	MLCD (%)	I/I_0 of MLCD
20.8	10	30	0.4761
	5	10	0.4875
	3	90	0.4719
29.54	2	70	0.4997
	2	20	0.5040

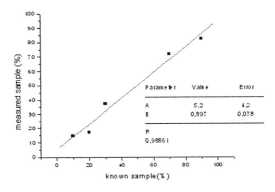

Figure 4 Plot of the percentagem of the measured sample for gamma transmission versus the percentage of the known sample.

For the validation of the theoretical predictions, measurements were performed with the energy lines 20.8 keV and 59.54 keV, with brine in paraffin using the epoxy box. The results are shown in the Table 8. Since the fraction of the sample was known, a comparison between the previously known percentages and those measured by gamma transmission were made. The calculations were done using equation 1. The results are shown in Figure 4.

4 Conclusion

According to the MLCD tables, it is possible to discriminate a substance in a biphasic fluid with a desirable minimum level of component discrimination, without the need of high counts number. It was also observed that the best discriminating energies, for all the mixtures, are the intermediary ones of 17.8, 20.8 and 26.35 keV, because of the 13.9 keV energy line is very attenuated, while the 59.54 keV is not very attenuated.

However, it is of extreme importance to include in this criterion the energy line of 59.54 keV, because if the thickness of the sample will be increased, the lower energies can be much attenuated. A good result was obtained from the comparison of the experimental data with the mathematical modeling data, so the mathematical modeling used is applicable to guide the planning of experiments to determine fractions of a substance in a biphasic fluid involving oil, natural gas or brine. It was obtained a good correlation between the previously known values of the biphasic mixture and the ones measured by transmission, as presented in Figure 4, showing that the technique presents good applicability.

Bibliography

[1] A. B. STRAPASSON. **Geologia do Petróleo**. Description of the Oil's basin. Available in: < http://geocities.yahoo.com.br/geologiadopetroleo/bacias.htm > . Access in: June 12, 2002.

[2] CEPETRO. **Petróleo**. Oil average composition. Available in: < http://www.cepetro.com.br/ > . Access in: February 25, 2001.

[3] EPOXI. **Silaex Química Ltda**. Chemical formula of the epoxy. Available in: < http://www.silaex.com.br/epoxi.htm > . Access in: January 20, 2003.

[4] G.A . JOHANSEN and P. JACKSON. **Salinity independent measurement of gas volume fraction in oil/gas/water pipe flows**. Applied Radiation and Isotopes. V.53, p. 595-601, 2000.

[5] GASPETRO. **Gás Natural**. Gas average composition. Available in: < http://www.gaspetro.com.br/propr.htm > . Access in: December 20, 2001.

[6] GERWARD, N. GUILBERT, JENSEN, K. B. and LEVRING, H. **X-ray absorption in matter. Reenginnering XCOM** – Radiation Phys. And Chem. V. 60, p. 23-24, 2001.

[7] S.A . TJUGUM, G.A . JOHANSEN and M.B. HOSTAD. **The use of gamma radiation in fluid flow measurements**. Radiation Physics and Chemistry, V. 61, Issus 3-6, p. 797-798, 2001.

[8] G.J. ROACH, J.S. WATT, H.W. ZASTAWNY, P.E. HARILEY, W.K. ELLIS. **Trials of a gamma-ray multiphase flow meter on oil production pipelines at Thevenard Island**. CSIRO. Division of mineral and process engineering, PMB 5, Menai 2234, Australia, 1994.

Aplicação de um Detector de Telureto de Cádmio (CdTe) para Espectrometria Gama "IN SITU"

Luiz Diego Marestoni*, Carlos Roberto Appoloni[†] e Jaquiel Savi Fernandes[‡]

Laboratório de Física Nuclear Aplicada, Departamento de Física / CCE
Universidade Estadual de Londrina, Campus Universitário
Cx.P.: 6001, CEP.: 86.051-990, Londrina-PR.

Resumo. Visto que cada detector apresenta características específicas, a determinação de alguns parâmetros, tais como a melhor temperatura de funcionamento, shaping time, eficiência, resolução, etc, permite-nos otimizar a aplicabilidade deste. Para o detector de CdTe, produzido pela Amptek, com geometria de $3 \times 3 \times 1$ mm^3, protegido por uma janela de berílio com 0,25 mm, tensão aplicada de 600 V, determinou-se que a melhor temperatura de operação é -22°C, com "Shaping Time" de 3 μs e "Rise Time Discrimination" (RTD) com valor 3. O detector de CdTe é portátil, devido ao seu baixo peso e à não necessidade de resfriamento com nitrogênio líquido, pois este utiliza resfriamento termoelétrico do tipo Peltier. Isso permite que as medidas sejam realizadas no próprio local de estudo, técnica esta que recebe o nome de *"técnica de medida in-situ"*. Foi determinada a curva de eficiência absoluta versus energia utilizando uma fonte certificada da Agência Internacional de Energia Atômica (IAEA) de Eu-152, obtendo a equação: *Eficiência = 0,15 [0,05].Energia* $^{-1,65[0.21]}$, e da curva de resolução (FWHM) versus energia obtendo a equação

$FWHM$ *(keV)* $= 0,178[0,051].Energia(keV)^{0,523[0.038]}$. Para a validação da curva de eficiência foram realizadas medidas com a fonte de ^{133}Ba. Houve boa concordância entre as atividades certificadas e as medidas.

---

1 Introdução

No presente trabalho investigamos a possibilidade da aplicação de um detector de CdTe (3x3x1 mm) para espectrometria gama de vários radionuclídeos de baixa

---

* lu_diegomarestoni@yahoo.com.br
[†] appoloni@uel.br
[‡] jaquielfernandes@yahoo.com.br

atividade, com o intuito de utilizá-lo em medidas ambientais "in situ". Este detector já é amplamente citado na literatura para casos envolvendo altas atividades, como em vazamento de reatores e verificação em depósitos de rejeitos radioativos. Em relação aos demais detectores semicondutores utilizados na espectrometria gama padrão, o detector de CdTe apresenta uma vantagem no que diz respeito ao resfriamento do cristal, já que este não necessita de temperatura criogênicas, pois é resfriado por efeito Peltier. Devido a este fato e ao da eletrônica associada ter dimensões reduzidas, este permite que as medidas sejam realizadas *in situ*, reduzindo tempo e custo da análise dos dados.

Visto que cada detector apresenta características específicas, a determinação de alguns parâmetros, tais como: melhor temperatura de resfriamento do cristal, shaping time, eficiência, resolução, etc, permite-nos otimizar a aplicabilidade deste. Para validação da metodologia foi determinada a atividade de uma fonte certificada de ^{133}Ba. Com uma pastilha de ^{137}Cs foi possível determinar a atividade mínima detectável para suas duas linhas de energia, 32 e 661,62 keV.

2 Materiais e Métodos

O detector, com geometria de $3\times3\times1$ mm^3, produzido pela Amptek com janela de berílio de 0,25 mm [1], e acoplado à eletrônica nuclear padrão. A eletrônica nuclear consiste em um sistema ROVER, também desenvolvido pela Amptek, com aproximadamente 2,0 kg e 21,5 cm × 5 cm × 7,6 cm de dimensões.

O detector foi posicionado na posição vertical, com o cristal voltado para baixo, sobre uma mesa de medidas *in situ*. Foi aplicada uma tensão de 600 V. O detector ficou envolto por uma blindagem de chumbo com 5 cm de espessura e 37 cm de altura. A Figura 1 mostra o equipamento e a geometria utilizada no trabalho.

Figura 1 Sistema de Espectrometria Gama "in-situ"

Foram utilizadas neste trabalho fontes pontuais padrão do tipo pastilha de [152]Eu, [137]Cs e [133]Ba, com atividade de 399,2 kBq, 377,5 kBq e 376 kBq, respectivamente, certificadas pela Agência Internacional de Energia Atômica (IAEA). As fontes foram posicionadas a 4,15 cm do cristal. Com a utilização destas fontes certificadas foi possível determinar a curva de eficiência do detector através da Equação 1,

$$\varepsilon\,(E) = \frac{N}{A.P_\gamma.t} \tag{1}$$

onde A é a atividade do radionuclídeo que chega no detector (Bq), N representa a contagem líquida do fotopico de interesse, ε é a eficiência do detector para a linha de energia analisada, P_γ é a probabilidade de transição pelo decaimento gama e t é o tempo de contagem (s).

O calculo da centróide e FWHM foi realizado automaticamente pelo programa PMCA. A centróide (P) é calculada depois de subtrair o fundo e é dada pela média ponderada dos canais sob o fotopico.

$$P = \frac{\sum i.C_i}{\sum C_i} \tag{2}$$

onde C_i é o numero de contagem sob o canal i. O FWHM é a medida da largura em energia do fotopico na metade da máxima contagem no fotopico [5].

3 Resultados

Determinou-se, através de um planejamento fatorial 2^3 [2,3], que a melhor temperatura de operação é $-22°$ C, "Shaping Time" de 3 µs e "Rise Time Discrimination" (RTD) com valor 3. As Figuras 2 e 3 apresentam a curva de resolução em energia (FWHM) e a curva de eficiência, respectivamente, para o detector de CdTe, utilizando a fonte de [152]Eu, com ganho 02 na eletrônica e tempo de medida de 900 segundos.

A Figura 4 apresenta o espectro medido para a fonte de [133]Ba, onde se verifica que os fotopicos são assimétricos. Tal fato se deve a, durante a criação do par eletron/buraco pela radiação que interage perto do final do contato do detector, resultar em flutuações no tempo de coleção de carga. Como resultado, o espectro adquirido tem um aumento no fundo de radiação e degradação da energia de resolução. Para reduzir tal efeito é que estudamos o melhor RTD possível. Com a curva de eficiência podemos calcular a atividade e comparar com os dados certificados pela IAEA, realizando assim a validação da metodologia.

A Tabela 1 apresenta os dados da fonte de [133]Ba, no que diz respeito às linhas de energia analisadas e respectivas atividades calculadas. Verificamos que a atividade medida, com 95 % de confiança, é igual a certificada.

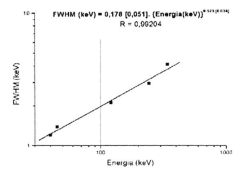

Figura 2 Curva de resolução para a fonte de ^{152}Eu a 4.15 cm do detector, ganho 02 e tempo de medida de 900 segundos.

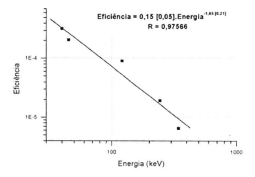

Figura 3 Curva de eficiência do detector de CdTe a 4.15 cm da fonte, com ganho 02 e tempo de medida de 900 segundos.

Tabela 1 Atividade medida de uma fonte de ^{133}Ba certificada 376±11 kBq (68 % de confiança).

Energia Literatura (keV)	Energia Determinada (keV)	Atividade Medida (kBq)
80,32	80,79	395
53,16	53,29	361
276,4	278,76	383
302,8	301,91	367
356,01	357,98	378
383,85	384	345
Atividade Média (95 % de confiança) = 373 ± 20 kBq		

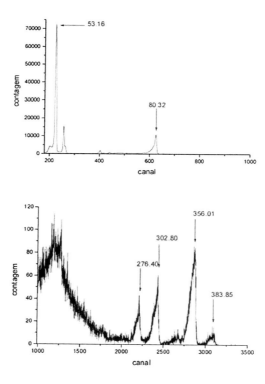

Figura 4 Espectro da fonte de ^{133}Ba com atividade certificada de 376 kBq (a) até o canal 1000 (b) do canal 100 até o canal 3500.

Como o presente trabalho tem como objetivo verificar a aplicação do detector de CdTe, é de grande importância conhecer sua Atividade Mínima Detectável (AMD) para as situações de interesse. A Tabela 2 apresenta a AMD para a fonte de ^{137}Cs, para as energias de 32 e 662 keV e respectivas geometrias.

Tabela 2 Atividade Mínima Detectável (AMD).

Fonte/Amostra	Distância (cm)	Tempo de Medida	Energia (keV)	AMD (kBq)
^{137}Cs/pastilha	4,15	900 s	32	15,09
^{137}Cs/pastilha	4,15	22 h	662	5,88

4 Conclusão

Conclui-se que a calibração em energia do sistema é boa, já que foi possível determinar as energias do fotopico do ^{133}Ba com boa precisão. Utilizando a curva de eficiência de detecção medida, pode-se determinar a atividade de qualquer radionuclídeo que contenha energia entre 40 e 344 keV, que corresponde à região estudada. Com o gráfico de FWHM podemos a capacidade de discriminação em energia do sistema, que é bastante boa para um sistema portátil. Analisando a AMD, conclui-se que o detector não é adequado para detectar atividades muito baixas, como no caso de traços radioativos naturais, devido ao fato de o cristal de CdTe, disponível comercialmente, possuir volume muito pequeno.No entanto, pode ser empregado para verificação in situ de situações com contaminação / acidentes de materiais radioativos. Caso fiquem disponíveis detectores de CdTe com cristais de maior volume, então esta conclusão pode ser revista e o sistema poderá ser utilizado para medir atividades menores.

Referências Bibliográficas

[1] AMPTEK. The ROVER. **User's Manual: Portable X-Ray and Gamma Ray Spectroscopy System.** Revision 2.2, 2001.

[2] NETO, B. B.; SCARMINIO, I. S.; BRUNS, R. E. **Como Fazer Experimentos: Pesquisa e Desenvolvimento na Ciência e Industria.** Ed. Unicamp – SP, 2001.

[3] FERNANDES, J. S. **Aplicabilidade de um espectrômetro portátil de CdTe e NaI(Tl) para a medida da atividade de césio-137 (^{137}Cs) e Berílio-7 (^{7}Be).** Tese (Mestrado em Física Nuclear Aplicada) – Universidade Estadual de Londrina, Londrina. 106p.,2005.

[4] DUSI, W.; AURICCHIO, N.; BRIGLIADORI, L.; DONATI, A.; LANDINI, G.; MENGONI, D.; PERILLO, E.; VENTURA, G. **A study of the spectroscopy response of planar CdTe detectors when irradiated at various angles of incidence.** Nuclear Instruments and Methods in Physics Research A. v. 506, p. 119 – 124, 2003.

[5] **http://www.amptek.com/mcasoft.htm** acessado em 20/07/2006.

Anais da XXVIII Reunião de Trabalho sobre Física Nuclear no Brasil
SP, Brasil, 2005

Calibração do Detector LR-115 II em Monitor de Placas Paralelas sob Condições de Ventilação Forçada

H. Javier Ccallata*, E. M. Yoshimura

Laboratório de Dosimetria, Instituto de Física, Universidade de São Paulo.
Rua do Matão, Travessa R., 187, CEP 05508-900, SP, Brasil.
Phone:+55(11)3091 6990, Fax: +55(11)3031 2742

Resumo. Neste trabalho apresentamos a calibração do detector LR-115 II, na geometria de Monitor de Placas Paralelas (MPP) coberto com filme fino de PVC em atmosfera de ^{222}Rn e filhos em duas condições ambientais: com e sem ventilação forçada. No ambiente estudado foram expostos simultaneamente os monitores NRPB/SSI com CR-39 e o MPP com três LR-115 (dois internos cobertos com folhas de alumínio de 22 e 25 μm de espessura, e um externo descoberto). A resposta do LR-115 II externo em relação aos internos, no MPP, foi avaliada nessa mesma atmosfera durante 10, 18 e 35 dias. A variação da densidade de traços ρ nos detectores internos e externo ao MPP indicou que a concentração de emissores alfa é bem maior no exterior que no interior do volume de detecção do MPP. Verificou-se também uma resposta independente do detector LR-115 II coberto com alumínio de 25 μm, frente à mudança da taxa de ventilação. A calibração relativa forneceu o fator $f_0 = (0,70 \pm 0,09)$ traços.m^3.cm^{-2} kBq^{-1}.h^{-1} para a concentração de radônio no local estudado.

1 Materiais e Métodos

No processo de calibração foi utilizada uma sala do IFUSP onde estão armazenadas fontes de ^{226}Ra em desuso. A metodologia baseia-se na exposição simultânea de dois tipos de monitores ambientais: o NRPB/SSI e o MPP (figura 1), cada um deles montado com seus respectivos detectores, e posicionados nas paredes a 1,70 m do solo. As exposições foram feitas durante 10,5 dias em duas condições ambientais: com e sem ventilação forçada. A ventilação do ambiente foi feita com ajuda de um ventilador comum, com uma taxa de fluxo de ar homogênea e constante. A relação entre a densidade de traços ρ no LR-115 II, o tempo de exposição t_e e a concentração

---

* hjavier@dfn.if.usp.br

média de emissores alfa A foi determinada através de $\rho = A.t_e\, f$, obtendo se o fator de calibração f do detector nessas condições ambientais. Para os detectores expostos (CR-39 e LR-115 II) foram usadas condições padrão de revelação e os traços neles formados foram contados de forma semi-automática com ajuda do software Leica Q-Win e um microscópio Leitz Diaplan.

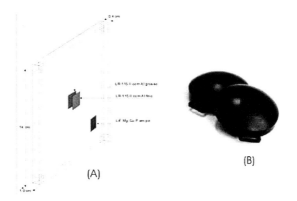

Figura 1 Monitores ambientais. A) Monitor de placas paralelas, MPP, com detectores LR-115 II. B) Monitor NRPB/SSI com detectores CR-39 no interior.

2 Resultados

O comportamento da densidade de traços ρ dos diferentes detectores em relação ao tempo de exposição mostra que o detector externo é sensível a variações ambientais

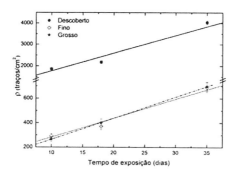

Figura 2 Relação de densidades de traços dos LR-115 II no MPP para os tempos de exposição

que os internos não podem detectar (figura 2). Como a concentração de radônio no ar foi constante no tempo, pode-se dizer que a concentração de emissores alfa é bem maior no exterior que no interior do volume de detecção ao MPP. Como a resposta dos dois detectores internos foi equivalente, a escolha do detector interno ao MPP foi feita levando em conta a maior probabilidade de detecção de traços considerando a energia residual de uma partícula alfa proveniente do decaimento do ^{222}Rn e filhos. Foi escolhido o LR-115 II coberto com alumínio de 25,0 μm.

2.1 Calibração relativa

Foram colocados seis monitores NRPB/SSI e 4 MPP em diversas posições dentro da mesma sala a 1,70 m do solo, alguns pendurados e outros colados na parede durante 10,5 dias com e sem ventilação forçada. Como era esperado, notou-se uma variação na concentração de radônio na sala devido à ação da ventilação (tabela 1). Este resultado mostra que a resposta do MPP é independente das taxas de ventilação no ambiente, ao menos na faixa estudada. A média do fator f nas duas condições ambientais foi de $f_o = 0,70 \pm 0,09$ traços.m^3.cm^{-2} kBq^{-1}.h^{-1}.

Tabela 1 Resultados da calibração relativa.

Condição de calibração	Não ventilado	Ventilado
Concentração de ^{222}Rn (A) [kBq.m^{-3}]*	$1,54 \pm 0,21$	$1,00 \pm 0,15$
Fator para o LR-115 II (f) [traços.m^3.cm^{-2} kBq^{-1}.h^{-1}]	$0,68 \pm 0,10$	$0,73 \pm 0,18$

* A concentração foi medida através do CR-39 no monitor NRPB/SSI. O fator de calibração deste conjunto para o ^{222}Rn é de $(2,8 \pm 0,2)$ traços.m^3.cm^{-2} kBq^{-1}.h^{-1}.

3 Conclusões

Na calibração relativa do MPP em relação ao monitor NRPB/SSI, verificou-se uma resposta independente do LR-115 II (coberto com alumínio de 25 μm) frente a mudanças de ventilação forçada no ambiente. Notou-se também um comportamento linear na densidade de traços do LR-115 II interno em relação ao tempo de exposição. O valor do fator f_o para o LR-115 II no MPP será utilizado no futuro para obter as concentrações de ^{222}Rn nos diferentes ambientes de estudo sob condições de ventilação diversas.

Referências Bibliográficas

[1] ICRP, Protection against ^{222}Rn at home and at work,. ICRP Publication 65, Annals of the ICRP 23, 1993.

[2] C. Orlando *et al.*, A passive radon dosimeter suitable for workplaces, *Radiat. Prot. Dosim.* 102(2): 163-168, 2002.

[3] M. Godoy *et al.*, Effects of environmental conditions on the radon daughters spatial distribution, *Rad. Meas.* 25(3), 213-221, 2002.

[4] I. Mäkeläinen, Calibration of bare LR-115 film for radon measurements in dwellings, *Radiat. Prot. Dosim.* 7(1-4): 195-197, 1984.

Anais da XXVIII Reunião de Trabalho sobre Física Nuclear no Brasil
SP, Brasil, 2005

Study of DNA and Protein Structures Through Perturbed Angular Correlations

J. C. de Souza[1],*, N. Carlin[1], A. W. Carbonari[2], M. S. Baptista[3], E. M. Szanto[1] and F. J. de Oliveira Filho[1]

[1] Departamento de Física Nuclear, Instituto de Física, Universidade de São Paulo, Caixa Postal 66318, CEP 05315-970, São Paulo, SP, Brazil

[2] Centro do Reator de Pesquisas, Instituto de Pesquisas Enérgicas e Nucleares, Av. Prof. Lineu Prestes, 2242, CEP 05508-000, São Paulo, SP, Brazil

[3] Departamento de Bioquímica, Instituto de Química, Universidade de São Paulo, Av. Prof. Lineu Prestes, 748, CEP 05508-900, São Paulo, SP, Brazil

Abstract. The Perturbed Angular Correlation (PAC) technique is largely used in Condensed Matter Physics experiments. Recently it has also been used in applications on biological problems, as the determination of structural characteristics of DNA and Protein molecules. We are implementing a PAC Spectrometer at the University of São Paulo Pelletron Laboratory. The main motivation is the study of DNA and Protein structures. We have already 6 BaF_2 detectors for gamma rays, which are adequate due to the good time resolution. The spectrometer mechanical structure is ready and tests have been performed in order to choose the best configuration for the data acquisition electronics, composed by a multiparameter CAMAC system. At the same time, we have performed, at IPEN-CNEN/SP, measurements to study the structure of organic systems like lipids and surfactants. If some characteristics of these molecules are known, then some mechanisms of interaction between them and protein molecules can be understood.

1 Introduction

The application of the PAC technique to Biological and Biochemical problems has brought very interesting results in the study of DNA and protein molecules structures [1, 2]. One of the advantages of this technique is the possibility of performing measurements with samples in solution and observing dynamical characteristics [3].

---

* jairocav@if.usp.br

We are implementing the PAC technique at the University of São Paulo Pelletron Laboratory with the objective of studying problems involving DNA and protein structure and functions. Allied to the charged particle external beam facility constructed by our research group [4], this technique can provide a versatile system. It is possible to produce structural modifications in protein or DNA molecules and by using the PAC technique, investigate them through the monitoring of the samples before and after irradiation.

2 Theory of Perturbed Angular Correlation of γ-Rays

The theory and the method of perturbed angular correlations of γ-rays are described in detail in [5]. We will show here only relevant points.

A γ-ray emitted from a radioactive nucleus carries angular momentum away from the nucleus in the nuclear decay. The total angular momentum is conserved in the nuclear decay, and this is the origin of the angular correlation of γ-rays. Experimentally, the technique consists in inserting a radioactive nucleus, like ^{111}In, in a specific medium and then observing the temporal variation or perturbation of the distribution of the γ-rays emitted in the nuclear decays, due to the electric field gradient (EFG) generated by the structure of the external medium. Figure 1 shows an example of the decay of ^{111}In and ^{111m}Cd.

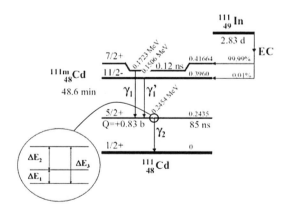

Figure 1 ^{111}In and ^{111m}Cd decays. The inset shows the three possible energy levels for the 5/2 spin intermediate state due to the influence of a electric field gradient (EFG). ΔE_1, ΔE_2 and ΔE_3 are the energy differences among the states.

The distribution of the second γ-ray with respect to the first is given by:

$$W(\theta, t) = \frac{e^{-t/\tau_N}}{4\pi\tau_N}(1 + A_2 G_2(t) P_2(\cos\theta)) \qquad (1)$$

where τ_N is the intermediate state mean life, A_2 describes the anisotropy of the angular correlation between the two γ-rays. P_2 is given by: $P_2(\cos\theta) = \frac{1}{2}(3\cos^2\theta - 1)$. $G_2(t)$ describes the temporal evolution of the probability to detect the second γ-ray in an interval t. For the case when the EFG is static we have:

$$G_2(t) = a_0 + \sum_{n=1}^{3} a_n \cos(\omega_n t) \qquad (2)$$

The angular frequencies ω_1, ω_2 and ω_3 above correspond to the energy differences ΔE_1, ΔE_2 and ΔE_3 (Figure 1). It is possible to obtain them directly through a $G_2(t)$ Fourier transform. These frequencies are, in the static case, directly proportional to the quadrupolar frequency, ω_Q.

The EFG can be described by two parameters [6]. In the Quantum Theory, the EFG is given by a 3×3 tensor, represented here by V. If an adequate choice of coordinate system is made, in which $|V_{zz}| \geqslant |V_{yy}| \geqslant |V_{xx}|$, the EFG is generally described by the component $|V_{zz}|$ (z component of V), which is directly proportional to the quadrupolar frequency ω_Q, and by the asymmetry parameter η, given by:

$$\eta = \frac{V_{xx} - V_{yy}}{V_{zz}} \qquad (3)$$

If there is a temporal dependence of the EFG, i.e. the orientation of the EFG changes in the time between the emissions of the two γ-rays, Equation (2) must be modified. This can be the case for molecules which are submitted to rotational diffusion due to Brownian motion in the sample, for molecules that are not rigid but change conformation, or undergo other dynamic processes in the time period between the emission of the two γ-rays. The term $G_2(t)$ passes to have an exponential dependence with time, by the factor e^{-t/τ_c}, for slow reorientation, and $e^{-2.8\omega_0^2\tau_c(1+\eta^2/3)t}$, in the fast limit. τ_c is denominated rotational correlation time, which for spherically symmetric molecules depends on the viscosity of the medium (ξ), the temperature (T) and the molecule volume (V), of the following mode:

$$\tau_c = \frac{V.\xi}{T.k_B}, \qquad (4)$$

where k_B is the Boltzmann constant.

3 Experimental Facility

It is possible to measure the distribution $W(\theta, t)$ through coincidences between pairs of detectors. With the 6 detector PAC spectrometer we are constructing (see Figure 2), we have 6 combinations for $\theta = 180°$, and 24, for $\theta = 90°$

Figure 2 new PAC spectrometer structure with BaF_2 detectors at the Pelletron Laboratory.

The experimental perturbation function is given by:

$$A_2 G_2(t) = 2\frac{W(180°, t) - W(90°, t)}{W(180°, t) + 2W(90°, t)} \qquad (5)$$

W(180°,t) and W(90°,t) represent the geometric averages of the coincidences measurements for 180° and 90°, respectively.

Generally, PAC spectrometers work with data acquisition electronics with a single channel analyzer to each detector and a router module [7] to separate, electronically, the coincidences of the two γ-rays and the energy spectra are not acquired.

Recently a new PAC spectrometer was constructed in which all the process are computational [8]. In this case the energy spectra are acquired.

Finally, the PAC spectrometer we are constructing uses a different electronics concept, in which the energy spectra are also acquired and the coincidences are also selected computationally, but through a multiparameter acquisition CAMAC system. The electronics schematic for two detectors is shown in Figure 3. The generalization of this example to 6 detectors is direct.

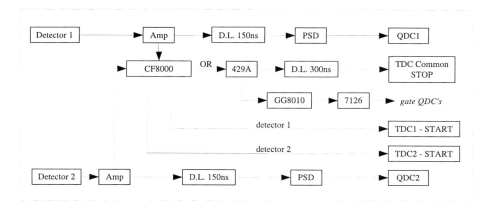

Figure 3 scheme of the data acquisition electronics system to the new PAC spectrometer at the Pelletron Laboratory, for two detectors. Amp is a 612AM LeCroy amplifier; D.L. is a cable delay line; PSD is a pulse shape discrimation module, used to attenuate the energy signal; QDC is a charge to digital converter Phillips 7166H. CF8000 is an ORTEC constant fraction discriminator; 429A is a LeCroy logic fan in fan out; GG8010 is a ORTEC gate & delay generator; 7126 is a Phillips level translator. The TDC module used is a time to digital converter Phillips 7186.

4 Measurements at IPEN/SP

We performed experiments at the *Instituto de Pesquisas Energéticas e Nucleares* (IPEN-CNEN/SP), where a 4 BaF_2 detectors PAC spectrometer [7] is available. Cardiolipin:phosphatidylcoline 1:1 vesicles [9] and *sodium dodecyl sulfate* (SDS) [10] samples were used.

For the experiments with vesicles the objective was to verify if the PAC technique could sense differences related to variation of their dimensions, provoked by adding a surfactant (SDS) to a vesicles sample, tranforming them into micelles. We have done measurements with vesicles at 295K and 77K. Also, with vesicles plus SDS at 295K and 77K.

After that, other measurements were performed with SDS samples in different enviroments (water and water plus methanol) in which the SDS concentrations of the solutions were varied to verify if the PAC parameters were sensitive to the molecular agregation or even to critical micellar concentration (CMC).

4.1 Materials and Methods

For the solution of vesicles, first it was diluited 5mg of cardiolipin and 5mg of phosphatidylcoline-98% in 0.5mL of chloroform. After that, the solution was dried with argon gas and then diluted in 2mL of mQ water. This solution was passed through a 100nm filter to form vesicles with this dimension. The sample, to be used in the spectrometer, was prepared by depositing 4.0μL of acqueous solution traced with $InCl_3$ in 40.0μL of vesicles solution. Two measurments were performed, the first one, at 295K and the second, at 77K.

The vesicles plus SDS sample was prepared adding 0.57mg of SDS to 40,0μL of vesicle solution (similar to the previous one), forming a micelles solution. In this manner, we would have smaller (approximately 10nm) molecules than in the first kind of sample. As in the previous sample, 4.0μL of acqueous solution traced with $InCl_3$ was added. The objective was to verify if the PAC thechnique would be capable to sense differences in the dimensions of the molecules.

In order to verify the sensitivity of PAC technique to the critical micellar concentration, we have performed several measurments with SDS samples, all at 295K. In these cases, we varied the SDS concentration. The $InCl_3$ solution to be added was diluted either in methanol or in water.

For all SDS series of measurments first it was prepared a 50.0mM acqueous stock solution, by adding 0.361g of SDS in 25.0mL of water. Each specific concentration of SDS samples was obtain through the dilution of one small volume of the stock solution in one separated recipient with a specific water volume, as the final SDS concentration was smaller than 50.0mM.

In the case of $InCl_3$ diluted in water, measurements were performed with 0.05mM, 0.2mM, 1.0mM, 2.67mM, 4.0mM, 8.0mM and 10.0mM concentrations of SDS.

When $InCl_3$ was diluted in methanol, the volumes of SDS solutions and the volume of methanol were different for each sample, according to the ^{111}In activity and concentration os SDS.

4.2 Results and Discussion

4.2.1 *Vesicles Samples*

The graphs in figures 4 and 5 refer to vesicles samples, the first one at 295K and the second, at 77k.

Following, Figures 6 and 7 are about vesicles plus 50mM SDS samples, at 295K and 77K. Visually, these graphs are similar to the previous ones.

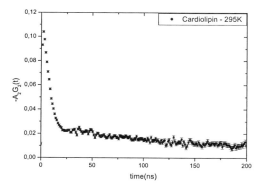

Figure 4 Perturbatoin function for vesicles sample at 295K.

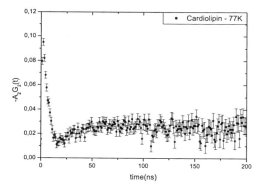

Figure 5 Perturbation funciton for vesicles sample at 77K.

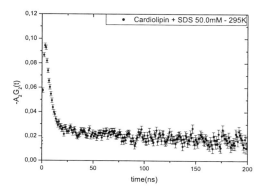

Figure 6 Perturbation function for vesicles plus SDS 50mM at 295K.

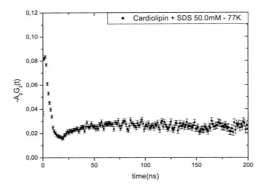

Figure 7 Perturbation function for vesicles plus SDS 50mM at 77K.

The most relevant parameters obtained by fitting the perturbation functions of the four graphs above are presented in Table 1:

Table 1 Results for vesicles at 295K (sample 1) and 77K (sample 2) and vesicles plus SDS at 295K (sample 3) and 77K (sample 4). The asymmetry parameter (η) is zero for the four samples. ν_Q is the quadrupolar frequency in MHz, δ is the width of ν_Q distribution and λ is the inverse of τ_c in Equation (4).

Sample	ν_Q (MHz)	δ	λ (ns^{-1})
1	81.4(5)	0.770(10)	15.6(7)
2	127.6(11)	0.570(10)	0
3	82.1(10)	0.620(20)	22.9(19)
4	127.7(10)	0.620(10)	0

At 295K there are dynamical interactions in the two samples, indicated by the values of λ (inverse of τ_c in Equation (4)), different from zero, in Table 1. This effect can also be noticed in the shape of the curves, which present exponential dampings.

The quadrupolar frequency is better determined at 77K, when the EFG doesn't change, and consequently, the value of λ is compatible with zero. This means we are dealing with a static situation.

If we compare the results for the two samples, we can notice that the significant difference between them is the value of λ. The quadrupolar frequencies don't change after adding SDS to the vesicles solution, indicating that the neighbourhood of ^{111}In didn't change conformation when the molecules passed from vesicle form to micellar form. In this manner, it is possible to believe that the EFG changes at

295K, due to rotational diffusion caused by the Brownian movement in the samples and the value of λ is bigger for the second sample because its molecules are smaller.

4.2.2 SDS Samples

For the experiments with SDS samples it is not possible to notice through the PAC technique the critical micellar concentration, which for acqueous solution at 295K, occurs around 7mM [11]. As previously mentioned, the concentration was varied from 0.05mM to 10.0mM. In the fits of the respective perturbation curves no significant changes were observed. Nevertheless, it was possible to observe that the medium, water or water plus methanol, modifies the behavior of the parameters.

In Figures 8 and 9 are presented two examples, 1.0mM that correspond to a sample with SDS concentration under CMC, and 10.0mM, that is above CMC. The shape of the perturbation curves don't change significantly, neither for the other samples in water solution. This indicates that there are no big changes in behavior of the molecules.

It was possible to adjust two binding sites for the ^{111}In (due to the charge of ^{111}In), and then the values of the quadrupolar frequencies fluctuated around 60MHz for the most populous site. The samples don't present dynamic interactions. Also, the quadrupolar frequencies were not well defined. It was not possible to adjust a free value of the parameter δ, that represents the width of the distribution of frequencies.

In Figure 10 it is shown a graph of a sample in which it was added 10μL of methanol to 1.0mM (10μL) acqueous solution sample. Then, we could obtain a SDS concentration, in such manner, intermediate between the two previous, for which we did not expect big changes if the solution was acqueous.

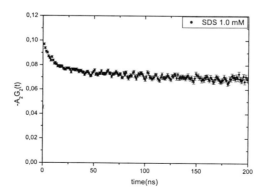

Figure 8 Perturbation function for SDS 1.0mM in water at 295K.

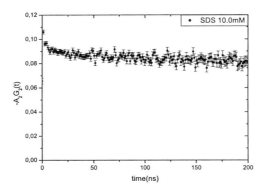

Figure 9 Perturbation function for SDS 10.0mM in water at 295K.

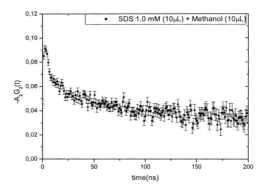

Figure 10 Perturbation function for SDS plus metanol at 295K.

As we can see, the shape of the curve in Figure 10 is clearly different from the perturbation function in Figures 8 and 9, what indicates that the medium, water or methanol, influences the SDS molecules bahavior. This can be explained by the changing of viscosity of the medium.

In the case of the samples that contained methanol, it was possible to fit one single site for ^{111}In. As in the acqueous samples, the frequencies in general were not well defined. This didn't allow us to find some systematics in the behaviors with the SDS concentration.

5 Conclusion

In this work we have presented a new PAC spectrometer being constructed at the University of São Paulo Pelletron Laboratory with an acquisition system based on multiparameter CAMAC system. We have also presented some results with cardiolipin:phosphatidylcoline vesicles and SDS samples from measurments at IPEN-CNEN/SP. For the vesicle samples, we can observe that the PAC technique is sensitive to differences between two samples for which the dimension of the molecule changes due to the presence of SDS surfactant, which tranforms one vesicle into a micelle, with out changes in the probe radioactive nucleus neighbourhood. On the other hand, the technique is not capable of sensing the changes caused by agregation of SDS molecules while the concentration of the solution is varied until forming micelles, and is not sensitive to the critical micellar concentration. Finally, we could observe that the behavior of SDS molecules changes due to the exchange of the medium, water or methanol. One of the possibilities to explain this bahavior is the changing of solvent viscosity.

Acknowledgements

We would like to thank FAPESP and CNPq for financial support.

Bibliography

[1] M. Ceolín, J. Biochem. Biophys. Methods **45**, 117 (2000).

[2] C. Damblon et al., J. Biol. Chem. **278**, 29240 (2003).

[3] L. Hemmingsen, K. N. Sas, E. Danielsen, Chem. Rev., **104**, 4027 (2004).

[4] N. Carlin, J. C. de Souza, E. M. Szanto, J. C. Acquadro, E. Okuno, J. Takahashi, N. K. Umisedo, F. J. de Oliveira Filho, J. A. C. Vasconcelos, Nuclear Instruments and Methods in Physics Research A, **540**, 215 (2005).

[5] H. Frauenfelder, R. M. Steffen, Alpha-,Beta and Gamma-Ray Spectroscopy, North-Holland, Amsterdam, p 997 (1965).

[6] G. Schatz, A. Weidinger,J. A. Gardner, Nuclear Condensed Matter Physics - Nuclear Methods and Applications, John Wiley & Sons (1996).

[7] C. Domienikan, Uma interface eletrônica para aquisição de 12 espectros de coincidências gama-gama atrasadas, São Paulo, MS Dissertation, CNEN-IPEN/SP (2001).

[8] C. H. Herden, M. A. Alves, K. D. Becker, J. A. Gardner, Hyperfine Interactions (Mar 2006).

[9] T. H. Haines, N. A. Dencher, FEBS Letters, **528**, 35 (2002).

[10] S. Micelli, D. Meleleo, V. Picciarelli, M. G. Stoico, E. Gallucci, Biophysical Journal, **87**, 1065 (2004).

[11] D. E. Otzen, Biophysical Journal, **83**, 2219 (2002).

Anais da XXVIII Reunião de Trabalho sobre Física Nuclear no Brasil
SP, Brasil, 2005
Artigo originalmente publicado no Brazilian Journal of Physics, Vol 36 – nº 4B,
Special Issue: XXVIII Workshop on Nuclear Physics in Brazil, pp.1353–1356 (2006).

Low-Energy Levels Calculation for ^{193}Ir

Guilherme Soares Zahn[1], Cibele Bugno Zamboni[1], Frederico Antonio Genezini[1], Joel Mesa-Hormaza[2] and Manoel Tiago Freitas da Cruz[3]

[1] Instituto de Pesquisas Energéticas e Nucleares, Caixa Postal 11049, 05422-970, São Paulo, SP, Brazil

[2] Centro Regional de Ciências Nucleares - CRCN-CNEN/PE

[3] Instituto de Física da Universidade de São Paulo, Caixa Postal 66318, 05315-970, São Paulo, SP, Brazil

Abstract. In this work, a model based on single particle plus pairing residual interaction was used to study the low-lying excited states of the ^{193}Ir nucleus.

In this model, the deformation parameters in equilibrium were obtained by minimizing the total energy calculated by the Strutinsky prescription; the macroscopic contribution to the potential was taken from the Liquid Droplet Model, with the shell and paring corrections used as as microscopic contributions. The nuclear shape was described using the Cassinian ovoids as base figures; the single particle energy spectra and wave functions for protons and neutrons were calculated in a deformed Woods-Saxon potential, where the parameters for neutrons were obtained from the literature and the parameters for protons were adjusted in order to describe the main sequence of angular momentum and parity of the bandheads, as well as the proton binding energy of ^{193}Ir. The residual pairing interaction was calculated using the BCS prescription with Lipkin-Nogami approximation.

The results obtained for the first three bandheads (the $3/2^+$ ground state, the $1/2^+$ excited state at E $\approx$ 73keV and the the $11/2^-$ isomeric state at E $\approx$ 80keV) showed a very good agreement, but the model so far greatly overestimated the energy of the next bandhead, a $7/2^-$ at E$\approx$ 299keV.

Keywords. Ir-193; beta decay; spectroscopy

---

1 Introduction

This work is a theoretical support for the experimental analysis of the excited states in ^{193}Ir populated by the β^- decay of ^{193}Os [1].

The low-lying states in ^{193}Ir have been thoroughly studied and, although this nucleus is usually described as triaxially-deformed [2, 3, 4, 5], it has also been recently described as a prolate rotor [6]. Most of these theoretical studies describe rather well the binding energies of the $3/2^+$ ground state and the $1/2^+$ excited state at 73keV, but fail to describe the negative-parity bandheads; in this regard, the best effort to this date seems to be the one by Drissi [7], where a variation of the *Interaction Boson-Fermion Model* (IBFM) describes to some degree the $11/2^-$ isomeric state at 80keV and has the next bandhead, a $7/2^-$ state, quite shifted from its original 299keV excitation energy, but still below 1MeV.

In this work, the low-lying states of ^{193}Ir were evaluated using a single-particle model in an axially-deformed potential, with pairing interaction corrections applied. The deformation parameters in equilibrium were obtained by minimizing the total energy calculated by the macroscopic-microscopic method [8]. The single particle energy spectra and wave functions for protons and neutrons were calculated in a deformed Woods-Saxon potential [9], with the potential parameters for neutrons obtained from the literature [10] and the parameters for protons adjusted in order to describe the main sequence of angular momentum and parity of the bandheads, as well as the proton binding energy. The residual pairing interaction was calculated using the BCS prescription with the Lipkin-Nogami approximation [11].

2 The Nuclear Model

2.1 Nuclear Deformation

The nuclear deformation was calculated using the macroscopic-microscopic method. Within this method, the total energy is given by

$$E_{tot}(\epsilon, \hat{\alpha}) = E_{macr}(\epsilon, \hat{\alpha}) + E_{micr}(\epsilon, \hat{\alpha}) \tag{1}$$

where ϵ and $\hat{\alpha}$ are the deformation parameters.

The macroscopic term is given by the liquid drop model. The microscopic portion can be divided into two components: the contribution associated with the shell correction energy and the pairing contribution. In order to obtain the equilibrium deformation the total energy is minimized with respect to the deformation parameters. The Cassini ovaloid shape parametrization was used, taking into consideration the terms associated with the quadrupole (ϵ) and hexadecapole (α_4) moments.

2.2 Single Particle Energies

The single particle states were calculated using the Woods-Saxon (W-S) potential. To determine the W-S potential twelve constants should be given, six for neutrons and six for protons:

V_0 depth of the central potential;

a_{so} diffuseness parameter of the spin-orbit part;

R_0 radius parameter;

r_{0-so} radius parameter of the spin-orbit potential;

a diffuseness nuclear parameter;

λ strength of spin-orbit interaction.

Several parameter sets have been proposed for the Woods-Saxon potential, usually determined by a global fit to various ground state nuclear properties of β-stable nuclei in a mass number range.

The Woods-Saxon potential consists of the central part V_{cent}, the spin-orbit part V_{so} and the Coulomb potential V_{Coul} for protons:

$$V^{WS}(r, z, \epsilon, \hat{\alpha}) = V_{cent}(r, z, \epsilon, \hat{\alpha}) + V_{so}(r, z, \epsilon, \hat{\alpha}) + V_{Coul}(r, z, \epsilon, \hat{\alpha}) \qquad (2)$$

where (r, z) are cylindrical coordinates.

The central part is defined in order to describe the density distribution function

$$V_{cent}(r, z, \epsilon, \hat{\alpha}) = \frac{V_0}{1 + e^{\frac{dist(r, z, \epsilon, \hat{\alpha})}{\alpha}}} \qquad (3)$$

where $dist$ is equal to the distance of a given point to the nuclear surface.

The depth of the central potential is parameterized by:

$$V_0 = V_0[1 \pm 0.063(N - Z)/(N + Z)] \qquad (4)$$

with positive signal for protons and negative for neutrons.

The spin-orbit term is defined by

$$V_{so}(r, z, \epsilon, \hat{\alpha}) = \lambda \left(\frac{h}{2Mc} \right)^2 \nabla V(r, z, \epsilon, \hat{\alpha}) \cdot (\vec{\sigma} \times \vec{p}) \qquad (5)$$

where M is the nucleonic mass, the vector operator $\vec{\sigma}$ stands for the Pauli matrices and $\vec{p}$ is the linear operator.

The Coulomb potential is assumed to be that corresponding to the nuclear charge $(Z - 1)e$, uniformly distributed inside the nucleus. For the Hamiltonian diagonalization, the eigenfunctions of an harmonic oscillator with axial symmetry in the cylindrical coordinates were used as a base.

The pairing energy was evaluated as in the commonly used prescription of the BCS approach. The hamiltonian operator in the BCS model contains two parts: the first corresponding to the single particle states and the second corresponding to the pairing interaction. If the particle term is diagonal, the BCS operator can be written in the formalism of the second quantization as

$$\hat{H} = \hat{H}_{sp} + \hat{H}_{pair} = \sum \epsilon_{\Omega i} \left(\alpha_i^\dagger \alpha_i + \alpha_{\bar{i}}^\dagger \alpha_{\bar{i}} \right) - \sum G_{ji} \alpha_j^\dagger \alpha_{\bar{j}}^\dagger \alpha_{\bar{i}} \alpha_i \qquad (6)$$

Finally, the system of equations of this model is

$$E_\Omega^\nu = \sum_i^{nu} \epsilon_{\Omega i} + \sum_i^{nu} \left(1 - \frac{\epsilon_{\Omega i} - \lambda_\nu}{\sqrt{(\epsilon_{\Omega i} - \lambda_\nu)^2 + \Delta_\nu^2}} \right) - \frac{\Delta_\nu^2}{G} \qquad (7)$$

where:

- ϵ Energy of single particle states;
- ν Number of quasi-particles;
- λ Chemical potential; and
- G Pairing constant.

3 Results

The values of the deformation parameters that minimize the total energy are $\epsilon = 0.1102$ and $\alpha_4 = -0.0404$; the representation of the total energy dependence with these parameters is shown in Fig. 1.

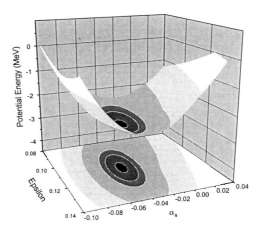

Figure 1 Dependence of the total energy with the deformation parameters.

Table 1 Parameters of the Woods-Saxon potential

	V_0 (MeV)	r_0 (fm)	a (fm)	r_{0-so} (fm)	a_{so} (fm)	λ
Universal Parameters [12]	49.6	1.275	0.70	1.32	0.70	36.0
Present Work	54.612	1.18992	0.70	1.27498	0.932	36.899

The parameters of the W-S potential for neutrons were obtained from ref. [10], and the parameters for protons were adjusted in order to describe the main sequence of angular momentum and parity of the three lowest-lying single particle states, as well as their binding energy; from these states it was then possible to calculate the quasi-particle states using the LINDEN code [11]. The values obtained were compared to the parameters suggested by Cwiok *et al.* [12], called *Universal Parameters* (see Table 1. In Figure 2, the present single particle states calculation is compared to the experimental results from [1] and to the results obtained using the *Universal Parameters*.

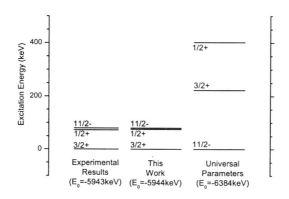

Figure 2 Comparison of the experimental bandheads of ^{193}Ir to the calculations performed using the present set of parameters and to the calculations using the Universal Parameters from Cwiok [12].

4 Conclusions

Our calculations reproduced very well the ground state and the two lowest-lying states in ^{193}Ir, while the *Universal Parameters* couldn't even reproduce the ground state. The $11/2^-$ state at 80keV, which has presented some problems in other theoretical explanations of this nucleus, in particular, was perfectly reproduced, showing that it can indeed be explained in terms of an axially-symmetric potential.

The next step in this work should be to try to reproduce the next two bandheads in ^{193}Ir, a $7/2^-$ state at 299keV and a $3/2^+$ state at 460keV, both of which were not included in the present analysis and presented serious trouble in other calculations found in the literature [2, 3, 4, 5, 6, 7].

Bibliography

[1] Zahn, G. S. et al., Bras. Journ. Phys. **35** (3B), 1 (2005).

[2] Bezakova, E. et al., Nucl. Phys. A **669**, 241 (2000).

[3] McGowan, F. K. et al., Phys. Rev. C **35**(3), 968 (1987).

[4] Stuchbery, A. E., Nucl. Phys. A **700**, 83 (2002).

[5] Stuchbery, A. E., J. Phys. G **25**, 611 (1999).

[6] Roussière, B. et al., Phys. Atom. Nucl. **64** (6), 1048 (2001).

[7] S. E. Drissi, Nucl. Phys. A **621**, 655 (1997).

[8] Garcia, F. et al., Comput. Phys. Commun. **120**, 57 (1999).

[9] Garcia, F. et al., Eur. Phys.J. A **6**, 49 (1999).

[10] Lojewski, Z. et al, Phys. Rev. C **51**, 601 (1995).

[11] Rodriguez, O. et al., Comput. Phys. Commun. **137**, 405 (2001).

[12] S. Cwiok et al. *Comput. Phys. Commun.* **46** (1987) 379.

Anais da XXVIII Reunião de Trabalho sobre Física Nuclear no Brasil
SP, Brasil, 2005

Evaluation of X-Ray Sensitive Image Plates as Detectors to Acquire Neutron Radiography Images

Maria Ines Silvani[1,], Gevaldo L. de Almeida[1], Rosanne C. A. A. Furieri[1], Ricardo Tadeu Lopes[2,†], Marcelo J. Gonçalves[2] and John Douglas Rogers[3,‡]*

[1] Instituto de Engenharia Nuclear - CNEN - C.P. 68550, Ilha do Fundão, 21945-970 Rio de Janeiro, RJ, Brazil

[2] UFRJ - COPPE

[3] RSC - RadSci Consultancy, Derby, U.K.

Abstract. This work presents preliminary results obtained from the utilization of X-ray sensitive Image Plates as detectors for thermal neutron radiography. The image plates have been made sensitive to thermal neutrons by attaching Gadolinium foils to them, acting as a neutron-to-charged particle converter. Images of several objects acquired with this kind of detector have been compared with those obtained by the conventional method using radiographic films. The response of the image plate and radiographic film to radiation dose, as a measure of their efficiencies, have been compared through their specific luminescence and optical density respectively, while their resolutions have been evaluated through the Modulation Transfer Functions - MTF. The Argonauta research reactor installed at the Instituto de Engenharia Nuclear-IEN/CNEN Brazil has been used as a source of thermal neutrons, furnishing a flux of 4.46×10^5 n $\cdot$ cm$^{-2} \cdot$ s^{-1} at its main channel outlet.

1 Introduction

The utilization of an Image Plate, henceforth called IP in this work, as detector in image-acquisition systems employing gamma and X-rays is a very attractive approach thanks to its high sensitivity to these radiations, that results in short exposure times. Besides its capability of being re-utilized, this kind of device,

---

* msouza@ien.gov.br

† ricardo@lin.ufrj.br

‡ jonh.d.rogers@btopenword.com

unlike the conventional radiographic film, does not require chemical processing for its development. This is performed by an interrogation laser beam in the scanning to produce a digital image. An image-acquisition system employing an IP is made up of four components: the radiation source, the IP itself, acting simultaneously as detector and temporary storage medium, the reader which converts the latent image stored in the IP into electrical signals, and a processor incorporating a proper software to translate those signals into a digital image. For systems employing thermal neutrons as the radiation source, it is necessary to add a neutron-to-ionizing radiation converter which can be homogeneously mixed with the phosphor contained in the IP or attach the converter to it externally as a separate piece. This work evaluates the feasibility of employing commercially available X-ray sensitive IPs, as devices to obtain neutron radiography images. To accomplish this task, the response of an IP having a Gadolinium foil attached to it has been analyzed when exposed to a thermal neutron field. The acquired images have then been compared with those obtained by conventional neutron radiography using radiographic films. The properties of the neutron field in which the experiments have been carried out are shown in table 1.

Table 1 Neutron field parameters at the outlet of the main channel, where a moderator block and a divergent collimator have been installed.

Thermal Flux	$4.46 \times 10^5 \, \mathrm{n} \cdot \mathrm{cm}^{-2} \cdot \mathrm{s}^{-1}$
Epithermal Flux	$6.00 \times 10^3 \, \mathrm{n} \cdot \mathrm{cm}^{-2} \cdot \mathrm{s}^{-1}$
Neutron/Gama Ratio	$3.00 \times 10^6 \, \mathrm{n} \cdot \mathrm{cm}^{-2} \cdot \mathrm{mR}^{-1}$
L/D Ratio	63.25
Cadmium Ratio	25
Beam Divergence	$1°16'$

2 Working Principle

The operating principle of an IP resembles that applied to electrostatic photocopy machines, where a luminous image of the object is projected onto a previously electrified plate of a suitable material. This light exposure causes a photoelectric effect, partially discharging regions of the plate in a non-homogeneous fashion and thus creating an *electrostatic image* of the object on that plate. This *invisible* image is then *developed* by using fine coal powder which is attracted by the electrified regions reproducing thus the original object pattern, which is finally *fixed* on a sheet of paper by heating. The electrical pattern is afterwards erased, providing a clean plate for re-use. Instead of visible light as in photocopy machines, an IP uses non-visible radiation such as neutrons or X-rays [1,2,3]. Its utilization involves exposure, development and erasure as follows.

2.1 Exposure

An IP is constituted by a layer of phosphor deposited on a polyester plate. The active phosphor layer is made of BaFBr:Eu^{+2}. As illustrated in Fig. 1, the incident radiation ejects an electron from the valence band of Eu^{+2}, which is oxidized to Eu^{+3}, to a higher energy level just below the conduction band, where it can be captured by existing traps in the crystal lattice. These trapped electrons form a latent and relatively stable electrical image which reproduces the pattern of the radiation beam that hits the IP.

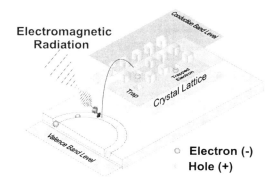

Figure 1 Exposure mechanism: the electrons are thrown from the valence band to existing traps in the crystal lattice by photoelectric effect.

When this radiation field is intercepted by an object, it casts a shadow on the Image Plate revealing its inner structure, as an attenuation map. Once exposed, the IP stores the information - i.e., the pattern of the radiation field that hit it during the exposition phase - for a period of time long enough to allow its manipulation and transfer from the irradiation to the *development* place.

2.2 Development

The IP is developed by scanning it with a laser beam that has enough energy to throw the trapped electrons into the conduction band. Once there, they migrate to the valence band, reducing Eu^{+3} back to Eu^{+2} and emitting visible light, as shown schematically in Fig. 2. The intensity of this light and the coordinates of its point of emission allow a reconstruction of the original image.

2.3 Erasure

After development, the residual electrical image in the IP is then erased by flooding it with intense white light, which sets the remaining trapped electrons free. They recombine with Eu^{+3} which is reduced to Eu^{+2}, allowing the use of the IP again.

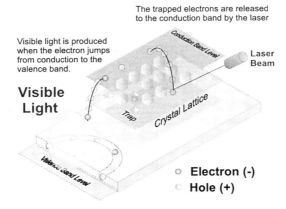

Figure 2 Development mechanism: a laser beam provides energy for the trapped electrons ejecting them into the conduction band. Their migration back to the valence band yields a memory mapped visible light distribution that reproduces the stored electrical pattern.

The whole process including exposure, development and cleaning of an IP is shown schematically in Fig. 3.

A neutron-sensitive IP must contain a material capable of converting neutrons into a kind of ionizing radiation. The conversion process can be accomplished using two approaches: attaching a conversion foil to an IP designed for X-ray operation or making the IP itself neutron-sensitive by introducing a converter material during its manufacture. Suitable converters emit photons or charged particles such as alpha, beta or protons after a nuclear reaction with the incoming neutron. Gadolinium, ^{6}Li-enriched Lithium, ^{10}B-enriched Boron are commonly employed as thermal neutron converters. In this work, an X-ray sensitive IP attached to an external converter foil has been used.

3 Methodology

Images of objects of different shapes and compositions have been acquired with the IP as well as with conventional radiographic films as benchmarks. The film used in these measurements is the double-sided *Industrex MX125* manufactured by Kodak.

The sensitivity of the IP to gamma-rays, always present in a reactor-produced neutron field, has been evaluated by taking images without the converter. Two kind of converters, both containing Gadolinium, have been tested. The first of this was a 50 µm-thick Gadolinium foil attached to the IP. The second one was a neutron-sensitive scintillation screen commonly used to obtain real-time neutron radiography images. Among the Gadolinium isotopes, ^{155}Gd and ^{157}Gd differ from the other in their high absorption cross sections - 61,000 and 254,000 barns

EVALUATION OF X-RAY SENSITIVE IMAGE PLATES AS DETECTORS... 231

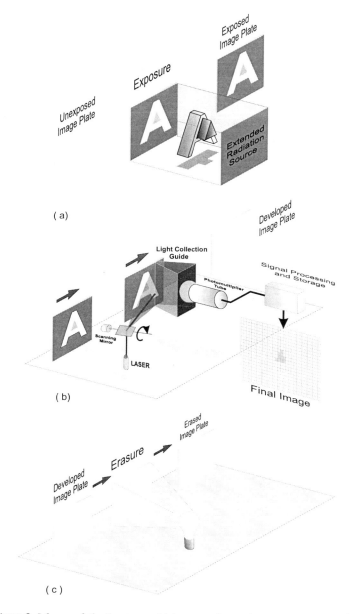

Figure 3 Scheme of the image acquisition procedure using an Image Plate. (a) Exposure (b) Development (c) Erasure

respectively - for thermal neutrons, being thus specially suited for the detection of these particles. The interaction of thermal neutrons with Gadolinium produces gamma-rays from the de-excitation of the composite nucleus as well as internal conversion electrons.

When used as an external converter, the Gd foil should be thick enough to absorb a substantial fraction of the incoming neutrons in order to produce sufficient internal conversion electrons, but thin enough to allow transmission of a reasonable fraction of them, and hence, a compromise has to be made. Fig. 4 shows the detection efficiency for internal conversion electrons - released by transmission and back-scattering - as a function the foil thickness [4]. Based on this behavior, the Gadolinium foil has been placed behind the IP with reference to the incoming neutron beam, operating thus in the back-scattering mode.

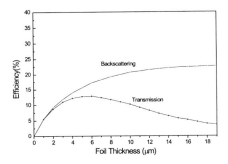

Figure 4 Detection efficiency for internal conversion electrons escaping from a gadolinium foil through transmission and back-scattering as a function of the foil thickness.

The optimal exposure time required to obtain images using the IP has been determined by comparison with the images acquired with conventional radiographic films. The performance of the IP has been evaluated through the *Modulation Transfer Function - MTF* which is a proper tool to express the behavior of an image-acquisition system on a quantitative basis. All IPs used in this work were manufactured by *Fuji-SR* and were designed for operation with X-rays. Read-out has been performed on the *Cyclone Storage Phosphor System (PERKIN-ELMER)* provided with a 83 micra spot size laser beam. The final images have been processed with the *Optiquant* software that has been specifically designed to optimize the image display and storage.

4 Quantitative Results

A full characterization of an image-acquisition system requires, among other parameters, knowledge of its sensitivity, linearity, dynamic range and resolution.

A comparison between the IP and radiographic film for the first three parameters is depicted in Fig. 5. Both ordinates express an arbitrary optical density measured from the digitalized images. These densities have been obtained by taking the average values at several regions of the images where the neutron beam arrived undisturbed. Since the responses of the radiographic film and the IP, to the exposure time, are inverted - the film becomes darker while the IP becomes brighter - it was necessary to invert the image from the IP, in order to perform a comparison with the image furnished by the radiographic film.

Such a procedure naturally introduces a somewhat arbitrary character into both ordinates precluding therefore an absolute comparison between them. However, in spite of this constraint, valuable information can still be inferred. Indeed it can be observed that the IP response to the radiation dose - proportional to the irradiation time - is reasonably linear, exhibiting additionally a high dynamic range as compared with the radiographic film. It can also be deduced that the IP requires a much shorter exposure time, thanks to its higher sensitivity.

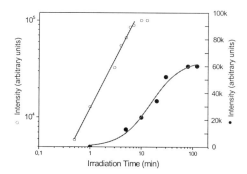

Figure 5 Response of the IP (left) and radiographic film (right) to exposure time. The IP exhibits better dynamic range, linearity and sensitivity.

Two types of resolution play important roles in an image-acquisition system: density and spatial resolution. The first one expresses the capability of the system to differentiate between materials with close attenuation properties for the employed radiation, while the second defines its ability to resolve closely-spaced features characterizing a high spatial frequency. As the spatial frequency increases, it becomes more and more difficult to resolve the individual features. This is quantitatively expressed by the *Modulation Transfer Function - MTF* as a decrease in the modulation with the frequency [5]. It is generally agreed [6] that an image has an acceptable quality when the modulation surpasses 10 %.

Fig. 6 exhibits the results arising from the measurements that were made to determine the MTF for systems employing IP and radiographic film as detectors. It can be figured out that, for IP, the 10 % *cutoff value* only would be reached at a very high spatial frequency, a fact which is not corroborated by the qualitative results which will be presented later in this work.

This phenomenon was caused by the high sensitivity of the IP to the gamma radiation present in the neutron field produced by the reactor. Indeed, this radiation spoils the light-shadow pattern cast on the IP by the neutron-tailored collimators used to obtain the MTF, as these are not opaque to gamma-rays.

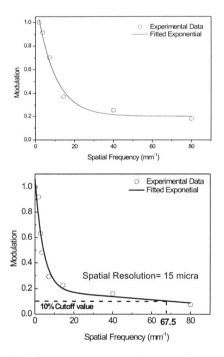

Figure 6 Data obtained from measurements to determine the *Modulation Transfer Function - MTF* for image-acquisition systems using image plates (top) and scanned radiographic films (bottom) as detector.

These difficulties do not arise in the case of radiographic films, as they are less sensitive to the local gamma-ray field. However, as the film has been digitalized to quantify the optical density distribution, the measured MTF actually reflects the combined performance of the film itself and the analog-to-digital transformation procedure. This procedure has a much poorer spatial resolution than the film itself, as inferred from Fig.6 where the 10 % *cutoff value* for the MTF occurs at a frequency of 67.5 mm^{-1} equivalent to a resolution of about 15 μm.

5 Qualitative Results

In addition to the images acquired with a converter, an image was taken without it to assess the response of the IP to the gamma-rays field. The results given in Fig. 7 are images of an automobile fuel injector taken with an IP (left) and with a radiographic film (right), both without converter. It can be seen that IPs are far more sensitive to gamma-rays than are conventional radiographic films. Hence in neutron fields from nuclear reactors where gamma radiation is always present, these detectors would produce hybrid-character images.

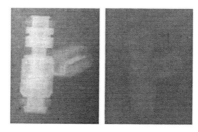

Figure 7 Images of an automobile fuel injector taken with an IP (left) and radiographic film (right) both without converter.

The acquisition of images using a converter has been performed using two different configuration. In the first one - where a 50 μm-thick Gadolinium foil was placed behind the IP - the neutrons transmitted through the object under inspection pass through the IP and interact with the Gd. The back-scattering reaction products then hit the IP. In the second configuration, the converter in the form of Gd_2O_3 was incorporated to a scintillation screen designed for operation with thermal neutrons, and the IP was placed behind it, catching therefore its feeble light. Images of several objects have been taken using the above described configurations and were compared with those taken with a radiographic film as follows.

Fig. 8 shows the conventional photograph of a bullet (a), which was sectioned after the acquisition of the radiographic images taken with the IP + Gd foil (b), the IP + scintillation screen (c) and radiographic film + Gd foil (d). The exposure times for acquiring (b), (c) and (d) were 5, 7 and 30 min respectively. The horizontal strips observed in (b) and (c) were caused by a malfunction of the development system, only later on recognized.

It can be observed in (d) that the lead projectile has a feeble attenuation property which is even lower than that for the powder grains and the fuse clearly visible towards the top of the images. The set IP + Gd (b) foil is somewhat more sensitive to gamma-rays as inferred from the brighter tonality of the lead projectile and, corroborated by the darker tonality of the powder chamber. As observed in (c),

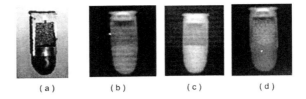

Figure 8 Images of a sectioned bullet taken with a conventional photographic-camera (a). Images of the original bullet with: (b) IP + Gd foil, (c) IP + neutron designed scintillation screen, (d) radiographic film + Gd foil.

the scintillation screen has an even higher sensitivity to gamma-rays, in spite of its neutron-designed character.

These conclusions are ratified by the images of an automobile fuel injector shown in Fig. 9, after the same detector combination sequence adopted in Fig. 8.

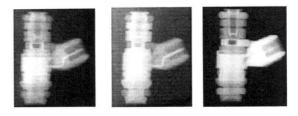

Figure 9 Images of an automobile fuel injector taken with: (left) IP + Gd foil, (centre) IP + scintillation screen, (right) radiographic film + Gd foil.

The unique capability thermal neutron have of passing through some heavy metals such as lead, steel, titanium, among others, make them a valuable tool for non-destructive assay of these materials. An example of this application, is the inspection of an aero-engine turbine stator-blade, designed to align the flow of gases exiting the combustion chamber at very high speed improving its efficiency. This blade is provided with channels through which cooling air is pumped to keep the blade metal temperature at an acceptable level. In the manufacturing process, the lost-wax technique is used in castings that contain complex internal channels. A ceramic material forms the shape of the channels in the mould and this must be completely removed by chemical means to ensure that there is no possibility of blockages to air flow. One of the techniques used to check for blockages in each turbine stator-blade is neutron radiography. Each turbine stator-blade is soaked in a gadolinium nitrate solution so that any ceramic material retained in a cooling channel absorbs this contrast agent. After thorough washing, the blade is radiographed. Any blades with the signs of blockages are returned for chemical treatment again in an attempt to recover that component.

Images of a turbine stator-blade obtained using the three image acquisition techniques previously described are shown in Fig. 10. Although making a cut in the stator-blade could reveal this type of flaw, it would destroy a high-cost component which could still be in good condition.

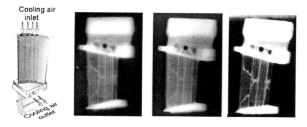

Figure 10 Radiographic images of an aero-engine turbine stator-blade acquired with: (left) IP + Gd foil, (centre) IP + scintillation screen, (right) radiographic film + Gd foil.

The images acquired with IP yield a poorer resolution than that obtained using a radiographic film. However, it is still possible to observe the retained ceramic material in the turbine stator-blade cooling channels.

The Fig. 11 shows the neutron radiography images of an oil pressure probe for automotive vehicles. A conventional photograph of the component is also shown for illustration purposes. Here again, it is possible to see the reduced contrast of the IP images compared with that obtained using the radiographic film.

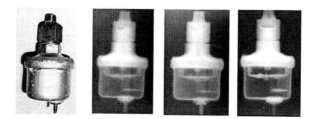

Figure 11 Conventional photograph of an oil pressure probe and neutron radiographies using: (left) IP + Gd foil, (centre) IP + scintillating screen, (right) radiographic film + Gd foil.

6 Conclusions

It is feasible to use an X-ray sensitive Image Plate - IP as a neutron detector, provided that a suitable converter is somehow incorporated into it. In this work a Gadolinium foil attached to the IP has been used as neutron-to-charged particle converter.

The IP is sensitive to the gamma-rays associated with the neutron field produced by a nuclear reactor. This feature should not be regarded as a handicap, as it allows - in one single exposure - the acquisition of hybrid gamma-neutron transmission images. These images of hybrid character can provide a richer information regarding the inner structure of the inspected object, otherwise achievable only by carrying out two different assays. However, images obtained with an IP exhibit poorer quality than those acquired with radiographic films. Yet, they do have the advantage that they can be re-utilized indefinitely (or at least a great number of times). Furthermore, they require shorter exposure and development times.

The development of an IP is simpler and cleaner than films, as they do not require any chemical processing. Moreover, its development yields digitalized images directly, whilst films would require scanning after the chemical development.

Bibliography

[1] J. Miyahara et al, A new type of X-ray area detector utilizing laser stimulated luminescence, Nuclear Instruments and Methods in Physics Research A246 (1986) pp.572-578.

[2] F. Cipriani, J. Castagna, M. S. Lehmann, C. Wilkinson,A large image-plate detector for neutrons, Physica B, 213&214(1995), pp.975-977.

[3] M. Schalapp, H. Conrad, H. von Seggern, Pixelated neutron image plates, Journal of Physics D: Applied Physics, 37 (2004)2607-2612.

[4] G. Melchart, G.Charpak, F. Sauli, The Multi-Step Avalanche Chamber as a Detector for Thermal Neutrons, Nuclear Instruments and Methods, (1981) 186, pp 613- 620.

[5] G. L. Almeida, M. I. Silvani, R. T. Lopes, A Hybrid Technique to evaluated the line spread function, Research and Development Brazilian Journal, vol. 4- no 3(2002) 798-803.

[6] ASTM E1441-95 and 1570-95a, Non-Destructive Testing, Radiation Methods, Computed Tomography. Guide for Imaging and Practice for Examination, ISO/TC 135/SC, N118 USA (1996)

Anais da XXVIII Reunião de Trabalho sobre Física Nuclear no Brasil
SP, Brasil, 2005

Development of Neutrongraphic Image-Acquisition Systems at the Instituto de Engenharia Nuclear

Maria Ines Silvani[1,], Gevaldo L. de Almeida[1],*
Rosanne C. A. A. Furieri[1], Ricardo Tadeu Lopes[2,†]
and Marcelo J. Gonçalves[2]

[1] Instituto de Engenharia Nuclear - CNEN
   C.P. 68550, Ilha do Fundão,
   21945-970 Rio de Janeiro, RJ, Brazil

[2] UFRJ - COPPE

Abstract. The Argonauta Reactor, at the *Instituto de Engenharia Nuclear*/CNEN - Brazil, is being used as a source of thermal neutrons to acquire transmission-generated radiographic images. Three image acquiring systems are currently being used for this purpose, each of them with their intrinsic advantages and constraints. The first of them is a conventional one, which utilizes a radiographic film as *detector* and *recording device*, while the remaining systems, both still under development, employ respectively scintillating screens - which convert the neutrons into photons of visible light - and image plates as detecting devices. This work presents neutrongraphic images of some objects acquired with these three systems aiming at a comparison between their performances, limitations and potential capabilities. All these systems use the Argonauta research reactor as source of thermal neutrons with a flux of de $4.46 \times 10^5 \mathrm{n} \cdot \mathrm{cm}^{-2} \cdot \mathrm{s}^{-1}$ at its main channel outlet where they have been installed. Taking into account the non-destructive character of the assays methods performed by these systems, the objects submitted to the inspections have been chosen in such a way that their features could only be properly visualized with thermal neutrons. Besides that, specifically designed test-objects have been as well utilized for the quantitative evaluation of the systems.

---

1 Introduction

Radiographic images acquired by transmission of thermal neutrons through some types of objects is sometimes the only tool capable to perform a proper

---

* msouza@ien.gov.br
† ricardo@lin.ufrj.br

non-destructive assay of them. Indeed, since this radiation ($E \sim 0.025$ eV) is strongly attenuated by hydrogen bearing materials, it is very powerful to detect them, with the production of high contrast images. Generally, the attenuation coefficient depends upon both nature and energy of the radiation. It is easily figured out that the higher the difference between them, the better the contrast will be. The contrast is an important image property which appears as a difference between gray tonalities of neighbor regions. Unlike the attenuation coefficients for gamma or X-rays, which grow steadily with the atomic number, those for thermal neutrons exhibit a somewhat erratic behavior [1]. Such a feature does not constitute a disadvantage at all, for it sometimes allows the differentiation of materials with close atomic numbers, otherwise difficult or even impossible using other kind of radiations.

2 Neutron Radiography Facility

A neutrongraphic image-acquisition system is comprised basically of a suitable neutron beam, a material to convert the incident neutrons into a kind of detectable radiation such as alpha, beta, gamma or visible light, and a device to detect, record and visualize the image.

2.1 Neutron Field

The neutron beam should be intense enough to yield a reasonable counting statistics after its transmission through the object under inspection. It should be well thermalized, i.e., to possess a high cadmium ratio: $R_{Cd} = (\phi_{th} + \phi_{ep})/\phi_{ep}$ where R_{Cd}, ϕ_{th} and ϕ_{ep} are the cadmium ratio, thermal and epithermic flux respectively. Its gamma-ray contents should be as low as possible, a requirement which could demand a proper filtration with bismuth or lead in order to reduce the gamma radiation ordinarily occurring in reactor-produced neutron fields.

For an extended source, as that virtual one assigned to the main outlet channel of a reactor, the divergence of its emitted neutron beam should be as low as possible to minimize the image degradation caused by penumbra effects. This divergence can be improved by using proper collimators but they concomitantly reduce the counting statistics, degrading thus the image. Therefore, a compromise has to be made based on an reasonable exposure time. No attempt has been made in this work to reduce the beam divergence since the graphite collimator ($L/D = 63.25$) installed within the main channel already limited it to an acceptable width (FWHM) of its related *Rocking Curve* of $1° 16'$ [2]. Further beam properties are given as follows [3] : $\phi_{th} = 4.46 \times 10^5 \text{n} \cdot \text{cm}^{-2} \cdot \text{s}^{-1}$, $\phi_{ep} = 6.00 \times 10^3 \text{n} \cdot \text{cm}^{-2} \cdot \text{s}^{-1}$, $\text{n}/\gamma = 3.00 \times 10^6 \text{n} \cdot \text{cm}^{-2} \cdot \text{mR}^{-1}$.

2.2 Neutron converters and Visualization

Neutrons interact with some elements in a nuclear reaction yielding photons or charged particles. Materials containing elements of this type - neutron converters - are usually added to the detectors or recording devices (the conventional radiographic film for instance perform both tasks) in order to make them sensitive to thermal neutrons. These converters can be mixed to the detector material already during its manufacture turning it into a neutron-designed detector or attached to it later on as an individual piece.

The mode to incorporate the converter into the system depends on the kind of device employed to record the images. For neutron, a Gadolinium foil acting as converter is usually attached to a radiographic film, eventually replaced by special screens named *Image Plates*, which store the information as an *electrical image* that can be later on *developed* by a proper reader. Although these devices yield a poorer image quality than radiographic films, they are reusable, demand a much shorter exposure time and do not require a cumbersome chemical development process. Moving images are taken with systems employing scintillating screens coupled to video-cameras. These screens emit visible light photons resulting from the energy loss of the electrons produced by the nuclear reaction between the neutron and the converter present in the scintillating screen. A brief scheme of the mechanisms involved in these three kind of systems currently utilized at the *Instituto de Engenharia Nuclear* using its Argonauta research reactor as thermal neutron source is depicted in Fig. 1.

3 Materials and Methods

An *Industrex MX125* radiographic film from Kodak, with a 50 μm thick gadolinium foil attached to it acting as converter, has been used as detector and recording device in the first system. It furnishes high quality images but requires about 30 min exposure time and 10 additional minutes for the chemical processing.

The second system utilizes SR-type image plates manufactured by Fuji. Since they are designed for X-rays, it was necessary to incorporate a Gadolinium foil converter as well. Typically, 7 minutes are required to get an image including 5 minutes exposure time and 2 minutes for readout. This task, which does not require any chemical processing is performed by a *Cyclone Storage Phosphor System* manufactured by Perkin-Elmer, which scans the *electrical image* stored in the image plate with a 83 μm spot-size laser. The final image is then recovered with the *Optiquant* software, written for an optimized image display, analysis, quantification, and data storage.

As the employed image plates are also sensitive to gamma-rays, they produce hybrid character pictures. Indeed, for a mixed neutron-gamma field the contrast would be improved for materials exhibiting a feeble difference between their thermal neutron absorption cross-sections, but a reasonable difference between their gamma attenuation coefficients. On the other hand, the contrast would be degraded for materials possessing inverse properties. Such a behavior is not necessarily a handicap because for objects with a complex inner structure involving different materials, a single radiography could reveal all the features at once, otherwise detectable only by carrying out two independent exposures.

The third neutrongraphic system being developed is comprised of a scintillation screen containing materials capable of converting neutrons into charged particles and their energy into visible light, and an optical system containing a first surface mirror coupled to a CCD video-camera. As the neutron beam disturbed by the features of the object being analyzed hits that screen, a light-shadow pattern is drawn, furnishing a neutron attenuation image of the object cross-section perpendicular to the beam.

This image is then reflected by the mirror of the optical system placed outside the neutron beam and perpendicular to it, avoiding thus radiation damages to the CCD camera. All components are enclosed within an aluminum box eliminating thus the ambient light.

The box is lined with a cadmium sheet acting as neutron shielding eliminating thus the scattered neutrons.

The CCD video-camera is coupled to a computer by means of a proper interface and its analogic image is converted to a digital format using a software developed for this purpose. The main advantage of this system is its capability to acquire real-time images allowing hence, the study of dynamic structures.

Alternatively, the video-camera can be replaced by a conventional photographic camera, as done in this work. This replacement precludes the image acquisition of moving objects, but produces better images.

The 35mm photographic camera employed in this work using an ASA 400 film, was equipped with a 50 mm f1.8 objective. A Panasonic video-camera model AW-E650 has been used in this work. This camera incorporates an optical system with a $1/2''$ prism, F1.4, a pick up device constituted by a $1/2''$ interline-type CCD with 811x494 pixels. Its standard illumination is 2,000 lx at F14, accepting a minimum value of 5×10^{-5} lx at F1.4.

The 15x10 cm scintillation screen is composed of Silver-activated ^{6}LiF(ZnS), - with ^{6}Li acting as a neutron-to-charged particle converter - being commercially designated as NE426.

4 Quantitative Results

Although a visual inspection is the ultimate criterion to evaluate the quality of images, it is not precise nor tight enough to assure a reliable and consistent

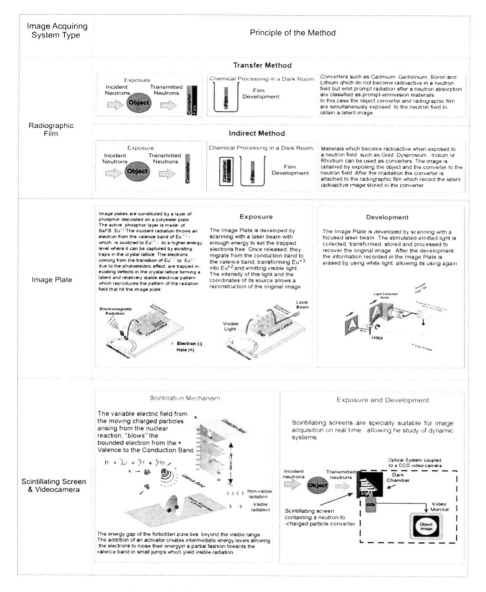

Figure 1 Methods to acquire neutrongraphic images.

comparison and classification between them. Therefore, it is necessary somehow to *quantify* their quality by using adequate criteria and tools.

The response of an image-acquisition system to a line-shaped image is a more or less blurred strip with an optical density profile resembling a Gaussian function, which is called *Line Spread Function*-LSF. As shown in Fig. 2, these curves overlap each other causing a convolution that reduces the amplitude-to-average value ratio known as *modulation*.

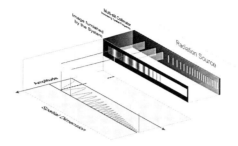

Figure 2 The impact of the Spatial Frequency on the outcome of an Image-Acquisition System.

As the spatial frequency increases this convolution becomes more and more intense reducing the modulation. The curve modulation x spatial frequency is called *Modulation Transfer Function*-MTF, and expresses how the capability of a system to resolve individual features is affected by that frequency. An ideal system would have a flat MTF but in a real one this function would drop with the frequency: slowly for good systems but swiftly for poorer ones.

Due to the high resolution of the radiographic film, its related Line Spread Function-LSF, is very narrow. Therefore, only at very high spatial frequencies those curves begin to overlap each other. Until there, the modulation remains close to 1. The used collimator has a maximum frequency at 80 mm^{-1}, which is not high

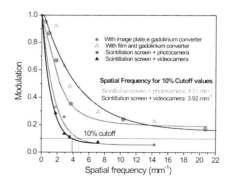

Figure 3 MTF for the thermal neutron systems. An estimation of the spatial resolution is given by the 10 % cutoff value in the modulation.

enough to cause a substantial convolution among the LSFs. Hence the Modulation Transfer Function remains flat along the measured range.

The MTFs for the other systems shown in Fig.3 have been measured using special slit collimators of different apertures made of aluminum with their slits filled up with metallic Gadolinium. These materials have been chosen due to high difference between their thermal neutron cross-sections, which provides an aimed high contrast image, necessary to reduce the uncertainty in the modulation measurements. It is generally agreed that an image has an acceptable quality when the modulation reaches at least 10 % [4].

5 Qualitative Results

The performance of an Image-Acquisition System engulfs aspects beyond the image quality itself, and hence, some desirable characteristics could be mutually exclusive. Therefore, a compromise should be made based on the priority level assigned for the specific application and on cost/benefit considerations. This approach is briefly illustrated in Fig.4 where a comparison - as a figure of merit - between the systems treated in this work is done.

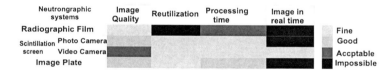

Figure 4 Performance of the three neutrongraphic systems treated in this work, after some important parameters and criteria.

Some neutrongraphies obtained in this work using three different systems are presented in the following pictures demonstrating the potential capability of the thermal neutrons as a non-destructive assay tool.

It would be feasible for instance to carry out researches on the soil physics by analyzing the water contents in the plant roots, for they act as an interface between the plant itself and the soil. Similarly, biological aspects such as moist transport and the growth of the roots could be studied through the neutrongraphic technique thanks to the high attenuation coefficient of hydrogen for thermal neutrons. Fig. 5 shows neutrongraphies of the plant *"Scindapsus aureus"*, (popularly known in Brazil as *Jiboia*- the Portuguese name for the huge snake Boa Constrictor) obtained with a radiographic film after 30 minutes exposure time.

In spite of their feeble thickness, it is possible as shown also in Fig.6, to visualize the leaf blade, its petiole and nervrures thanks to the strong neutron attenuation caused by their water contents,

Figure 5 Plant *"Scindapsus aureus"*, popularly known in Brazil as *Jiboia* a), *zoom* of the delimitated region shown in b).

Figure 6 Neutrongraphies of leaves 1) *"Tibouchina granulosa"*, popularly known in Brazil as *quaresmeira*, 2) leaf of a non-identified plant.

The inner structure of oily fruits or seeds can be as well visualized using thermal neutrons due to the high hydrogen contents of the oil, as shown in Fig. 7, taken with a radiographic film and a Gadolinium converter. It is possible to visualize details of the interior of a peanut and a walnut and even to differentiate between a normal walnut and a dried up one.

Figure 7 Neutrongraphies acquired with a radiographic film: (a) Peanut (b) Normal Walnut (c) Dried up Walnut.

A comparison between neutrongraphic images of an oil pressure probe for automotive vehicles acquired with radiographic films, scintillating screens + photo or video-cameras as well as with image plates is presented in Fig. 8. One can infer from the pictures that the systems possess different spatial resolution but for all cases it is possible to observe the construction details of the device through its iron case which is practically transparent to thermal neutrons. The exposure time for the acquisitions (a), (b), (c) and (d) were 30 min, 30 seg, 2 seg e 5 min respectively. The horizontal strips observed in (d) were caused by a malfunction of the readout system, only later on identified.

Non-destructive assays are essential whenever devices, pieces of equipment or their components should not undergo invasive or destructive inspection due to

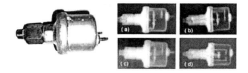

Figure 8 Conventional photograph of an oil pressure probe and neutrongraphies using: (a) Radiographic film + Gd foil, (b) photo camera + scintillating screen, (c) video camera + scintillating screen, (d) IP + Gd foil.

safety or high-cost reasons. A typical example is the case of the *stator blade* of an aeronautic turbine. This blade, designed to align the flow of exiting gases improving thus the thrust, is provided with air-cooling channels to prevent temperature rise beyond the acceptable limit. Its manufacturing process involves the application of the lost-wax technique where a ceramic material forms the shape of the channels in the mould. This material should be afterwards removed by chemical means to preclude any possibility of air flow blockage. A cut could reveal such a kind of flaw, but it would destroy a high-cost piece which could still be recovered. Therefore, a suitable technique to detect an eventual blockage is neutron radiography. After this technique, the stator-blade is soaked in a Gadolinium nitrate solution which is absorbed by the ceramic material. The stator-blade is then radiographed after a thorough washing which eliminates all the contrast agent, except that remaining in the ceramic material. The high difference between the thermal neutron cross-sections of *Monel* (a nickel-copper alloy) and Gadolinium allows the acquisition of high contrast images. A scheme of the stator-blade and its positioning in the engine is shown in Fig.9.

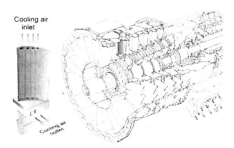

Figure 9 Stator blade and its position in the aircraft engine. The component is provided with channels - cooled by pressed air - to prevent temperature rise. The engine drawing was picked up from [5].

A component of this type has been inspected by neutron radiography addressing the eventual detection of this kind of fail. The results obtained with radiographic film (a), image plate (b) and a scintillating screen (c) coupled to a video-camera are shown in Fig 10. The exposure time for these acquisitions were 30, 5 minutes and 2 seconds, respectively.

Figure 10 Neutrongraphies of a stator-blade from an aircraft turbine. Some cooling channels are still partially obstructed by the ceramic material used in the manufacturing process. Radiographic film + Gd foil (left). IP + Gd foil (center) video camera + scintillating screen (right).

A remarkable difference between X-rays and neutrons is seen in Fig.11 which depicts radiographic images of a conventional bullet taken with X-rays and thermal neutrons, as well as a conventional photograph of the same bullet cut *after* the acquisition of the radiographic images. The X rays, have been produced by an ampoule operating at 80 kV. Both images have been acquired with a radiographic film.

While X-ray is strongly attenuated by the lead projectile, and very feebly by the powder and the percussion fuse, the neutron behaves in the inverse way, as seen in the neutrongraphy, where the dark tonality means a high neutron flux. These opposite features, make X-rays and neutrons complementary tools for non-destructive assays.

Figure 11 Conventional photograph of the sectioned bullet (left) its radiography (centre) and neutrongraphy (right). The powder grains and the percussion fuse are clearly visible.

6 Conclusions

The performance of two different image acquisition systems employing scintillating screens and image plates - still being developed or optimized - have been compared with the conventional one using radiographic films. The results have shown the superior quality of the images furnished by the last one. Yet, the remaining systems, in spite of their poorer image quality - but still at a good or acceptable level -

possesses so many advantages that the employ of the radiographic film system only would be worthwhile when the resolution plays a crucial role in the process. These advantages include the reutilization of the detectors, shorter exposure and development times, no need for a cumbersome chemical processing in a darkroom and the possibility to get digital images directly. All these features result in lower costs and efforts. Besides that the system equipped with a scintillating screen optically coupled to a video-camera is the only one capable to deal with moving images.

Neutron-sensitive image plates have the converter mixed to the phosphorescent material during the manufacturing process. Since these devices were not available in the laboratory, a Gadolinium foil acting as an *external converter* was attached to an Image Plate designed for X-ray operation. It has been found that this arrangement works quite well, but it is sensitive to the gamma radiation present in the reactor-produced neutron field.

The hybrid sensitivity of this arrangement constitutes in many situations an advantage rather than a flaw, as it allows to get a neutron-gamma radiography at once, otherwise achievable only by carrying out two independent exposures.

Bibliography

[1] J.C. Domanus, 1992, Practical Neutron Radiography, Kluwer Academic Press, London

[2] G. L. Almeida, M. I. Silvani, R. C. A. A. Furieri, M. J. Gonçalves and R. T. Lopes, Brazilian J. of Physics, 35,771(2005)

[3] R.C. A. A. Furieri and M. I. Silvani, Mapeamento do Fluxo de Nêutrons do Reator Argonauta na nova Configuração do Núcleo, XII ENFIR Brazil (2000)

[4] ASTM E1441-95 and 1570-95a, Non-Destructive Testing, Radiation Methods, Computed Tomography. Guide for Imaging and Practice for Examination, ISO/TC 135/SC, N118 USA (1996)

[5] Barnes& Nobel Books, The Lore of Flight, New York (1996)

Anais da XXVIII Reunião de Trabalho sobre Física Nuclear no Brasil
SP, Brasil, 2005

Systematics of Half-lives for Alpha-Emitter Nuclides

M. M. N. Rodrigues*, S. B. Duarte, O. A. P. Tavares and E. L. Medeiros

Centro Brasileiro de Pesquisas Físicas - CBPF/MCT
Rua Dr. Xavier Sigaud 150, 22290-180, Rio de Janeiro, Brazil

Abstract. A systematic study of partial alpha decay half-lives using a one -parameter, effective model for the potential barrier is reported. The tunneling probability is calculated for the whole set of nuclides having experimental half-life values. The model parameter is adjusted in order to minimize the standard deviation of the ratio (in $\log_{10}$-scale) of calculated results to experimental values.

After the discovery of the cluster radioactive decay of nuclei [1-2] a significant effort has been done in order to present a general systematization of the alpha decay and cluster emission half-lives within a same theoretical framework [3-5]. The present work is a contribution to this direction by means of a model systematization of alpha decay half-life, representing a first step to extend the procedure to cluster emission process. We call attention also to the recent detection of proton emission process from nuclei far away the beta stability line, being a topic of great interest in the recent literature [5-7]. The model systematization proposed here can be immediately applied to discuss this topic.

According to the quantum mechanical theory for alpha emission from nuclei [8], the decay constant λ can be calculated as:

$$\lambda = \lambda_O P, \qquad P = e^{-G}, \qquad G = \frac{2}{\hbar} \int \sqrt{2\mu(s)\,[V(s) - Q_\alpha]}\, ds, \qquad (1)$$

where

$$\lambda_o = \frac{\upsilon}{2a} = \frac{\sqrt{2}}{2a} \left(\frac{Q_\alpha}{\mu_o} \right)^{\frac{1}{2}} \quad \text{and} \quad Q_\alpha = m_p - (m_d + m_\alpha) \ . \qquad (2)$$

---

* Fellow, CNPq Brazilian Agency, under contract N° 117373/2004-4, e-mail: nicke@cbpf.br

The G-factor in Eq. (1) is calculated over the entire barrier region, $a \leqslant s \leqslant b$ (see Fig. 1) and it is the sum of two parts. The first one is calculated in the overlapping region $a-c$ in Fig 1, by using a given potential barrier model for this molecular form of the dinuclear system. The second one is calculated in the region $c-b$ in Fig. 1 by using the conventional centrifugal plus coulomb barrier. In the overlapping region both the effective mass, $\mu(s)$ and the potential energy, $V(s)$, are not known exactly. Duarte and Gonçalves [9] have studied the behavior of $\mu(s)$ for different cases of nuclear hadronic decay. For alpha particle emission their results for the effective mass can be describe quite well by a parametrization given by

$$\mu(s) = \mu_o \left(\frac{s-a}{c-a} \right)^p, \quad \text{for} \quad a \leqslant s \leqslant c \tag{3}$$

and

$$\mu(s) = \mu_o = \frac{m_d m_\alpha}{m_d + m_\alpha} \quad \text{for} \quad s > c \tag{4}$$

with $p > 0$ and m_d, m_α being the masses of the daughter nucleus and alpha particle, respectively.

The expression for the potential barrier in the overlapping region (shaded area in Fig. 1) is well parametrized by

$$V(s) = Q_\alpha + (V_c - Q_\alpha) \left(\frac{s-a}{c-a} \right)^q, \quad \text{for} \quad a \leqslant s \leqslant c \tag{5}$$

and

$$V_c = \frac{2 Z_d e^2}{c} + \frac{l(l+1)\hbar^2}{2\mu_o c^2} \quad \text{for} \quad s > c, \tag{6}$$

with $q > 0$, as shown in references $[3, 5, 9 - 11]$. The values of Q_α have been calculated from the most recent experimental atomic mass-excess evaluation [12].

The half-lives calculated with the present model were compared with experimental data for the whole set of alpha emitters (327 cases) of alpha transitions of angular moment $l = 0$. The single model parameter g is a combination of the exponents p and q in Eq. (3, 5) defined by $g = \frac{2}{2+p+q}$. This parameter can assume values in the interval $0 < g \leqslant \frac{2}{3}$, and it was adjusted in order to minimize the standard deviation,

$$\sigma = \left\{ \left(\frac{1}{N} \right) \sum \left[\log_{10} \left(\frac{T_{(i)}^{calc}}{T_{(i)}^{exp}} \right) \right]^2 \right\}^{\frac{1}{2}}. \tag{7}$$

The obtained adjusted value for parameter g was $g = 0.122$, as shown in Fig. 2-a, and the deviation of the calculated values with respect to experimental data is shown (Fig. 2(b)). This result shows that the present systematics is able to give half-life values for α-emitter nuclides within one order of magnitude.

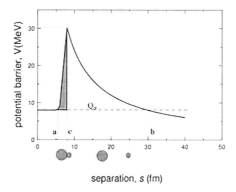

Figure 1 Potential barrier showing potential energy values for the molecular nuclear configuration of the separating system (shaded area). The turning points a and b are also shown, and the touching configuration is for separation between centers of fragments at $s = c$.

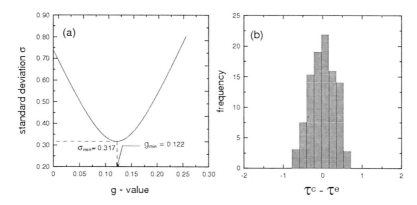

Figure 2 The standard deviation σ given by Eq. (6) as a function of g-value is shown in part (a). Note that a minimum value for σ is reached at $g = 0.122$. Part (b) shows the frequency distribution of the quantity $\tau^c - \tau^e = \log T^c_{1/2} - \log T^e_{1/2}$.

Bibliography

[1] H. J. Rose and G. A. Jones, Nature (London) **307**, 245(1984).

[2] H. G. de Carvalho, J. B. Martins and O. A. P. Tavares, Phys. Rev. C **34**, 2261 (1986).

[3] D. N. Poenaru, M. Ivascu, A. Sāndulescu and W. Greiner, Phys. Rev. C **32**, 572 (1985).

[4] K. P. Rykaczewski, E. Phys. J. A **15**, 81 (2002).

[5] S. B. Duarte, O. A. P. Tavares, F. Guzmán, A. Dimarco, F. García, O. Rodríguez and M. Gonçalves, Atom. Data Nucl. Data Tables **80**, 235 (2002).

[6] M. Karny et al., Phys. Rev. Lett. **90**, 012502 (2003).

[7] F. Guzmán, M. Gonçalves, O. A. P. Tavares, S. B. Duarte, F. García and O. Rodríguez, Phys. Rev. C **59**, R2339 (1999).

[8] P. Marmier and E. Sheldon: Physics of Nuclei and Particles vol I (New York, Academic Press, 1969) p. 303.

[9] S. B. Duarte and M. G. Gonçalves, Phys. Rev. C **53**, 2309 (1996).

[10] O. A. P. Tavares, E. L. Medeiros and M. L. Terranova, J. Phys. G: Nucl. Part. Phys. **31**, 129 (2005).

[11] O. A. P. Tavares, M. L. Terranova and E. L. Medeiros, Nucl. Inst. Meth. Phys. Res. B, **243**, 256 (2006).

[12] G. Audi, A. H. Wapstra, and C. Thibault, Nucl. Phys. A **729**, 337 (2003).

Anais da XXVIII Reunião de Trabalho sobre Física Nuclear no Brasil
SP, Brasil, 2005

Effects of the Mixed Thermal Neutron-Gamma Radiation on Plasmid DNA From IEA-R1 Reactor of IPEN

Maritza R. Gual[1,*], Airton Deppman[2], Felix Mas[1], Paulo R. P. Coelho[3], A. N. Gouveia[4], Oscar R. Hoyos[1] and A. C. Schenberg[4]

[1] Instituto Superior de Tecnologías y Ciencias Aplicadas, InSTEC, Havana, Cuba;

[2] Instituto de Física, Universidade de São Paulo, IF-USP, Brazil

[3] Instituto de Pesquisas Energéticas e Nucleares, IPEN-CNEN/SP, Brazil

[4] Instituto de Ciencia Biomédicas,ICB, São Paulo, Brazil

Abstract. The experimental results of the mixed thermal neutron-gamma radiation-DNA interaction are presented in this paper. The experiments were conducted at a research Boron Neutron Capture Therapy (BNCT) facility using the IEA-R1 reactor of IPEN-CNEN, Sao Paulo. This paper also includes the method used for neutron and gamma dose calculations. The fractions of supercoiled (undamaged), circular (resulted from single-strand breaks (SSBs)) and linear (resulted from double-strand breaks (DSBs)) molecules of plasmid DNA produced by the mixed thermal neutron-gamma radiation from the IEA-R1 reactor are presented.

1 Introduction

Nowadays, there is a growing interest in neutron interaction of DNA molecule. This is due to the fact that neutrons are used in the radiation treatment of cancer patients.

The interaction of ionizing radiation with living matter involves direct and indirect effects. In the direct effect, the ionizing radiation deposits energy directly in the DNA (ex. heavy particles, protons and neutrons). The indirect effect consists on the interaction of DNA molecules with the products of water radiolysis: hydroxyl radicals (OH), hydrogen atoms and aqueous-electrons (ex. X or gamma rays and

---

* mrgual@yahoo.es

electrons). The interaction is studied through the observation of single-strand and double-strand breaks of the DNA molecule. Radiation damage to DNA is evaluated by electrophoresis through agarose gels, which shows the conformation of the DNA after irradiation. The method used for neutron and gamma doses measurements is shown.

2 Materials and Methods

2.1 DNA sample specification

Plasmid DNA has been used for studying how radiation can induce breakage of DNA molecules [1-5]. Plasmids are a convenient system, because they are well-defined size and breakage detection is readily accomplished by gel electrophoresis.

DNA plasmid samples containing $+2.961$ kb of pBsKS were isolated from Escherichia coli using the lise alkali method; wich involves the extraction using organic solvents and separation by centrifuge in Cesium Chlorid. 15 μl-samples were placed in small eppendorf for the irradiation. A support was built to place each eppendorf. Also, a mechanism was introduced to aid the placement and removal of the eppendorf while the reactor was in operation.

The DNA sample at concentration of 50 ng/μl was prepared. One sample was used as a control and the others as irradiated. The DNA solutions were irradiated, in 0.5 ml eppendorf tubes (polypropylene material). The plasmid was purified according to [6]. In this irradiation, the plasmid DNA used was of approximately 10 % supercoiled form. This plasmid was originally 85 % supercoiled and degraded as time went by.

2.2 Neutron irradiation

Neutron irradiation was performed in the BH-3 at the research Boron Neutron Capture Therapy (BNCT) facility at the IEA-R1 reactor of IPEN-CNEN. See figure 1. The optimal filters sets were obtained by means of Monte Carlo simulations in a previous work[7].

The reactor was operated with 3.5 MW. The neutron field is compound for: 48 % of the thermal neutrons, 13 % of epithermal neutrons and the 19 % of the fast neutrons. For this cause prevail the (n,γ) radiative capture in this irradiation field. The contribution of gamma photons (originating from the reactor) was about 20 % of the fluence.

Gold activations foils were used to the thermal and epithermal neutron fluxes measurements, while applying the Cadmium rate technique in each mixed-field radiation. The LiF Thermoluminescent dosimeter was used to estimate the gamma exposure of the DNA, which constitutes itself an unavoidable contamination of the

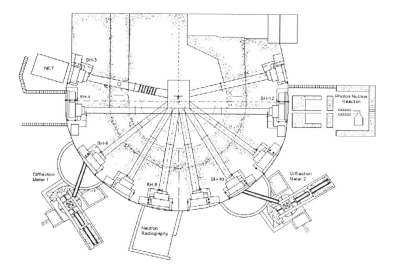

Figure 1 Structural layout of the IEA-R1 reactor at IPEN-CNEN.

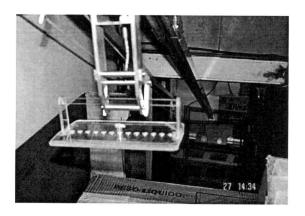

Figure 2 Mechanism to place and remove the sample in the irradiation beam-hole

thermal neutron field. The count rate was measured with a High Pure Germanium (HPGe) Detector. The neutron doses rate was of 2.8 Gy/h.

To place and remove the sample in the irradiation beam hole with the reactor in operation, a mechanism was construed. This mechanism is represented in fig. 2.

The system is formed by an Aluminum rail 10 m long, over which runs a 4x4 cm^2 wagon. In the experimental installation, the rail has a declined of 25 degrees and the wagon is manually moved through a flexible cable. Another similar cable is used for moving a mechanical hand which is used to grasp the sample holders. The construction was carried out in the facilities of IF–USP (Brazil).

2.3 Electrophoresis and DNA quantification

Supercoiled DNA (form I) consists of the undamaged plasmid, circular (form II) resulting from single- strand breaks (SSBs), and linear (form III) resulting from double-strand breaks (DSBs).

The agarose gel was prepared with 0.7% of agarose, and 5 µL de ethidium bromide. 10 µl samples were placed in each well. The run time lasted 2 h under 75 volts potencial.

The gel images were obtained in the EAGLE EYE II transilluminator, using a CCD camera connected to a computer. This way, the images were stored in BMP format with a 640x480 pixel resolution. Agarose gel electrophoresis (figure 3) show a decrease in the amount of the supercoiled form, an increase in that of the circular form, and the appearance and increase in the amount of the linear form upon irradiation. The damage quantification of the DNA bands was obtained with the program Gel Analysis[8]. We used pictures with three integrations, wich presented a low background. Also, the mass ladder was used in the mass calibrations of the samples. Furthermore, a factor of 1.7 was applied to the adherence correction of the Ethidium Bromide in the supercoiled DNA.

2.4 Neutron and gamma doses measurements

The characteristics of the mixed radiation field were determined with gold activation foils and TLD-100 (7.5% 6Li) at a reactor power of 3.5MW.

The photon and thermal neutron doses must be determined separately:

$$D = D_n + D_\gamma$$

Neutron measurements were performed using an Au-197 activation foils. After neutron irradiation, the Au-197 foil is activated. The neutron flux is obtained after activation rate calculations. Gold foils were irradiated both bare and covered by 0.6 cm thick of Cadmium (Cd) foils. In order to measure the epithermal neutron flux the Au foils is covered with Cd. With the Cd foil in place, all neutron contributions with energies less than 0.4 eV are neglected. Neutron flux to kerma conversion factors were used to determine the thermal and/or epithermal (with Cd) neutron dose.

The gamma doses were carried out with TLD-100 dosimeters. It is more attractive for gamma measurements. The Li natural (TLD-100) thermoluminescent detectors are most used in mixed neutron and gamma fields, due to the high thermal neutron capture cross section (941 barns).

3 Results and Discussions

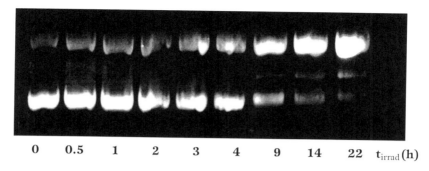

Figure 3 Ethidium fluorescence for samples with 50 ng/μl irradiated at the BH-3

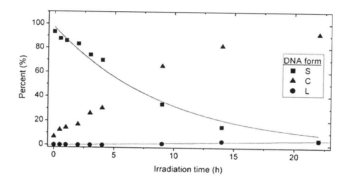

Figure 4 Percent of the forms of plasmid: supercoiled(S), circular(C) and linear (L) for thermal neutron-gamma mixed radiation as a function of irradiation time.

We fit an exponential curve to the data of the supercoiled conformations. The fraction of supercoiled form(S) is the decrease decay function of the irradiation time and can be well fitted by the following equation: $S = A*\exp(B*t)$, where t is the irradiation time:

Parameter	Value	Error
A	97.4	3.1
B	−0.111	0.011

The linear conformations were fit to a linear equation, $L = A^*t$,

Parameter	Value	Error
A	0.180	0.016

The three sets of data show the expected behavior. During irradiation, the amount of supercoiled form decreases, while the circular- and linear forms increase, due to the the production of SSB and DSB.

4 Conclusions

1. A better understanding of the neutron radiation effects in the DNA molecule is reached.

2. When the DNA molecules are irradiated in a mixed thermal neutrons and photons field the damage is produced due to both the neutrons and the gamma rays.

3. It is illustrated the constructed mechanism for positioning and removing of the sample in the irradiation beam hole with the reactor in operation.

Acknowledgements

The authors are thankful to the Latin-American Center of Physics(CLAF), Fundação de Amparo á Pesquisa do Estado de São Paulo(FAPESP), Conselho Nacional de Desenvolvimento Cientifico e Tecnológico (CNPq) and Petróleo Brasileiro SA(PETROBRAS) for support this work. We are grateful to Italo Salzano, Julio Benedito and Edno Aparecido from IPEN for their kindest help.

Bibliography

[1] A.A.Stankus, et al., Energy deposition events produced by fission neutrons in aqueous solutions of plasmid DNA, Int. J.Radiat.Biol., 1995, vol. 68, No.1, 1-9.

[2] M.Stothein-Maurizot, M.Charlier and R.Sabattier, DNA radiolysis by fast neutrons, Int. J.Radiat.Biol., 1990, vol. 57, No.2, 301-313.

[3] Cowan,R, et al., Breakage of double-stranded DNA due to single-stranded nicking, J. of Theorical Biology, 127,229-245(1987).

[4] Dudley T. Goodhead, The initial physical damage produced by ionizing radiations, Int. J.Radiat.Biol., 1989, vol. 56, No.5, 623-634.

[5] Dalong Pang, et. al., Spacial Distribution of Radiation Induced Double-Strand Breaks in Plasmid DNA as Resolved by Atomic Force Microscopy. Radiation Research 164, 755-765(2005).

[6] Andreia N. Gouveia, et. al., Evaluation of plasmid DNA purification for studies of radiation damage, XXVI RTFNB, 2002.

[7] M. R. Gual,O. Rodriguez, F. Guzman, A. Deppman, J.D.T. Arruda Neto, V.P. Likhachev, Paulo R. P. Coelho and Paulo T. D. Siqueira, Study of neutron-DNA interaction at the IPEN BNCT research facility, Brazilian Journal of Physics, vol.34, no.3A, september 2004, p 901-903.

[8] Felix Mas Milian, Airton Deppman, Andreia N.Gouveia, Jorge O. Echeimberg, Maritza R. Gual, João D.T. Arruda-Neto, Oscar Rodríguez, Ana Clara Schenberg and Fernando Guzmán, Validation of methodology for study the DNA-radiation interaction, Accepted for publication in the Revista Nucleus.

Anais da XXVIII Reunião de Trabalho sobre Física Nuclear no Brasil
SP, Brasil, 2005
Artigo originalmente publicado no Brazilian Journal of Physics, Vol 36 – nº 4B,
Special Issue: XXVIII Workshop on Nuclear Physics in Brazil, pp.1357–1362 (2006).

Simple Analytical Expression for Vector Hypernuclear Asymmetry in Nonmesonic Decay of $^5_\Lambda$He and $^{12}_\Lambda$C

César Barbero[1], Francisco Krmpotić[2], Alfredo P. Galeão[3]

[1] Departamento de Física, Facultad de Ciencias Exactas,
Universidad Nacional de La Plata, C.C. 67, 1900 La Plata, Argentina.

[2] Instituto de Física, Universidade de São Paulo,
C. P. 66318, 05315-970 São Paulo, SP, Brazil,
Instituto de Física La Plata, CONICET, 1900 La Plata, Argentina.

[3] Instituto de Física Teórica, Universidade Estadual Paulista,
Rua Pamplona 145, 01405-900 São Paulo, SP, Brazil.

Abstract. We present general explicit expressions for a shell-model calculation of the vector hypernuclear parameter in nonmesonic weak decay. We use a widely accepted effective coupling Hamiltonian involving the exchange of the complete pseudoscalar and vector meson octets (π, η, K, ρ, ω, K*). In contrast to the approximated formula widely used in the literature, we correctly treat the contribution of transitions originated from single-proton states beyond the s-shell. Exact and simple analytical expressions are obtained for the particular cases of $^5_\Lambda$He and $^{12}_\Lambda$C, within the one-pion-exchange model. Numerical computations of the asymmetry parameter, a_Λ, are presented. Our results show a qualitative agreement with other theoretical estimates but also a contradiction with recent experimental determinations. Our simple analytical formulas provide a guide in searching the origin of such discrepancies, and they will be useful for helping to solve the hypernuclear weak decay puzzle.

Keywords. hypernuclear decay; asymmetry parameter; one-meson-exchange model

The free decay of a Λ hyperon occurs almost exclusively through the mesonic mode, $\Lambda \rightarrow \pi N$, emerging the nucleon with a momentum of about 100 MeV/c. Inside the nuclear medium ($p_F \sim 270$ MeV/c) this mode is Pauli blocked and, for all but the lightest Λ hypernuclei ($A \geqslant 5$), the weak decay is dominated by the nonmesonic channel, $\Lambda N \rightarrow NN$, with enough kinetic energy to put the two emitted nucleons above the Fermi surface. The nonmesonic hypernuclear weak decay (NMHWD) offers us a very unique opportunity to investigate strangeness-changing weak interaction between hadrons. Assuming that the NMHWD is saturated by the

263

$\Lambda N \rightarrow NN$ mode, the transitions receive contributions either from neutrons ($\Lambda n \rightarrow nn$) and protons ($\Lambda p \rightarrow np$), with rates Γ_n and Γ_p, respectively ($\Gamma_{NM} = \Gamma_n + \Gamma_p$, $\Gamma_{n/p} = \Gamma_n/\Gamma_p$). Over the last three decades amount of theoretical and experimental effort has been invested in solving an interesting puzzle: the theoretical models cannot reproduce simultaneously the experimental values of $\Gamma_{n/p}$ and a_Λ.

From the experimental side there is actually an intense activity, as can be seen in the light of the experiments under way and/or planned at (KEK) [1], (FINUDA) [2] and Brookhaven National Laboratory (BNL) [3]. The preliminary results give a $\Gamma_{n/p}$ ratio value very close to 0.5 [4, 5, 6] and the measurements of a_Λ favor a negative value for $^{12}_\Lambda C$ and a positive value for $^5_\Lambda He$ [5, 6, 7, 9, 8]. On the other hand, from the pioneering work of Block and Dalitz [10] there have been many theoretical attempts dedicated to solving the puzzle. The earlier studies were based on the simplest model of the virtual pion exchange [11]. This model naturally explains the long range part of the two body interaction, and it reproduces reasonably well the total decay rate but fails badly in reproducing the other observables. In order to achieve a better description, some improvements have been introduced: (i) models that include the exchange of different combinations of other heavier mesons, like η, K, ρ, ω and K* [12]–[19], to reproduce the short range part of the interaction ; (ii) analysis of the two nucleon (2N) stimulated process $\Lambda NN \rightarrow NNN$ [20, 21]; (iii) inclusion of interaction terms that violate the isospin $\Delta T = 1/2$ rule [22]–[24]; (iv) description of the short range baryon-baryon interaction in terms of quark degrees of freedom [25, 26]; (v) correlated (in the form of σ and ρ mesons) and uncorrelated two-pion exchanges [27]–[31]. We emphasize that none of these models give a fully satisfactory description of all the NMHWD observables simultaneously, even though a consistent (though not sufficient) increase of $\Gamma_{n/p}$ has been found. All such models reproduce quite well Γ_{nm}, but seem to strongly underestimate the experimental n/p branching ratio, Γ_n/Γ_p. Besides, the measurements favor a negative value of a_Λ for $^{12}_\Lambda C$ and a positive one for $^5_\Lambda He$, meanwhile all existing calculations based on strict one-meson-exchange (OME) models [26, 14, 15, 16, 18, 32, 33, 19] find values between -0.73 and -0.19 for $^5_\Lambda He$ [34, 35] and, when results are available in the same model, very similar values for $^{12}_\Lambda C$.

Most calculations of the asymmetry parameter make use of an approximate formula (Eq. 18, below) which, however, is valid only for s-shell hypernuclei. Since an essential aspect in the asymmetry puzzle presented above concerns the comparison of its values for $^5_\Lambda He$ and $^{12}_\Lambda C$, it would be of great interest to have an expression that is applicable to both cases. Besides, simple formulas containing kinematical factors and nuclear matrix elements separately, will provide a manageable and useful tool. They will be a guide in searching the origin of discrepancies between theoretical calculations and experimental data. Thus, we present in this paper a general formalism for the vector asymmetry parameter in nonmesonic decay and derive exact and simple analytical expressions, within the simplest one-pion exchange model (OPEM), for the particular cases of $^5_\Lambda$He and $^{12}_\Lambda$C hypernuclei.

Single-Λ hypernuclei produced in a (π^+, K^+) reaction end up with considerable vector polarization along the direction normal to the reaction plane, $\hat{n} = (\mathbf{p}_{\pi^+} \times \mathbf{p}_{K^+})/|\mathbf{p}_{\pi^+} \times \mathbf{p}_{K^+}|$. The angular distribution of protons emitted in $\Lambda p \to np$ decay is given by [34]

$$\frac{d\Gamma(J_I M_I \to \hat{p}_2 t_p)}{d\Omega_{p_2}} = \frac{\Gamma_p}{4\pi} \left(1 + P_V A_V \, \hat{p}_2 \cdot \hat{n}\right), \tag{1}$$

where Γ_p is the full proton-induced decay rate, and

$$A_V = 3 \sqrt{\frac{J_I}{J_I + 1}} \frac{\sigma_1(J_I)}{\sigma_0(J_I)}, \tag{2}$$

is the *vector hypernuclear asymmetry*, with

$$\sigma_0(J_I) = \sum_{M_I} \sigma(J_I M_I),$$

$$\sigma_1(J_I) = \frac{1}{\sqrt{J_I(J_I + 1)}} \sum_{M_I} M_I \sigma(J_I M_I), \tag{3}$$

being

$$\sigma(J_I M_I) = \int d\Omega_{p_1} \int dF \sum_{s_1 s_2 M_F} |\langle \mathbf{p}_1 s_1 t_n \, \mathbf{p}_2 s_2 t_p \, J_F M_F | V | J_I M_I \rangle_{\mathcal{A}}|^2. \tag{4}$$

Here V is the OME potential involving the complete pseudoscalar and vector meson octets $(\pi, \eta, K, \rho, \omega, K^*)$ which has been extensively discussed in Refs. [17, 18], $\mathbf{p}_1 s_1$ $(\mathbf{p}_2 s_2)$ are the momenta and spin projections of the emitted neutron (proton) and

$$\int dF \ldots = 2\pi \sum_{J_F} \int \frac{p_2^2 \, dp_2}{(2\pi)^3} \int \frac{p_1^2 \, dp_1}{(2\pi)^3} \, \delta\left(\frac{p_1^2}{2M} + \frac{p_2^2}{2M} + \frac{|\mathbf{p}_1 + \mathbf{p}_2|^2}{2M_F} - \Delta_{J_F}\right) \ldots, \tag{5}$$

with M and M_F being the nucleon and residual nucleus masses and Δ_{J_F} the liberated energy.

We assume the hyperon to stay in the $j_\Lambda = 1s_{1/2}$ single-particle state, and the initial hypernuclear state is built as $|J_I\rangle \equiv |(j_\Lambda J_C) J_I\rangle$, where $|J_C\rangle$ is the ^{A-1}Z ground state. Thus, the intrinsic asymmetry parameter is defined as

$$a_\Lambda = \begin{cases} A_V & \text{for} \quad J_I = J_C + 1/2, \\ -\frac{J_I + 1}{J_I} A_V & \text{for} \quad J_I = J_C - 1/2. \end{cases} \tag{6}$$

Rewriting the transition amplitude in (4) in the total spin (S, M_S) and isospin (T, M_T) basis we get

$$\sigma(J_I M_I) = \int d\Omega_{p_1} \int dF \sum_{S M_S M_F} \left| \sum_T (-)^T \langle \mathbf{p} P S M_S T J_F M_F | V | J_I M_I \rangle_\mathcal{A} \right|^2, \quad (7)$$

where $\mathbf{p} = \frac{1}{2}(\mathbf{p}_2 - \mathbf{p}_1)$, $\mathbf{P} = \mathbf{p}_1 + \mathbf{p}_2$ and

$$\langle \mathbf{p} P S M_S T J_F M_F | V | J_I M_I \rangle_\mathcal{A}$$
$$= \frac{1}{\sqrt{2}} \left[\langle \mathbf{p} P S M_S T J_F M_F | V | J_I M_I \rangle - (-)^{S+T} \langle -\mathbf{p} P S M_S T J_F M_F | V | J_I M_I \rangle \right]. \quad (8)$$

For the integration in (7) we take the z-axis as indicated in Fig. 1 and choose p_2, θ_{p_1} and ϕ_{p_1} as independent variables. Thus, expanding the final state in terms of relative and center-of-mass partial waves of the emitted nucleons [17], after integration on ϕ_{p_1} and recoupling of angular momentum, we get

$$\sigma(J_I M_I) = \frac{1}{2}(4\pi)^4 \int d\cos\theta_{p_1} \int dF \sum_{STT'} (-)^{T+T'}$$
$$\times \sum_{lL\lambda J} \sum_{l'L'\lambda'J'} i^{-l'-L'-l-L} (-)^{\lambda+S+J+J'+J_I+J_F} \hat{l}\hat{l}'\hat{L}\hat{L}'\hat{\lambda}\hat{\lambda}'\hat{J}\hat{J}'$$
$$\times \sum_{kK\kappa} \hat{\kappa}\hat{J}_I (J_I M_I \kappa 0 | J_I M_I)(l0l'0|k0)(L0L'0|K0) [Y_k(\theta_p, \pi) \otimes Y_K(\theta_P, 0)]_{\kappa 0}$$
$$\times \begin{Bmatrix} J_I & \kappa & J_I \\ J & J_F & J' \end{Bmatrix} \begin{Bmatrix} \kappa & J' & J \\ S & \lambda & \lambda' \end{Bmatrix} \begin{Bmatrix} l & l' & k \\ L & L' & K \\ \lambda & \lambda' & \kappa \end{Bmatrix}$$
$$\times (plPL\lambda SJTJ_F; J_I |V|J_I)(pl'PL'\lambda'SJ'T'J_F; J_I |V|J_I)^*. \quad (9)$$

where $(plPL\lambda SJTJ_F; J_I |V|J_I) = \frac{1}{\sqrt{2}} \left[1 - (-)^{l+S+T} \right] \langle plPL\lambda SJTJ_F; J_I |V|J_I \rangle$.
From [17] we have

$$\langle plPL\lambda SJTJ_F; J_I |V|J_I \rangle$$
$$= (-)^{J_F+J-J_I} \hat{J}_I^{-1} \sum_{j_p} \langle J_I \| \left(a_{j_\lambda}^\dagger a_{j_p}^\dagger \right)_J \| J_F \rangle \mathcal{M}(plPL\lambda SJT; j_\lambda j_p), \quad (10)$$

where

$$\mathcal{M}(plPL\lambda SJT; j_\lambda j_p) = \frac{1}{\sqrt{2}} \left[1 - (-)^{l+S+T} \right] (plPL\lambda SJT|V|j_\lambda j_p J). \quad (11)$$

The two-particle spectroscopic amplitudes are cast as [17]

$$\langle J_I \| \left(a_{j_\lambda}^\dagger a_{j_p}^\dagger \right)_J \| J_F \rangle = (-)^{J+J_I+J_F} \hat{J}\hat{J}_I \begin{Bmatrix} J_C & J_I & j_\lambda \\ J & j_p & J_F \end{Bmatrix} \langle J_C \| a_{j_p}^\dagger \| J_F \rangle. \quad (12)$$

To continue, we will adopt the extreme shell model and restrict our attention to cases where the single-proton states are completely filled in $|J_C\rangle$. This is so, for the cores of both $^5_\Lambda$He ($J_I = 1/2, J_C = 0$), and $^{12}_\Lambda$C ($J_I = 1, J_C = 3/2$) *. In such cases, the final nuclear states take the form $|J_F\rangle \equiv |(j_p^{-1}J_C)J_F\rangle$. This leads to

$$\langle J_C \| a^\dagger_{j_p} \| J_F\rangle = (-)^{J_F+J_C+j_p} \hat{J}_F. \tag{13}$$

Putting all this together and performing the summation on J_F, we obtain

$$\sigma_\kappa(J_I) = \frac{4}{\pi} \hat{J}_I^3 \hat{\kappa}^{-1} \sum_{j_p} \int d\cos\theta_{p_1} \int dF_{j_p} \sum_{STT'} (-)^{T+T'}$$

$$\times \sum_{lL\lambda J} \sum_{l'L'\lambda'J'} i^{-l'-L'-l-L} (-)^{\lambda+S+J_I+J_C-j_p+\kappa} \hat{l}\hat{l}'\hat{L}\hat{L}'\hat{\lambda}\hat{\lambda}'\hat{J}^2\hat{J}'^2$$

$$\times \sum_{kK} (l0l'0|k0)(L0L'0|K0) [Y_k(\theta_p,\pi) \otimes Y_K(\theta_p,0)]_{\kappa 0}$$

$$\times \begin{Bmatrix} j_\Lambda & J_I & J_C \\ J_I & j_\Lambda & \kappa \end{Bmatrix} \begin{Bmatrix} \kappa & j_\Lambda & j_\Lambda \\ j_p & J & J' \end{Bmatrix} \begin{Bmatrix} \kappa & J' & J \\ S & \lambda & \lambda' \end{Bmatrix} \begin{Bmatrix} l & l' & k \\ L & L' & K \\ \lambda & \lambda' & \kappa \end{Bmatrix}$$

$$\times \mathcal{M}(plPL\lambda SJT; j_\Lambda j_p) \mathcal{M}^*(pl'PL'\lambda'SJ'T'; j_\Lambda j_p), \tag{14}$$

where

$$\int dF_{j_p} \ldots = \int p_2^2\, dp_2 \int p_1^2\, dp_1\, \delta\left(\frac{p_1^2}{2M} + \frac{p_2^2}{2M} + \frac{|\mathbf{p}_1+\mathbf{p}_2|^2}{2M_F} - \Delta_{j_p}\right) \ldots \tag{15}$$

The intrinsic asymmetry parameter can finally be written as [31]

$$a_\Lambda = \frac{\omega_1}{\omega_0}, \tag{16}$$

where

$$\omega_\kappa = (-)^\kappa \frac{8}{\sqrt{2\pi}} \hat{\kappa}^{-1} \sum_{j_p} \int d\cos\theta_{p_1} \int dF_{j_p}\, Y_{\kappa 0}(\theta_p,0)$$

$$\times \sum_{TT'} (-)^{T+T'} \sum_{LS} \sum_{l\lambda J} \sum_{l'\lambda'J'} i^{l-l'} (-)^{\lambda+\lambda'+S+L+j_p+\frac{1}{2}}$$

$$\times \hat{l}\hat{l}'\hat{\lambda}\hat{\lambda}'\hat{J}^2\hat{J}'^2 (l0l'0|\kappa 0)$$

$$\times \begin{Bmatrix} \kappa & 1/2 & 1/2 \\ j_p & J & J' \end{Bmatrix} \begin{Bmatrix} \kappa & J' & J \\ S & \lambda & \lambda' \end{Bmatrix} \begin{Bmatrix} l' & l & \kappa \\ \lambda & \lambda' & L \end{Bmatrix}$$

$$\times \mathcal{M}(plPL\lambda SJT; j_\Lambda j_p) \mathcal{M}^*(pl'PL\lambda'SJ'T'; j_\Lambda j_p), \tag{17}$$

---

* The decay of the $^{11}_\Lambda$B hypernuclei will be analyzed in a forthcoming paper.

with $L = 0$ for the $1s_{1/2}$ state, and $L = 0$ and 1 for the $1p_{3/2}$ state. It has been verified that $\omega_0 = \Gamma_p$, in such way that the new information is carried by ω_1. The presence of the Clebsch-Gordan coefficient in Eq. (17), for $\kappa = 1$, ensures that l and l' have opposite parities. Since the initial state in the two matrix elements has a definite parity, this implies that all contributions to ω_1 come from interference terms between the parity-conserving and the parity-violating parts of the transition potential. Furthermore, the antisymmetrization factor in Eq. (11) shows that the two final states have $T \neq T'$. These are general properties of the asymmetry parameter.

We have used these general expressions for the numerical computation of a_Λ within several OME models. The values of ω_0 and ω_1 are shown in Tables 1 and 2, for $^5_\Lambda$He and $^{12}_\Lambda$C respectively, in units of the free Λ decay constant, $\Gamma^{(0)}_\Lambda = 2.50 \times 10^{-6}$ eV. In the case of $^5_\Lambda$He, we have also included, between parentheses, the values obtained with the approximate formula usually adopted in the literature [34], which is strictly valid only for s-shell hypernuclei:

$$A_V(^5_\Lambda\text{He}) \approx \frac{2\,\Re\left[\sqrt{3}\,ae^* - b\,(c^* - \sqrt{2}\,d^*) + \sqrt{3}\,f\,(\sqrt{2}\,c^* + d^*)\right]}{|a|^2 + |b|^2 + 3\,(|c|^2 + |d|^2 + |e|^2 + |f|^2)}, \tag{18}$$

where

$$
\begin{aligned}
a &= \langle np,^1S_0|V|\Lambda p,^1S_0\rangle, & b &= i\langle np,^3P_0|V|\Lambda p,^1S_0\rangle, \\
c &= \langle np,^3S_1|V|\Lambda p,^3S_1\rangle, & d &= -\langle np,^3D_1|V|\Lambda p,^3S_1\rangle, \\
e &= i\langle np,^1P_1|V|\Lambda p,^3S_1\rangle, & f &= -i\langle np,^3P_1|V|\Lambda p,^3S_1\rangle.
\end{aligned} \tag{19}
$$

Table 1 Results for the asymmetry parameter, a_Λ, for nonmesonic decay of $^5_\Lambda$He.

Model/Calculations	$\omega_0(1s_{1/2})$	$\omega_1(1s_{1/2})$	a_Λ
π	0.6492	-0.2913	$-0.4487\ (-0.4456)$
(π, η, K)	0.3920	-0.2412	$-0.6153\ (-0.6384)$
$\pi + \rho$	0.5937	-0.1776	$-0.2991\ (-0.3155)$
$(\pi, \eta, K) + (\rho, \omega, K^*)$	0.5526	-0.2974	$-0.5382\ (-0.5388)$
Experiment	KEK-PS E278 [9]		0.24 ± 0.22
$a_\Lambda = A_V(^5_\Lambda\text{He})$	KEK-PS E462 [7] (preliminary)		$0.11 \pm 0.08 \pm 0.04$

We find values ranging from -0.62 to -0.29, in qualitative agreement with other theoretical estimates [26, 14, 15, 16, 18, 32, 33, 19, 34, 35] but in contradiction with some recent experimental determinations [5, 6, 7, 9, 8]. We remark here that the approximate formula ignores the fact that the final state of nonmesonic decay is a three-body one and does not include the full contribution of the

Table 2 Results for the asymmetry parameter, a_Λ, for nonmesonic decay of $^{12}_\Lambda$C.

Model/Calculations	$\omega_0(1s_{1/2})$	$\omega_0(1p_{3/2})$	$\omega_1(1s_{1/2})$	$\omega_1(1p_{3/2})$	a_Λ
π	0.5206	0.5954	−0.2400	−0.2596	−0.4477
(π, η, K)	0.3336	0.3817	−0.2057	−0.2217	−0.5975
$\pi + \rho$	0.4922	0.5514	−0.1461	−0.1596	−0.2930
$(\pi, \eta, K) + (\rho, \omega, K^*)$	0.4619	0.5083	−0.2546	−0.2599	−0.5303
Experiment	KEK-PS E160 [8]				-0.9 ± 0.3
$a_\Lambda = -2A_V(^{12}_\Lambda C)$	KEK-PS E508 [7] (preliminary)				-0.44 ± 0.32

transitions coming from proton states beyond the s-shell, being therefore of only limited validity, and should not be used for p-shell hypernuclei such as $^{12}_\Lambda$C, or, even worse, for heavier ones. This being said, comparison of the corresponding values for a_Λ in Table 1 shows that the formula works well within its range of validity. Also, from Table 2 we observe that the p-shell contributions to ω_0 and ω_1 for $^{12}_\Lambda$C are by no means negligible. However, they are also in approximately the same ratio, so that the effect on a_Λ is much smaller.

In order to obtain simple and manageable analytical expressions for the asymmetry parameter, we adopt now the simplest OME model, which takes into account only the contribution of one pion exchange. As can be seen form our numerical results, the main characteristics of the nonmesonic hypernuclear weak decay asymmetry are well described within this OPEM. Then, as a starting point, this model will be enough to obtain such analytical formula*. After a straightforward analytical evaluation of the summation terms indicated in (17), together with the calculation of the nuclear matrix elements indicated in Eqs. (11)-(13) within the strict OPEM, we can write

$$\omega_0 = \frac{24}{\pi} \int d\cos\theta_{p_1} \int dF_{j_p} \left[3\left(S_\pi^0\right)^2 + 18\left(T_\pi^{20}\right)^2 + \left(P_\pi^{10}\right)^2 \right],$$

$$\omega_1 = \frac{32\sqrt{3}}{\sqrt{\pi}} \int d\cos\theta_{p_1} \int dF_{j_p}\, Y_{10}(\theta_p, 0)\left(S_\pi^0 - 2T_\pi^{20}\right) P_\pi^{10}, \tag{20}$$

for $^5_\Lambda$He, and

$$\omega_0 = \frac{24}{\pi} \int d\cos\theta_{p_1} \int dF_{j_p} \left\{ (1+R)[3(S_\pi^0)^2 + 18(T_\pi^{20})^2 + (P_\pi^{10})^2] \right.$$
$$\left. + 7(S_\pi^1)^2 + \frac{1}{5}(T_\pi^{11})^2 + \frac{9}{5}(T_\pi^{31})^2 + \frac{7}{3}(P_\pi^{21})^2 \right\},$$

---

* An analytical expression including the contribution of the complete octets meson will be deduced and analyzed in a forthcoming paper.

$$\omega_1 = \frac{32\sqrt{3}}{\sqrt{\pi}} \int d\cos\theta_{p_1} \int dF_{j_p} \, Y_{10}(\theta_p, 0) \left[(1+R) \left(S_\pi^0 - 2T_\pi^{20} \right) P_\pi^{10} \right.$$

$$\left. + \left(S_\pi^1 + \frac{1}{5}T_\pi^{11} - \frac{9}{5}T_\pi^{31} \right) P_\pi^{21} \right], \tag{21}$$

for $^{12}_{\Lambda}$C. Here S_π^l, T_π^{lJ} and P_π^{lJ}, and $R = (bP)^2/3$ are, respectively, the nuclear moments and the ratio between the $1p_{3/2}$ and $1s_{1/2}$ contributions defined in Section IV C from Ref. [17]. These formulas clearly show that the s-part of $^{12}_{\Lambda}$C results agrees with the $^{5}_{\Lambda}$He ones. This has also been tested numerically, as can be seen from Tables 1 and 2. There we observe that this conclusion is also valid within the complete OME model. Simultaneously, our numerical computation shows that the contribution of the p-shell of $^{12}_{\Lambda}$C has approximately the same magnitude that the s-one, both for ω_0 and ω_1 (this has been extensively discussed in Ref. [17] concerning the decay rate). On the other side, it is also well known that the main contribution comes from the lower quantum numbers and, as it has been discussed in Ref. [17], the tensorial T_π^{20} term is the dominant one in the hypernuclear weak decay case. Thus, approximating the p-shell contribution of $^{12}_{\Lambda}$C by the s-one, and retaining only the terms containing the tensorial moment T_π^{20} in our formula, we arrive to a very simple expression for a_Λ:

$$a_\Lambda \simeq -\frac{4\sqrt{3\pi}}{27} \frac{\int d\cos\theta_{p_1} \int dF_{j_p} \, Y_{10}(\theta_p, 0) T_\pi^{20} P_\pi^{10}}{\int d\cos\theta_{p_1} \int dF_{j_p} (T_\pi^{20})^2}. \tag{22}$$

This expression is valid both for $^{5}_{\Lambda}$He and C$^{12}_{\Lambda}$C hypernuclear decays, within the OPEM. Our results presented in Eqs. (20), (21) and (22) suggest that, in order to obtain a stronger theoretical dependence of a_Λ on the particular hypernucleus considered (as indicated by experimental data) future improvements of the model are required. We believe that a different decay mechanism should be introduced for s- and p-shell hypernuclei, beyond the Born approximation.

Summarizing, we have derived simple formulas for the evaluation of the asymmetry parameter, which exactly include the effects of three-body kinematics in the final states and correctly treat the contribution of transitions originated from proton states beyond the s-shell. Besides, we have deduced exact analytical expressions for the particular cases of $^{5}_{\Lambda}$He and $^{12}_{\Lambda}$C written, within the OPEM, in terms of nuclear moments. The numerical values of a_Λ, calculated within different OME models, variate from -0.62 to -0.29. The negative value systematically obtained for a_Λ for the two hypernuclei indicates that it will be hard to get a positive or zero value for it in the $^{5}_{\Lambda}$He case, at least within strict OME models. The puzzle posed by the experimental results for a_Λ in s- and p-shell hypernuclei remains unexplained. However, our results suggest that additional changes in the nuclear model and/or the decay mechanism should be introduced for sand p-shell hypernuclei, in such way that the improvements lead to a stronger theoretical dependence of a_Λ on the hypernucleus considered, which could reverse the asymmetry sign in the case of $^5_\Lambda$He, as required to reproduce the more recent experimental data on this observable.

Bibliography

[1] H. Outa *et al.*, KEK Report No. KEK-PS E462, 2000.

[2] A. Feliciello, Nucl. Phys. **A691**, 170c (2001); P. Gionatti, Nucl. Phys. **A691**, 483c (2001).

[3] R.L. Gill, Nucl. Phys. **A691**, 180c (2001).

[4] G. Garbarino, A. Parreño and A. Ramos, Phys. Rev. C **69**, 054603 (2004); *ibid*, Nucl. Phys. **A754**, 137c (2005).

[5] H. Bhang *et al.*, Nucl. Phys. **A754**, 144c (2005).

[6] H. Outa *et al.*, Nucl. Phys. **A754**, 157c (2005).

[7] T. Maruta *et al.*, nucl-ex/0402017; *ibid*, Nucl. Phys. **A754**, 168c (2005).

[8] S. Ajimura *et al.*, Phys. Lett. **B282**, 293 (1992).

[9] S. Ajimura *et al.*, Phys. Rev. Lett. **84**, 4052 (2000).

[10] M.M. Block and Dalitz, Phys. Rev. Lett. **11**, 96 (1963).

[11] J. B. Adams, Phys. Rev. **156**, 832 (1967) (*Cf.* correction pointed out in Ref. [12].)

[12] B. H. J. McKellar and B. F. Gibson, Phys. Rev. **C 30** (1984) 322.

[13] K. Takeuchi, H. Takaki and H. Bandō, Prog. Theor. Phys. **73**, 841 (1985).

[14] J. F. Dubach, G. B. Feldman, B. R. Holstein and L. de la Torre, Ann. Phys. (N.Y.) **249**, 146 (1996).

[15] A. Parreño, A. Ramos and C. Bennhold, Phys. Rev. **C 56**, 339 (1997) and references therein.

[16] A. Parreño and A. Ramos, Phys. Rev. **C 65**, 015204 (2001).

[17] C. Barbero, D. Horvat, F. Krmpotić, T. T. S. Kuo, Z. Narančić and D. Tadić, Phys. Rev. **C 66**, 055209 (2002).

[18] C. Barbero, C. de Conti, A.P. Galeao and F. Krmpotić, Nucl. Phys. **A726**, 267 (2003).

[19] C. Barbero, A.P. Galeao and F. Krmpotić, Phys. Rev. **C 72**, 035210 (2005).

[20] A. Ramos, E. Oset, and L. L. Salcedo, Phys. Rev. **C 50**, 2314 (1995).

[21] A. Ramos, M.J. Vicente-Vacas and E. Oset, Phys. Rev. **C 55**, 735 (1997), Erratum-ibid Phys. Rev. **C 66**, 039903 (2002).

[22] A. Parreño, A. Ramos, C. Bennhold and K. Maltman, Phys. Lett. **B435**, 1 (1998).

[23] W. M. Alberico and G. Garbarino, Phys. Lett. **B486**, 362 (2000).

[24] J-H. Jun, Phys. Rev. **C 63**, 044012 (2001).

[25] T. Inoue, M. Oka, T. Motoba and K. Itonaga, Nucl. Phys. **A633**, 312 (1998).

[26] K. Sasaki, T. Inoue and M. Oka, Nucl. Phys. **A669**, 331 (2000), Erratum-ibid **A678**, 455 (2000).

[27] M. Shmatikov, Phys. Lett. B322, 311 (1994); *ibid*, Nucl. Phys. **A580**, 538 (1994).

[28] K. Itonaga, T. Ueda and T. Motoba, Nucl. Phys. **A577**, 301c; (1994) *ibid*, Nucl. Phys. **A585**, 331c (1995); *ibid*, Nucl. Phys. **A639**, 329c (1998).

[29] E. Jido, E. Oset and J.A. Palomar, Nucl. Phys. **A694**, 525 (2001).

[30] K. Sasaki, M. Izaki and M. Oka, Phys. Rev. **C 71**, 035502 (2005).

[31] C. Barbero and A. Mariano, Phys. Rev. **C 73**, 024309 (2006).

[32] K. Itonaga, T. Ueda and T. Motoba, Phys. Rev. **C 65**, 034617 (2002); K. Itonaga, T. Motoba and T. Ueda, *Electrophoto Production of Strangeness on Nuand Nuclei* (Sendai03), K. Maeda, H. Tamura, S.N. nakamura and O. Hashimoto eds., World Scientific (2004) pp. 397-402.

[33] K. Sasaki, T. Inoue and M. Oka, Nucl. Phys. **A702**, 477 (2002).

[34] W. M. Alberico and G. Garbarino, Phys. Rep. **369**, 1 (2002).

[35] W. M. Alberico, G. Garbarino, A. Parreño and A. Ramos, Phys. Rev. Lett. **94**, 1 (2005).

Atualização das Energias dos Padrões Gama para a Escala do CODATA2002: Efeito Sobre as Correlações

Zwinglio O. Guimarães-Filho, Otaviano Helene

Instituto de Física da Universidade de São Paulo

Resumo. Energias de raios gama são correlacionadas com as constantes fundamentais, cujos valores recomendados são determinados periodicamente pelo CODATA. O efeito da atualização dos valores recomendados para as constantes fundamentais sobre as correlações entre as energias de um conjunto de raios gama padrões para calibração é estudado neste trabalho. Exemplos de energias de raios gama cujas correlações sofreram alterações significativas devido à mudança da escala do CODATA86 para o CODATA2002 são apresentados e interpretados qualitativamente.

1 Introdução

Para estabelecer um conjunto de valores recomendados para a energia de raios gama adequados para serem utilizados como padrões para calibração, Helmer e van der Leun[1] consideraram 4 tipos de informações (esquematicamente representados na Tabela 1): *(i) medidas absolutas do comprimento de onda; (ii) medidas relativas do comprimento de onda; (iii) medidas com detectores semicondutores; (iv) informações provenientes de relações baseadas em esquemas de decaimento.*

Destas fontes de informações, apenas o método *(i)* relaciona as energias dos raios gama com grandezas externas ao conjunto de padrões gama: o parâmetro de rede do Si, $a_{Si}^{\#}$; a constante de plank, h, e a carga elementar, e (estes últimos, envolvidos no fator de conversão de comprimento de onda para energia, $E_\lambda = \frac{h.c}{e}$). Os outros métodos apenas estabelecem relações entre as energias de diferentes raios gama entre si. Assim, as energias gama são proporcionais à relação

$$f = \frac{E_\lambda}{a_{Si}^{\#}} = \frac{h.c}{e.a_{Si}^{\#}} \tag{1}$$

e, portanto, dependem dos valores adotados para as constantes fundamentais.

Tabela 1 Classificação das informações consideradas no ajuste para determinar as energias dos raios gama padrões para calibração

Difração em cristais	**(i) Medidas absolutas de comprimento de onda** *(Determinação da razão entre o comprimento de onda da transição e o parâmetro de rede do Silício)* **(ii) Medidas relativas de comprimento de onda** *(Determinação da razão entre os comprimentos de onda de duas transições gama entre si)*
(iii) Medidas em detectores semicondutores	**Medidas da energia de raios gama com detectores semicondutores calibrados** **Medidas de diferenças de energia entre dois raios gama com detectores semicondutores**
(iv) Relações baseadas no esquema de decaimento	

Como conseqüência da origem das informações que determinam o conjunto de padrões de energias gama, estes padrões são correlacionados tanto entre eles quanto com as constantes fundamentais. Entretanto, estas correlações não foram determinadas na referência [1]. Para determiná-las, usamos um procedimento baseado no ajuste simultâneo, pelo Método dos Mínimos Quadrados, da energia dos 330 raios gama do conjunto a partir das ~850 informações consideradas para o estabelecimento do conjunto de padrões.

O número de dados e de parâmetros do ajuste para determinar as constantes fundamentais estão apresentados na Tabela 2 (classificados com respeito ao tipo de informação). O número de graus de liberdade deste ajuste é 396, o que permitiu avaliar a consistência das incertezas atribuídas às informações experimentais consideradas. As principais correlações, módulos superiores a 0,7, determinadas com o uso deste procedimento foram publicadas na referência [2].

2 Atualização dos valores recomendados para as constantes fundamentais

As referências [1] e [2] são consistentes com a escala dos valores recomendados para as constantes fundamentais da compilação do CODATA de 1986[3], porém, após esta já ocorreram 2 as outras compilações (em 1998[4] e 2002[5]) e, portanto, foi necessário atualizar as energias dos padrões gama (valores e respectiva matriz de covariância) para torna-las compatíveis com a escala mais recente das constantes fundamentais.

ATUALIZAÇÃO DAS ENERGIAS DOS PADRÕES GAMA PARA A ESCALA DO CODATA2002...

275

Tabela 2 Dados experimentais e parâmetros do ajuste para determinar os valores recomendados para as energias dos raios gama padrões para calibração

Dados		Parâmetros		
Origem	*Quantidade*	*Tipo*	*Quantidade*	*Obs*
Relação com as constantes fundamentais	1	Relação com as constantes fundamentais	1	
Difração (i) Medidas "absolutas"	79			
(ii) Medidas relativas	121	Energias gama	331	330 raios gama (há 2 parâmetros para o 722 keV da^{108}Ag)
(iii) Detectores semicondutores Calibrados	39			
Diferença de energia	367			
(iv) Relações baseadas nos esquemas de decaimento	248	Níveis nucleares excitados	127	26 nuclídeos
Número de dados	855	Número de parâmetros	459	

As alterações de E_λ e $a_{Si}^{\#}$, entre as compilações do CODATA de 1986 (CODATA-86[3]), de 1998 (CODATA98[4]) e de 2002 (CODATA2002[5]) estão representadas na Figura 1. As razões f correspondentes a cada uma destas compilações estão apresentadas na Tabela 3.

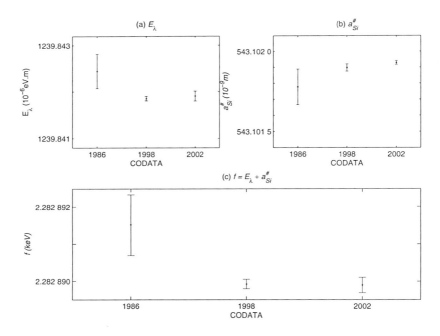

Figura 1 Evolução dos valores recomendados para as constantes fundamentais envolvidas na determinação das energias dos padrões gama. (a) fator de conversão de comprimento de onda em energia, E_λ; (b) parâmetro de rede do Si nas medidas absolutas de comprimento de onda, $a_{Si}^{\#}$; (c) relação $f = \frac{E_\lambda}{a_{Si}^{\#}}$

Tabela 3 Valores da relação entre as constantes fundamentais, f, de acordo com os valores recomendados pelas compilações CODATA de 1986 a 2002

Origem de E_λ e de $a_{Si}^{\#}$	Identificação	$f = \frac{E_\lambda}{a_{Si}^{\#}}$ (keV)	incerteza relativa
CODATA 1986 [3]	f_{86}	2,282 891 51 (83)	$0,36.10^{-6}$
CODATA 1998 [4]	f_{1998}	2,282 889 93 (13)	$0,06.10^{-6}$
CODATA 2002 [5]	f_{2002}	2,282 889 89 (23)	$0,10.10^{-6}$

As conseqüências sobre as correlações entre as energias dos padrões gama provocadas pela atualização da escala do CODATA86 para a do CODATA2002, obtidas com o uso do procedimento apresentado nas referências [6] e [7], podem ser observadas no histograma da Figura 2. A relação entre as variâncias, $V_x = \sigma_x^2$ e $V_y = \sigma_y^2$, a covariância, $V_{x,y}$, e a correlação $\rho_{x,y}$ de duas grandezas x e y é[8]: $\rho_{x,y} = \frac{V_{x,y}}{\sigma_x . \sigma_y}$.

As variâncias correspondem ao valor esperado do erro quadrático da grandeza, $V_x = \langle \varepsilon_x^2 \rangle$, enquanto as covariâncias ao valor esperado do produto dos erros das grandezas $V_{x,y} = \langle \varepsilon_x . \varepsilon_y \rangle$. Assim, as fontes de erro comuns a duas grandezas podem contribuir para a covariância entre elas. Estas contribuições serão positivas quando a superestimação (subestimação) de uma das grandezas devido à fonte de erro comum, implicar na superestimação (subestimação) da outra. As contribuições para a covariância serão negativas quando o efeito da fonte de erro comum for tal que, quando superestima uma, subestima a outra.

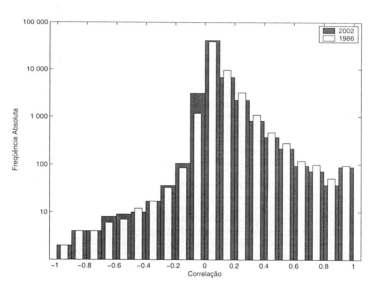

Figura 2 Histograma comparativo das correlações entre as energias gama dos padrões de calibração, obtidas considerando os valores recomendados para as constantes fundamentais do CODATA86 e do CODATA2002.

3 Interpretação das alterações nas correlações

Na Figura 2 pode-se observar a redução sistemática nas correlações positivas e o conseqüente aumento das negativas, evidenciando que as correlações entre as energias dos padrões gama se tornaram menos positivas com a atualização. Isto se deve ao fato das energias gama serem proporcionais a f e, portanto, a covariância entre duas energias devido apenas a f é sempre positiva. Logo, é compreensível que a atualização das constantes fundamentais (com a redução da incerteza de f) tenha tornado as covariâncias entre as energias gama menos positivas.

Tabela 4 Valores recomendados para a energia de alguns raios gama quando usando as escalas do CODATA86 e do CODATA2002. A correlação dessas energias gama com razão f entre as constantes fundamentais em cada uma destas escalas também é apresentada.

Nuclídeo	Energia (keV)		Correlação com f	
	1986	**2002**	**1986**	**2002**
[169]Yb	63.12043(4)	63.12039(3)	0,58	0,19
[169]Yb	93.61445(7)	93.61439(6)	0,48	0,15
[169]Yb	109.77923 (5)	109.77915(4)	0,75	0,30
[169]Yb	118.18940 (12)	118.18931(11)	0,37	0,11
[169]Yb	130.52293 (6)	130.52284(3)	0,85	0,41
[169]Yb	177.21303 (8)	177.21290(4)	0,85	0,40
[169]Yb	197.95672 (8)	197.95658(5)	0,87	0,44
[169]Yb	261.07708(11)	261.07689(6)	0,87	0,44
[169]Yb	307.73581(12)	307.73559(6)	0,92	0,56
[192]Ir	136.34281(22)	136.34272(21)	0,23	0,06
[192]Ir	205.79432(20)	205.79417(18)	0,38	0,11
[192]Ir	295.95655(15)	295.95634(10)	0,73	0,29
[192]Ir	308.45516(18)	308.45494(14)	0,63	0,22
[192]Ir	316.50609(15)	316.50587(10)	0,79	0,33
[192]Ir	416.4696(5)	416.4693(5)	0,29	0,08
[192]Ir	468.06875(23)	468.06841(17)	0,72	0,28
[192]Ir	484.5752 (4)	484.5749(4)	0,39	0,12
[192]Ir	588.5817(5)	588.5812(5)	0,42	0,13
[192]Ir	604.41120(28)	604.41077(18)	0,79	0,34
[192]Ir	612.46212(26)	612.46169(15)	0,85	0,41
[192]Ir	884.5372(6)	884.5366(5)	0,56	0,18
[198]Au	411.80198(17)	411.80169(9)	0,89	0,49
[198]Au	675.8836(7)	675.8832(7)	0,33	0,10
[198]Au	1087.6841(8)	1087.6833 (7)	0,49	0,15

Esta análise permite compreender, por exemplo, a redução na correlação entre as energias dos gamas de 307keV do [169]Yb e de 612keV do [192]Ir (de 0,82 para 0,33), pois esta correlação era basicamente devida a f, uma vez que estas energias foram determinadas com elevada precisão em medições absolutas de comprimento de onda (e não há relações baseadas nos esquemas de decaimento envolvendo estes gamas).

Os valores recomendados, e as correlações das energias de alguns raios gama determinados por medidas absolutas de comprimento de onda (fontes de [169]Yb, [192]Ir e [198]Au) com a razão entre as constantes fundamentais f nas escalas de 1986 e 2002 são apresentadas na Tabela 4. A Tabela 4, mostra que, na escala do CODATA86, a correlação entre o gama de 307keV do [169]Yb e f era de 0,92, reduzindo-se para 0,56 quando usando a escala do CODATA2002. Analogamente, para o 612keV do ^{192}Ir a redução foi de 0,85 para 0,41. Com isso, a covariância entre estas energias devido à fonte comum (valor adotado para f) se reduziu devido à diminuição da incerteza em f.

A redução da contribuição positiva na covariância entre duas energias gama devido à redução da incerteza de f permite entender, também, a alteração da correlação entre as energias dos gamas de 109keV e 197keV do ^{169}Yb (de 0,55 para -0,14). A correlação neste caso se tornou negativa por haver uma relação baseada no esquema de decaimento envolvendo estas duas energias gama (109keV + 197keV = 307keV + diferença das energias de recuo do núcleo), o que introduz uma contribuição negativa na covariância entre estas energias (como a soma das energias é conhecida com elevada precisão, a superestimação de uma tende a subestimar a outra). Assim, com a redução da contribuição positiva na covariância devido à f, esta se tornou negativa por causa das outras fontes de covariâncias. Isto deu origem à correlação negativa observada entre as energias destes gamas (o sinal da correlação é o mesmo da covariância).

Porém, há casos em que as correlações aumentaram (devido à redução das correspondentes variâncias), como, por exemplo, a correlação entre o 675keV e o 1087keV do ^{198}Au que aumentou de 0,980 para 0,993 e a correlação entre o 136keV e o 604keV do ^{192}Ir (de 0,57 para 0,65). O aumento das correlações nestes casos é devido às relações baseadas nos esquemas de decaimento envolvendo estes raios gama com outros determinados com elevada precisão em medidas absolutas de comprimento de onda que tiveram suas incertezas muito reduzidas com as alterações das constantes fundamentais (no caso do ^{198}Au, o raio gama de 411keV, e no caso do ^{192}Ir, o gama de 468keV). Portanto, nestes casos, o efeito da redução das variâncias superou o efeito da redução das covariâncias, tornando as correlações maiores.

Agradecimentos

Os autores agradecem o suporte parcial da FAPESP, do CNPq e da Agência Internacional de Energia Atômica (IAEA)

Referências Bibliográficas

[1] R.G. Helmer e C. van der Leun, *Recommended standards for gamma-ray energy calibration (1999)*, Nucl. Instr. and Meth. **A450** (2000) 35.

[2] O. Helene, Z.O. Guimarães-Filho, V.R. Vanin et al., *Covariances between gamma-ray energies*, Nucl. Instr. and Meth. **A460** (2001) 289.

[3] E.R. Cohen e B.N.Taylor, *The 1986 adjustment of the fundamental physical constants*, Rev. Mod. Phys 59 (1987) 1121.

[4] P.J.Mohr e B.N.Taylor, *CODATA recommended values of the fundamental physical constants: 1998*, Rev. Mod. Phys 72 (2000) 351.

[5] P.J.Mohr e B.N.Taylor, *The 2002 CODATA recommended values of the fundamental physical constants*, Rev.Mod.Phys 77 (2005) 1.

[6] Z.O. Guimarães-Filho, O. Helene, V.R. Vanin e N.L. Maidana, *Fitting and updating gamma-ray energies*, AIP Conference Proceedings **769** (2005) 382.

[7] Z.O. Guimarães-Filho, *Energias de raios gama padrões: suas covariâncias e relações com as constantes fundamentais*, Tese de doutoramento, IFUSP (2004).

[8] ISO – International Organization for Standardization, *Guide to the expression of uncertainty in measurement*, Geneva, Suíça (1993).

Anais da XXVIII Reunião de Trabalho sobre Física Nuclear no Brasil
SP, Brasil, 2005

Correlações Entre as Energias dos Raios Gama que Seguem o Decaimento do ^{56}Co e do ^{66}Ga

Zwinglio O. Guimarães-Filho, Otaviano Helene, Vito R. Vanin

Instituto de Física da Universidade de São Paulo

Resumo. Apresentamos as correlações entre as energias dos raios gama que seguem os decaimentos por captura eletrônica do ^{56}Co e do ^{66}Ga, determinadas com o uso de um procedimento baseado no ajuste simultâneo da energia dos raios gama do conjunto de padrões a partir de todas as informações consideradas para estabelece-lo. Há diversas correlações elevadas (superiores a 0,7) envolvendo as energias dos raios gama das fontes de ^{56}Co e ^{66}Ga, tanto entre gamas emitidos pela mesma fonte, quanto entre gamas de fontes diferentes. Portanto, estas correlações também precisam ser consideradas nas compilações de resultados envolvendo fontes diferentes.

---

1 Introdução

Energias de padrões gama são correlacionadas tanto entre elas quanto com as constantes fundamentais devido ao modo como seus valores foram determinados. Em 2000, Helmer e van der Leun[1] publicaram um conjunto de valores recomendados para as energias de ~350 raios gama adequados para serem utilizados como padrões para calibração. As correlações entre as energias destes raios gama, no entanto, não foram determinadas na referência [1]. Estas correlações são fundamentais para a utilização das informações quando o resultado final depende do valor adotado para a energia de mais de um raio gama.

2 Origens das correlações entre energias gama

As energias do conjunto de padrões da referência [1] foram determinadas por: *(i)* difração em cristais de elevada pureza; *(ii)* difração em cristais curvos; *(iii)* medida em detectores semicondutores; e *(iv)* relações baseadas nos esquemas de decaimento (relações cascata/cross-over). Destes métodos, apenas o item *(i)* relaciona diretamente a energia do raio gama com as constantes fundamentais h (constante de Plank), e (carga elementar), e a_{Si} (o parâmetro de rede do Si); enquanto os demais só permitem estabelecer relações entre energias de raios gama. No entanto, o método *(i)* só foi utilizado para determinar a energia de alguns raios gama do ^{198}Au, ^{192}Ir, ^{169}Yb, ^{170}Tm e ^{99}Mo de energias inferiores a 700 keV.

As principais correlações (correlações superiores a 0,7, em módulo) existentes entre energias de transições gama do conjunto de padrões da referência [1] foram apresentadas na referência [2]. O estudo das origens das correlações, efetuado na referência [2], mostrou que tanto o uso de relações baseadas nos esquemas de decaimento, quando de medidas com o uso de detectores semicondutores levam ao estabelecimento de correlações elevadas. No caso dos raios gama que seguem o decaimento do ^{56}Co e do ^{66}Ga, ambas as origens de grandes correlações estão presentes.

O conhecimento destas correlações é particularmente relevante no caso do ^{56}Co e do ^{66}Ga por serem as principais fontes de raios gama empregadas na calibração de detectores semicondutores na faixa de 2 a 5 MeV. Além disso, as correlações envolvendo os raios gama de energias superiores a 2 MeV emitidos por estas fontes são todas positivas.

3 Resultados

Os esquemas de níveis nucleares do ^{56}Fe e ^{66}Zn, que seguem os decaimentos por captura eletrônica do ^{56}Co e do ^{66}Ga, respectivamente, mostrando apenas os raios gama considerados no conjunto de padrões, estão apresentados nas Figuras 1 e 2.

Os valores recomendados para as energias dos raios gama que seguem os decaimentos do ^{56}Co e do ^{66}Ga, estão apresentados, respectivamente, nas Tabelas 1 e 2. Para obter estes resultados, as cerca de 850 informações selecionadas na

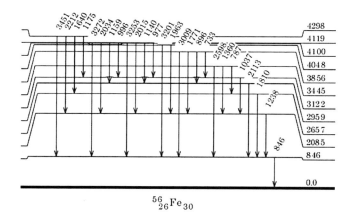

Figura 1 Esquema de níveis nucleares considerado para o ^{56}Fe, indicando a relação com os raios gama que seguem o decaimento por captura eletrônica do ^{56}Co e que foram utilizados nas relações cascata/cross-over.

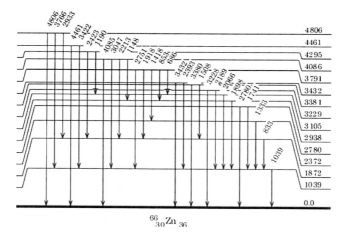

Figura 2 Esquema de níveis nucleares considerado para o ^{66}Zn, indicando a relação com os raios gama que seguem o decaimento por captura eletrônica do ^{66}Ga e que foram utilizados nas relações cascata/cross-over.

referência [1] foram utilizadas como dados de entrada em um ajuste pelo Método dos Mínimos Quadrados para determinar simultaneamente a energia dos cerca de 350 raios gama do conjunto de padrões. Detalhes sobre o procedimento podem ser obtidos nas referências [3] e [4].

Tabela 1 Valores recomendados para as energias dos raios gama que seguem o decaimento do ^{56}Co.

Gama(keV)	Energia(keV)	$\frac{\sigma_E}{E}$ (x10^6)
733	733,5089 (26)	3,6
787	787,7393 (27)	3,5
846	846,7631 (22)	2,6
896	896,497 (5)	6
977	977,364 (4)	3,7
996	996,938 (4)	4,4
1037	1037,8345 (24)	2,3
1140	1140,352 (6)	5
1159	1159,925 (6)	5
1175	1175,0853 (26)	2,2
1238	1238,2731 (23)	1,8
1335	1335,38 (3)	24
1360	1360,197 (3)	2,5
1640	1640,448 (5)	3,0
1771	1771,329 (4)	2,0

continua ...

Gama (keV)	Energia (keV)	$\frac{\sigma_E}{E}$ (x10^6)
1810	1810,726 (4)	2,4
1963	1963,715 (5)	2,6
2015	2015,179 (5)	2,3
2034	2034,752 (5)	2,4
2113	2113,099 (5)	2,5
2212	2212,896 (4)	1,6
2598	2598,438 (4)	1,6
3009	3009,560 (4)	1,5
3201	3201,942 (6)	1,8
3253	3253,405 (5)	1,7
3272	3272,977 (6)	1,7
3451	3451,117 (5)	1,5

Tabela 2 Valores recomendados para as energias dos raios gama que seguem o decaimento do ^{66}Ga.

Gama (keV)	Energia (keV)	$\frac{\sigma_E}{E}$ (x10^6)
686	686,082 (6)	9
833	833,5328 (20)	2,4
853	853,032 (7)	8
1039	1039,2198 (30)	2,9
1148	1147,893 (10)	9
1190	1190,283 (8)	7
1333	1333,115 (5)	3,5
1418	1418,756 (6)	3,9
1459	1458,661 (12)	8
1508	1508,156 (7)	4,9
1741	1740,906 (18)	11
1898	1898,835 (7)	3,9
1918	1918,334 (5)	2,7
2066	2065,782 (7)	3,6
2173	2173,326 (16)	8
2189	2189,623 (6)	2,8
2213	2213,189 (10)	4,5
2393	2393,128 (8)	3,3
2423	2422,525 (8)	3,3
2751	2751,841 (5)	1,9
2780	2780,096 (18)	7
2933	2933,359 (10)	3,4
2977	2977,08 (5)	16

continua...

Gama (keV)	Energia (keV)	$\frac{\sigma_E}{E}$ $(\times 10^6)$
2993	2993,21 (4)	12
3047	3046,692 (10)	3,3
3228	3228,806 (6)	1,9
3380	3380,849 (7)	2,1
3422	3422,038 (9)	2,7
3432	3432,307 (8)	2,3
3766	3766,852 (10)	2,7
4085	4085,860 (10)	2,6
4461	4461,200 (10)	2,3
4806	4806,008 (11)	2,3

Com o uso do procedimento das referências [3, 4], os valores recomendados para as energias dos raios gama do conjunto de padrões são obtidos diretamente dos parâmetros ajustados. As incertezas e correlações entre estas energias foram determinadas com o uso da matriz de covariância dos parâmetros do ajuste.

As Tabelas 3 e 4 apresentam as principais correlações (correlações maiores que 0,7 ou menores que −0,3) existentes entre os raios gama que seguem os decaimentos do ^{56}Co e do ^{66}Ga, respectivamente. Correlações importantes (correlações de módulo superior a 0,6) envolvendo pelo menos um raio gama de energia maior que 1 MeV das fontes de ^{56}Co e ^{66}Ga são apresentados na Tabela 5.

Tabela 3 Correlações entre energias de raios gama que seguem o decaimento do ^{56}Co.

$\gamma_1(^{56}\text{Co})$	$\gamma_2(^{56}\text{Co})$	$\rho(\gamma_1,\gamma_2)$
1140	896	0.717
1640	1360	−0.389
1771	733	0.723
1810	787	−0.386
"	1640	−0.392
2015	977	0.854
2034	996	0.865
2113	896	−0.664
"	1140	−0.541
"	1159	−0.510
2212	1175	0.745
2598	1360	0.833
"	1810	0.788
3009	1771	0.863

continua ...

$\gamma_1(^{56}\text{Co})$	$\gamma_2(^{56}\text{Co})$	$\rho(\gamma_1,\gamma_2)$
3201	1963	0.924
3253	977	0.771
"	2015	0.909
3272	996	0.784
"	2034	0.916
3451	1175	0.774
"	1238	0.773
"	2212	0.918
"	3009	0.734

Tabela 4 Correlações entre energias de raios gama que seguem o decaimento do ^{66}Ga.

$\gamma_1(^{66}\text{Ga})$	$\gamma_2(^{66}\text{Ga})$	$\rho(\gamma_1,\gamma_2)$
1418	1333	−0.467
1898	853	−0.714
"	1148	−0.340
2066	686	−0.709
"	1190	−0.408
2213	1148	0.709
2751	1918	0.928
2780	1741	0.987
3047	1148	0.727
"	2213	0.980
3228	2189	0.882
3380	1508	0.874
3432	2393	0.930
3766	2933	0.981
4085	2213	0.937
"	3047	0.958
4461	3422	0.955
4806	2933	0.943
"	3766	0.962

Tabela 5 Correlações importantes (módulo superior a 0,6) envolvendo ao menos um raio gama de energia superior a 1 MeV das fontes de ^{56}Co e ^{66}Ga.

γ_1	γ_2	$\rho(\gamma_1,\gamma_2)$
1175 (^{56}Co)	1121 (^{182}Ta)	0.693
"	1157 "	0.674

continua ...

γ_1	γ_2	$\rho(\gamma_1, \gamma_2)$
1175 (^{56}Co)	1189 (^{182}Ta)	0.693
"	1221 "	0.692
"	1231 "	0.680
"	1257 "	0.674
"	1273 "	0.688
"	1289 "	0.692
"	1373 "	0.687
"	1387 "	0.686
1238 (^{56}Co)	1121 "	0.789
"	1157 "	0.776
"	1189 "	0.789
"	1221 "	0.789
"	1231 "	0.786
"	1257 "	0.776
"	1273 "	0.787
"	1289 "	0.789
"	1373 "	0.787
"	1387 "	0.787
1963 (^{56}Co)	1898 (^{66}Ga)	0.634
"	1918 "	0.717
"	2189 "	0.714
"	2751 "	0.687
"	3228 "	0.795
2212 (^{56}Co)	3380 "	0.615
3201 (^{56}Co)	1918 "	0.698
"	2189 "	0.766
"	2751 "	0.693
"	3228 "	0.865
3451 (^{56}Co)	3380 "	0.670
"	1121 (^{182}Ta)	0.752
"	1157 "	0.736
"	1189 "	0.752
"	1221 "	0.752
"	1231 "	0.745
"	1257 "	0.736
"	1273 "	0.749
"	1289 "	0.752
"	1373 "	0.749
"	1387 "	0.749

4 Conclusão

Foram encontrados vários casos de correlações elevadas entre raios gama emitidos pela mesma fonte (como a correlação de 0,92 entre 2212keV e o 3451keV do ^{56}Co e a correlação de 0,96 entre o 3766keV e o 4806keV do ^{66}Ga), bem como correlações importantes entre raios gama do ^{56}Co com outros do ^{66}Ga de energias próximas (como a correlação de 0,87 entre os gamas de 3201keV do ^{56}Co e de 3228keV do ^{66}Ga), cujo conhecimento e correta consideração são necessários para que as compilações de resultados envolvendo detectores calibrados com gamas destas fontes possam ser efetuadas de maneira adequada.

Correlações tão grandes como estas, envolvendo as duas principais fontes de calibração de energias entre 2 e 5 MeV, influenciam a determinação dos resultados de cada experimento e correlacionam resultados de experimentos diferentes. Um exemplo do efeito das correlações em resultados de medidas de espectroscopia gama é apresentado na referência [5].

Agradecimentos

Os autores agradecem o suporte parcial da FAPESP, do CNPq e da Agência Internacional de Energia Atômica (IAEA).

Referências Bibliográficas

[1] R.G. Helmer e C. van der Leun, *Recommended standards for gamma-ray energy calibration (1999)*, Nucl. Instr. and Meth. **A450** (2000) 35.

[2] O. Helene, Z.O. Guimarães-Filho, V.R. Vanin et al., *Covariances between gamma-ray energies*, Nucl. Instr. and Meth. **A460** (2001) 289.

[3] Z.O. Guimarães-Filho, *Energias de raios gama padrões: suas covariâncias e relações com as constantes fundamentais*, Tese de doutoramento, IFUSP (2004).

[4] Z.O. Guimarães-Filho, O. Helene, V.R. Vanin e N.L. Maidana, *Fitting and updating gamma-ray energies*, AIP Conference Proceedings **769** (2005) 382.

[5] Z.O. Guimarães-Filho, O. Helene, *Energy of the 3/2 + state of ^{229}Th reexamined*, Phys. Rev. C **71** (2005) 044303

Anais da XXVIII Reunião de Trabalho sobre Física Nuclear no Brasil
SP, Brasil, 2005

Electric Field Interference in DNA Repair: Case Study and Perspectives to Improve Cancer Radiotherapy

J. D. T. Arruda-Neto[1-2], M. C. Bittencourt-Oliveira[3], E. C. Friedberg[4],
E. C. Silva[1-3], A. C. G. Schenberg[5], M. C. C. Oliveira[1], T. C. Hereman[3],
J. Mesa[1], T. E. Rodrigues[1], K. Shtejer[1], C. Garcia[1], F. Garcia[6]

[1] IF-University of São Paulo, São Paulo, SP, Brazil.

[2] UNISA-University of Santo Amaro, São Paulo, SP, Brazil.

[3] ESALQ-University of São Paulo, Piracicaba, SP, Brazil.

[4] University of Texas Southwestern Medical Center, Dallas, TX, USA.

[5] ICB-University of São Paulo, São Paulo, SP, Brazil.

[6] Medical Physics Group Santa Cruz State University, Ilhéus, BA, Brazil.

[7] InSTEC-Instituto Superior de Tecnologías y Ciencias Aplicadas, La Habana, Cuba.

Abstract. Cell mechanisms associated with repair of DNA strand breaks are not so far fully understood. Several seminal studies, dealing with diverse aspects of complex signaling transduction pathways have been published recently [1, 3]. However, a plethora of intriguing, key, and first principles questions, raised by all of these studies, remain unanswered [4]. In particular: how do repair proteins respond in a coordinated way, and how are a few DSBs within three billion base pairs in the cell reached and recognized ? Here we propose a biophysical approach which accounts for the elucidation of these questions, by reasoning that repair proteins find their route to the DNA damage sites by means of electric signaling and orientation of their dipole moments. To demonstrate this approach, we report an experiment in which cell-killing induced by both irradiation and exogenous electric fields in the prokaryote Microcystis panniformis was examined. Our results show that when cells exposed to 3KGy of radiation are immediately submitted to a weak electric field, cell death increases more than an order of magnitude compared to the effect of radiation alone, unambiguously demonstrating the electrical nature of the repair process. This finding propitiates a possibility to circumvent a major challenge in cancer radiotherapy: the delivering of a lethal dose of radiation to a tumor target volume while minimizing damage to the surrounding normal tissue.

1 Introduction

The mechanisms by which cells sense DNA strand breaks remain to be elucidated. For example, the fast induction of ATM kinase activity in mammalian cells, immediately after exposure to ionizing radiation, suggests that it acts at an early stage of signal transduction. Although many elegant experiments have dealt with important aspects of DNA repair [1, 3], several fundamental questions were left unresponded. In particular[4] : what is so specific about DNA breaks *vis-à-vis* chromatin changes?, how does ATM sense directly structure disruption in "relaxed"chromatin?, what factors determine the impressive speed and extent of the ATM response?, among others.

In this work we address and propose answers to three key questions recently raised and discussed by Bartek and Lukas, namely[4],

a) By means of what kind of interaction are repair proteins activated?

b) How do repair proteins and their recruiting complexes plan navigation routes as to reach the DNA damage site? In other words, what in these molecular motors works as a gyrocompass?

c) How do proteins throughout the cell nucleus respond in a coordinated way, and how are a few DSB within three billion base pairs recognized?

Here we propose a signaling mechanism for the activation of repair proteins and their recognition of a few DNA breaks within the entire genome, while accounting for the elucidation of the questions mentioned before. Our approach comprises the following sequence of reasoning:

1. electric signals generated by DNA damages are able to induce long-range activation of repair proteins almost instantly.

2. a static electric field established at the damage site functions as a "marker".

3. huge electric dipole moments, associated with proteins in general, explain how repair proteins find their navigation route to sense and reach the DNA damage site.

Such a possibility suitably complements the explanation provided by Banerjee et al.(2005) on how the human repair enzyme 8-oxoguanine glycosylase finds aberrant nucleotides (from oxidative damage) among the myriad of normal ones[5].

Finally, we performed an experiment on cell-killing induced by both irradiation and exogenous electric fields in the prokaryote Microcystis panniformis. The rationale underlying this idea was to destructively interfere with navigation routes and sensing of repair proteins, and strongly suggesting thus that these proteins are sensitive to the presence of weak electric fields (see item-3 above). Then, if the reasoning in item-3 is true the one in item-2 is automatically verified, since without an "electric marker"the orientation of the proteins dipoles is prevented. These two items run independently from item-1, which is proposed by means of arguments from the DNA electromechanical properties.

2 The proposed signaling mechanism

2.1 Repairing cascade activation

Let's start by recalling that DNA is wrapped around nucleosomes are arranged into ordered arrays of solenoid-like structures, which are then packaged into supercoiled loops. The action of a drug or ionizing radiation could give rise to localized excitations in such ordered, strongly nonlinear array, like typical *intrinsic localized modes* (ILM) [6]. In response to these ILM the system can undergo conformational changes and buckling. Actually, ILM excitations have been associated with the storage and transport of energy released during ATP hydrolysis, with the local opening of the DNA double-helix, and with the buckling, folding and collapse of a biopolymer chain. However, as it is used to occur in highly structured systems as DNA, mechanical deformations (resulting from conformational changes) automatically generate electrical disturbances, and vice-versa. In this sense, the DNA is an *electromechanical continuum*. Under this perspective, it is understandable why a sudden change in structure conformation induces an equally fast electric signal, a transient (by e.g. producing an instantaneous electric dipole). There are many observations of protein phosphorylation by weak electromagnetic fields [7] and, by the same reason, the same should occur with the head-protein of the repair reaction cascade.

2.2 The underlying static electric field

Strand breaks also generate static electric fields of quadrupolar nature, as revealed by experiments of perturbed angular correlations of γ rays, an alternative and elegant experimental technique to study the molecular dynamics of DNA [8, 9]. While electric transients have short durations, the static electric (quadrupole) field persists until the completion of DNA repair, being responsible for both the orientation of the repair proteins and their recognition of the damaged sites.

3 Case study

We used the prokaryote *Microcystis panniformis* to our approach, as it is resistant to radiation and heat [10]. Similar peculiarities led Imamura et al.(2002) to use Deinococcus radiodurans in the study of cell-killing induced by the combined action of radiation and heat [11]. In our case study, cultures of M. panniformis were exposed to 3KGy of γ radiation and immediately submitted to static electric fields of 20 v.cm^{-1}.

3.1 Methods Cell cultivation and density counting

M. panniformis BCCUSP100 strain (Cyanobacteria) was collected in the Barra Bonita reservoir located in the Tiête River, SP, Brazil ($22^0 29' S$, $48^0 34' W$). It was isolated by micromanipulation techniques using a microscope (Nikon E200, Melville, NY, USA). The isolated colony was washed by transferring it to several drops of water until all other microorganisms were removed, and was then placed into liquid medium. The non-axenic batch culture was grown in BG-11 medium [12] for a 14:10 hours (light: dark) photoperiod, at 22 ± 0.5 ^{0}C with an intensity of 30 ± 2 mol photons.m^{-2}.s^{-1}. Light intensity was measured using a spherical quantum photometer (LI-COR mod. LI-250, Lincoln, NE, USA) inside a culture flask containing the medium.

Prior to the beginning of the experiments a pre-culture of 2.2 L was prepared to obtain an inoculum in appropriate physiological condition. The pre-culture (exponential growth phase) was divided in amounts of 10 mL and inoculated in 30 mL of new medium in triplicate. Each total volume of 40 mL was housed in a glass tube. The cultures were initiated with 2.6 x 106 cells.mL^{-1}and maintained as previously described. The irradiations were carried out at the Nuclear Energy Center for Agriculture (CENA-USP, Piracicaba,SP,Brazil), using the γ source facility Gamma beam, model 650. All cell samples in glass tubes were simultaneously irradiated with 3KGy at a rate of 0.94 KGy/h. Three glass tubes were exposed for 2h to a 20 v.cm^{-1} static electric field between the plates of a capacitor, immediately after irradiation.

Cell densities were estimated by microscopic counts of cell samples stained with Lugol 4 percent solution in Fuchs Rosenthal hemacytometer. The average number of counted cells ranged from 6000 to 1000 in the first and second days after irradiation, respectively. We note that a total of 400 counted cells is required to obtain an error of $\sim$ 10 percent at a confidence level of 95 percent [13].

3.2 Nomenclature and definitions

N(t) is generically the average number of cells measured at time t (in days), obtained from the cell counting results in triplicate. The triplicate results were averaged, and it was found that their dispersion does not exceed 10 percent. Regarding the time scale, inoculation takes place at t = 0 and irradiation at t = 5d.

More specifically, $N_C(t)$ refers to control samples, while $N_\gamma(t)$ and $N_{\gamma E}(t)$ are the number of surviving cells after γ irradiation and after irradiation plus the application of an electric field, respectively. All results were normalized to N(5d), i.e., they were divided by N(5d). In this sense, each data point displayed in Fig. 1 represents a fraction of N(5d).

From the survival data shown in Fig.1 we calculated average *cell death rates*, defined as $r_{de}(t) \Delta N_{de}(t) / \Delta t$, where $\Delta N_{de}(t) = N(t) - N(t + \Delta t)$ is the number cells killed in the time interval Δt. Since we chose $\Delta t = 1$ day we have $r_{de}(t) = \Delta N_{de}(t)$, and our first data point (Fig. 2) corresponds to $N(5) - N(6) = r_{de}(5.5)$. Such a quantity is useful since it provides the speed by which cells are killed daily after the irradiation, and its measuring unity is cells per day, i.e. (d^{-1}).

4 Results and discussion

If a cell population is irradiated with a dose sufficient to induce DNA damage in most of its individuals, and additionally, if an external and static electric field (E-field) with suitable intensity is applied to such a population, the following sequence of events may occur:

(1) the repair proteins will tend to gradually align their electric dipoles with the direction of the E-field; (2) as a consequence, the majority of these proteins will be unable to reach damaged sites in DNA; and (3) lack of repair, particularly of DSB, may lead to cell death. In fact, cells have evolved complex signaling pathways to arrest the progression of the cell cycle in the presence of DNA damage, thereby providing increased time for repair mechanism to operate. But, when the burden of genomic insult is too large to be effectively met by the various cell responses, cells are able to initiate programmed cell death (apoptosis) [1].

As shown in Fig.1a, the application of a weak electric field (20v/cm) did not affect cell growth in control experiments. The average of the absolute differences between the two sets of data points is nearly zero.

On the other hand, the situation for irradiated cells is different (Fig.1b). The effect of the same electric field applied for 2 hours immediately after the completion of irradiation is dramatic (Fig.1c). In this figure, which shows the ratio of the two curves in Fig.1b, $N_\gamma(t)/ N_{\gamma E}(t)$, we observe that the number of surviving cells is approximately 12 times greater than that irradiated (γ) and submitted to an electric field $(\mathbf{E})$ on the second day after irradiation $(t = 7d)$.

This result is well appreciated in Fig.2, where we note that:

    i. While for irradiation alone (with 3 KGy) the average *cell death* rate is roughly 0.35N(5d), i.e., the initial cells, N(5d) are killed at a rate of about 35 percent per day in the first day, for $\gamma + \mathbf{E}$ we obtained $r_{de}(5.5) = 0.75N(5d)$, i.e., 75 percent of the total per day.

    ii. Besides its massive killing rate, the association $\gamma + \mathbf{E}$ is considerably faster. Actually, 3/4 of the entire cell population is killed in the first 12h

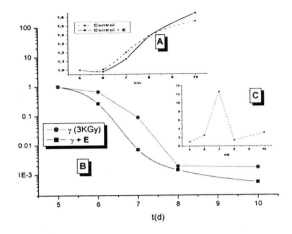

Figure 1 Results for the M. panniformis. (a) Growth curves for the control and control plus the application of an electric field. (b) Surviving curves following gamma irradiation, with and without the application of E. (c) Ratio between the two data sets shown in (b). In this and the next figure, the lines connecting the data points are only to guide the eyes.

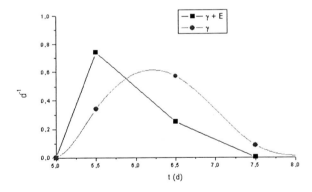

Figure 2 Cell death rates for the same two experimental situations shown in fig. 1b.

(t = 5.5d), explaining why the rate drops so abruptly in subsequent days. In contrast, when only γ is applied, the death rate peaks at about 24h–36h (t = 6–6.5d).

iii. Since the killing rate (r_{de}) is inversely proportional to the DNA damage repair rate (r_{da}), i.e., $r_{de} \propto 1/r_{da}$, the fact that in the first day we observe r_{de} for $\gamma + \mathbf{E}$ substantially higher than that for γ alone, demonstrates that the action of $\mathbf{E}$ is accentuated by repair mechanisms.

This last finding is corroborated by examination of the asymptotic behavior of $N_\gamma(t)$ and $N_{\gamma E}(t)$ at $t > 10$ days (Fig.1). These two sets of data run asymptotically $t_0 \sim t = 30d$. Since nearly all the initial cells were damaged by the 3KGy of γ radiation, the asymptotic values for both experiments, $N_\gamma(asym)$ and $N_{\gamma E}(asym)$ equal the number of successfully repaired cells (N_r), because all cells not repaired or deficiently repaired are already inviable. From the values obtained we estimate that $N_r(\gamma) > 3N_r(\gamma + E)$, showing that repair mechanisms work more efficiently in the absence of **E**.

These findings offer potential answer to the question: "Why does the E-field act only on DNA repair?"

An electric field of 20v/cm is very weak indeed, unlikely to inflict lethal cell damage or even disturbances in metabolism, as confirmed by our results (Fig.1a). The growth curves of *M. panniformis* for control and control + **E** are quite similar, indicating that the E-field of 20 v/cm has no effect on cellular metabolism. The possible redirecting of transport routes by the E-field lasts for only 2 hours and there are other route options to follow. The route for DNA repair proteins toward the DNA damaged site is, according to our model, single-optioned (single navigation route).

5 Conclusions

Our results from a case study show that:

1. Application of **E** plus irradiation increases cell death by over one order of magnitude.

2. Only application of **E** has no effect on control samples.

3. Combination of these two pieces of information leads one to conclude that **E** acts only on the DNA repair process, since it is too weak to induce damage toward apoptosis by itself.

From these facts we can infer that:

a. The only role a weak **E** can play in the repair process is to prevent mobility of repair proteins toward the damage site.

b. By the same reason it is reasonable inferring that through electric dipole orientation, repair proteins sense the static electric field at the damage and use it as a navigation cue.

Finally, our case study with a prokaryote shows that our approach is, at least, qualitatively correct (in the sense that electric fields do interfere destructively in the DNA repair process). However, it would be necessary to perform similar experiments with mammalian cells in order to get quantitative estimates envisaging possible applications. These studies were already launched by our multidisciplinary research group.

6 Final Remarks

All of these findings and reasoning constitute a fair answer to the long-standing question[5] enunciated in our *Introduction*: "Why do proteins throughout the cell respond in a coordinated way, and how is this achieved, e.g. how are a few DSB within *3 billion base pairs* of DNA localized and recognized?"

Moreover, our approach points to the possible existence of another link between DNA damage, changes in chromatin structure, and downstream signaling events, which could stimulate and orientate novel studies and applications in this area. It is ironic that radiation, a valuable agent in the treatment of cancer, can itself cause cancer. It is becoming more and more evident that the primary problem in cancer is not an exploration of such isolated issues as the connection between radiation and cancer. There are, in this regard, many other and much more stimulating problems to tackle; we pick the most recent one, namely the search to find synthetic compounds to displace p53 from Mdm2. In fact, the first sign that the p53-Mdm2 interaction might be *"druggable"* appeared recently [2]. We would like to add another item to the list: the study on how to interfere destructively in the DNA repairing process by means of exogenous physical agents, as a possible new strategy to drastically reduce radiation doses in cancer therapy.

Acknowledgements

This work was supported by grants from FAPESP, CNPq and CAPES, Brazilian agencies for the promotion of science.

Bibliography

[1] Bakkenist, C.J. and Kastan, M.B. DNA damage activates ATM through intermolecular autophosphorylation and dimer dissociation. Nature 421, 499-506 (2003).

[2] Petrini, J.H. and Stracker, T.H. The cellular response to DNA double-strand breaks: defining the sensors and mediators. Trends Cell Biol. 13, 458-462 (2003).

[3] van den Bosch, M., Bree, R.T.and Lowndes, N.F. The MRN complex: coordinating and mediating the response to broken chromosomes. EMBO Rep. 4, 844-849 (2003).

[4] Bartek, J. and Lukas, J. Damage alert. Nature 421, 486-488 (2003).

[5] Banerjee, A., Yang, W., Karplus, M. and Verdine, G.L. Structure of a repair enzyme interrogating undamaged DNA elucidates recognition of damaged DNA. Nature 434, 612-618 (2005).

[6] Campbell, D.K. and Flach, S. Kivshar, Y.S. Localizing energy through nonlinearity and discreteness. Physics Today 57, 43-49 (2004).

[7] Weisbrot, D., Lin, H., Ye, L., Blank, M. and Goodman, R. Effects of mobile phone radiation on reproduction and development in Drosophila melanogaster. Journal of Cellular Biochemistry 89, 48-55 (2003).

[8] Kalfas, C.A., Sideris, E.G. and Martin, P.W. Perturbed gamma-gamma angular correlation studies of 111In bound to double and single stranded DNA. Int.J.Appl.Radiat. Isot. 35, 889-893 (1984).

[9] Kalfas, C.A., Sideris, E.G., Loukakis, G.K. and Anagnostopoulou-Konsta, A, Rotational correlation times in irradiated DNA. J. of Non-Crystalline Solids 172-174, 1121-1124 (1994).

[10] Castenholz, R.W. and Garcia-Pichel, F., Cyanobacterial responses to UV-Radiation, in: The Ecology of Cyanobacteria, pp. 591-611, Whitton, B.A. and Potts, M. (eds.), Kluwer Academic Publishers, The Netherlands (2000).

[11] Imamura, M., Sawada, S., Kasahara-Imamura, M., Harima, K. and Harada, K. Synergistic cell-killing effect of a combination of hyper-thermia and heavy ion beam irradiation: In expectation of a break through in the treatment of refractory cancers. International Journal of Molecular Medicine 9, 11-18 (2002).

[12] Rippka, R., Deruelles, J., Waterbury, J. B., Herdman, M., Stanier, R. Y. 1979. Generic assigments, strain histories and properties of pure cultures of cyanobacteria. J. Gen. Microbiol. 111, 1-61.

[13] Guillard, RRL. Division rates. In Stein, J.R. Handbook of phycological methods: culture methods and growth measurements. London: Cambridge University Press, 1973. p. 289-311.

Anais da XXVIII Reunião de Trabalho sobre Física Nuclear no Brasil
SP, Brasil, 2005
Artigo originalmente publicado no Brazilian Journal of Physics, Vol 36 – nº 4B,
Special Issue: XXVIII Workshop on Nuclear Physics in Brazil, pp.1363–1365 (2006).

Single Neutron Pick-up on ^{104}Pd

M. R. D. Rodrigues[1], J. P. A. M. de André[1], T. Borello-Lewin[1],
L. B. Horodynski-Matsushigue[1], J. L. M. Duarte[1],
C. L. Rodrigues[1] and G. M. Ukita[1,2]

[1] Instituto de Física, Universidade de São Paulo - Caixa Postal 66318, CEP 05389-970,
São Paulo, SP, Brazil

[2] Faculdade de Psicologia, Universidade de Santo Amaro, Rua Professor Enéas da
Siqueira Neto, 340, CEP 04829-300, São Paulo, SP, Brazil

Abstract. Low-lying levels of ^{103}Pd have been investigated through the (d,t) reaction on ^{104}Pd, at an incident deuteron energy of 15.0 MeV. Outgoing particles were momentum analyzed by an Enge magnetic spectrograph and detected in nuclear emulsion plates, with an energy resolution of 8 keV. Previous (d,t) work suffered from a much worse resolution than that here achieved. A partial analysis of the data obtained is reported, referring to six out of the fourteen scattering angles for which data were obtained. Angular distributions associated with eight of the thirteen levels seen up to 1.1 MeV of excitation have been compared to DWBA one-neutron pick-up predictions. Both, the attributed excitation energy values and the transferred angular momenta are in excellent agreement with the results of other kind of experiments, as tabulated by the Nuclear Data Sheets. Some peculiar structure characteristics, associated with the yrast $5/2^+$, $3/2^+$ and $7/2^+$ states found in the Ru chain could be recognized also in ^{103}Pd, pointing to the possibility of a more global understanding of this transitional mass region.

1 Introduction

The transitional A $\sim$ 100 mass region presents several intriguing aspects, which so far have defied a complete understanding. Among those, several have been put forward by investigations performed by the S. Paulo Nuclear Spectroscopy with Light Ions Group [1, 2, 3, 4, 5, 6].

The nuclear emulsion technique, employed in association with the good characteristics of the Pelletron-Enge-Spectrograph system in São Paulo, puts the Group in a very competitive position with respect to one-neutron transfer studies, as are (d,t) [2, 6] and (d,p) [3, 4] reactions, where high energy resolution and low background are mandatory for best experimental results. One-neutron transfer is very conclusive for disclosing similarities between low-lying states in neighbouring nuclei, being thus ideal for comparative structure studies along nuclear chains. An exhaustive study has been performed by the Group on the Ru isotopic chain [2, 3, 4, 5, 6], which demonstrated that some low-lying, in particular the yrast levels, contain structure information which may be tracked along the whole A ~ 100 region [5]. This bibliographic research showed, however, that the quality of the data available throughout the region is well bellow what has been achieved for Ru.

The present investigation aims at filling the gap for the Pd isotopic chain. This paper concerns partial results on the ^{104}Pd(d, t)^{103}Pd reaction which were presented at the XXVIII Workshop on Nuclear Physics in Guarujá, S. Paulo, Brazil. The previous ^{104}Pd(d, t)^{103}Pd data of Scholten et al. [7], had a typical resolution of 65 keV, much worse than what can be achieved by the techniques in use by the S. Paulo Group. There is, in addition, an older (d,t) study on ^{104}Pd [8], which has not been published, but was taken by the Nuclear Data Sheets compilation [9] as reference for excitation energy values, with an estimated uncertainty of 2 keV, instead of the 15 keV typical for the results of Scholten et al. [7]

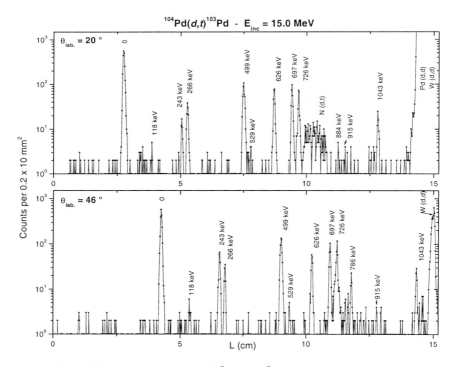

Figure 1 Triton spectra obtained at 20^0 and at 46^0 laboratory scattering angles. The excitation energies attributed to the levels populated in the reaction are indicated above each of the corresponding triton peaks.

SINGLE NEUTRON PICK-UP ON ^{104}PD 301

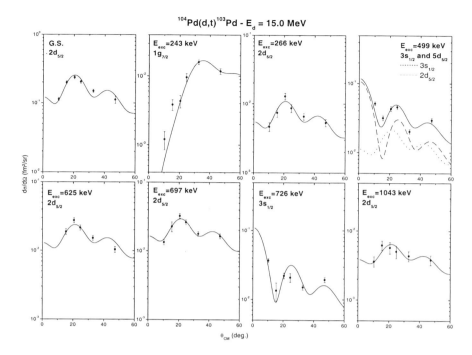

Figure 2 Experimental angular distributions of ^{104}Pd(d,t)^{103}Pd in comparison with DWBA one-neutron pick-up predictions.

2 Experimental Procedure

Spectra associated with a total of fourteen scattering angles, judiciously chosen, were measured for the ^{104}Pd(d,t)^{103}Pd reaction at an incident energy of 15.0 MeV, with the S. Paulo Pelletron-Enge-Spectrograph system. The target of ^{104}Pd, isotopically enriched to 98.98%, with a thickness of 20 μg/cm^2, was produced in São Paulo by the electron bombardment method. Only rather small amounts of the usual contaminations by C, N, O, Si, S, K e W were diagnosed.

The emerging tritons were momentum analyzed by the Enge Magnetic Spectrograph and detected in nuclear emulsion plates (Fuji 7D, 50 μm thick), which covered 50 cm along the focal plane. After processing, the exposed plates were scanned in strips of 200 μm across the plates. An energy resolution of 8 keV was achieved.

One great advantage of the nuclear emulsion detector is its insensibility to the abundant background, mostly X and γ rays from (n, γ) reactions in the spectrograph iron core after deuteron break-up. FIG.1 displays the triton spectra obtained at 20^0 and at 46^0 laboratory scattering angles.

3 Some Results and Discussion

Inspection of FIG.1, where the position, L, along the focal plane is proportional to the momentum of the emerging triton, demonstrates the very low background attained in this energy region and the excellent resolution. Some of the small peaks indicated in FIG.1 could be identified with the help of the other spectra already processed.

Only six of the fourteen spectra measured at different scattering angles have been analyzed up to now. Further work is in progress. Even so, experimental angular distributions have been obtained for eight of the thirteen levels identified up to 1.1 MeV. The respective excitation energies are presented in FIG.1, also for those peaks confirmed only by the whole set of analyzed spectra. Previously published (d,t) work on ^{104}Pd had identified, in this same energy region, only eight states [7]. That study, by Scholten et al. [7], was clearly unable to distinguish the 243 keV + 266 keV doublet, presenting only a level populated by $\ell = 4$ at 0.24 MeV. Furthermore, it did not see the 726 keV ($\ell = 0$) state and the three weakly excited levels. There is the unpublished study by Rickey et al. [8], which also reported a total of thirteen levels. However, their tentative state [8] at 905 keV is, up to now, not confirmed by the present data. There is no hint in those previous studies for the weakly excited 118 keV level. The excitation energies presented by this unpublished work are in rather good accord with those of the present research, for which an uncertainty of 1 keV is estimated.

The excitation energies attributed in the present study are moreover in excellent agreement with the precise gamma ray results tabulated in the Nuclear Data Sheet (NDS) compilation [9]. Of the levels there reported, only three have not been identified in the present nor in any of the former (d,t) studies [7, 8]. These are levels of spins $9/2^+$ and $11/2^+$, two of which have been associated, respectively, with the second members of bands built on the ground state and on the $7/2^+$ level. As for the 815(2) keV level, reported in NDS as corresponding to a $\ell = 2$ transfer in (d,t) [9], it was seen neither by Scholten et al. [7], nor in the present experiment.

The experimental angular distributions are shown in FIG.2, in comparison with DWBA predictions, calculated in the usual manner [6]. The orbitals attributed to the transferred neutron are indicated in the figure.

The transferred angular momentum (ℓ) values are very well discriminated by the angular distribution shapes and also in perfect agreement with informations gathered by previous experiments [9]. The nature of the $\ell = 0 + 2$ doublet detected at 499 keV, which corresponds to the 498.948 keV ($1/2^+$) and 504.24 keV ($3/2^+$) levels reported by NDS, could be more clearly assessed than in the former (d,t) work [7], due to the much improved resolution.

Also for ^{103}Pd, as had been stressed for the isotone ^{101}Ru in a previous publication by the S. Paulo Group [6], the yrast $\ell = 2$ ($J^{\pi} = 5/2^{+}$), and $\ell = 4$ ($J^{\pi} = 7/2^{+}$) levels carry most of the respective spectroscopic strengths. In particular, the $7/2_1^+$ state, which in ^{103}Pd is located at 243 keV, is part of an intriguing systematics of low-lying levels of similar characteristics found between 0.2 and 0.4 MeV throughout the region [5]. In ^{103}Pd this state has been identified as bandhead of a rotational structure which was seen up to $19/2^+$, at 2764.38 keV, in a gamma ray study [9].

Another aspect which is worth mentioning is that the peculiar low-lying $3/2_1^+$ level, which becomes ground-state in ^{103}Ru, maintains in this (d,t) study the characteristic of being very weakly excited, corresponding in ^{103}Pd to an excitation energy at 118 keV. In fact, the $3/2_1^+$ level is barely seen in the spectra shown in FIG.1, taken at two rather distinct scattering angles. This particular $3/2^+$ level may be followed throughout most part of the A $\sim$ 100 region, having been interpreted by the S. Paulo Group as indicative of a coexisting configuration [5].

Acknowledgements

This work was partially supported by FAPESP and CNPq.

Bibliography

[1] M. R. D. Rodrigues, C. L. Rodrigues, T. Borello-Lewin, L. B. Horodynski-Matsushigue, J. L. M. Duarte and G. M. Ukita, Brazilian Journal of Physics, vol. **34**, no. 3A, 777 (2004).

[2] M. R. D. Rodrigues, T. Borello-Lewin, L.B. Horodynski-Matsushigue, J. L. M. Duarte, C. L. Rodrigues, M. D. L. Barbosa, G. M. Ukita and G. B. da Silva, Phys. Rev. C **66**, 034314 (2002).

[3] L.B. Horodynski-Matsushigue,; C.L. Rodrigues, F.C. Sampaio and T. Borello-Lewin, Nucl. Phys. A **709** 73-84 (2002).

[4] M. D. L. Barbosa, T. Borello-Lewin, L. B. Horodynski-Matsushigue, J. L. M. Duarte, G. M. Ukita and L. C. Gomes, Phys. Rev. C **59** (1998) 2689-2702.

[5] T. Borello-Lewin, J. L. M. Duarte, L. B. Horodynski-Matsushigue, and M. D. L. Barbosa, Phys. Rev. C **57** (1998) 967-970.

[6] J. L. M. Duarte, L. B. Horodynski-Matsushigue, T. Borello-Lewin, Phys. Rev. C **50** (1994) 666-681

[7] O. Scholten et al., Nucl.Phys. A**348**, 301 (1980).

[8] F. A. Rickey, R. E. Anderson, J. R. Tesmer, Priv. Comm. (1973).

[9] D. de Frenne and E. Jacobs, Nucl. Data Sheets, **93**, 447 (2001).

Anais da XXVIII Reunião de Trabalho sobre Física Nuclear no Brasil
SP, Brasil, 2005
Artigo originalmente publicado no Brazilian Journal of Physics, Vol 36 – nº 4B,
Special Issue: XXVIII Workshop on Nuclear Physics in Brazil, pp.1366–1370 (2006).

The Nuclear Matter Effects in π^0 Photoproduction at High Energies

T. E. Rodrigues[1], J. D. T. Arruda-Neto[1,2], J. Mesa[1], C. Garcia[1], K. Shtejer[1,3], D. Dale[4] and I. Nakagawa[4].

[1] Instituto de Física da Universidade de São Paulo, P.O. Box 66318, CEP 05315-970, São Paulo, Brazil

[2] Universidade de Santo Amaro / UNISA, São Paulo, Brazil

[3] Center of Applied Studies for Nuclear Developments (CEADEN), Havana, Cuba

[4] Department of Physics and Astronomy, University of Kentucky, Lexington KY, 40506, USA

Abstract. The in-medium influence on π^0 photoproduction from spin zero nuclei is carefully studied in the GeV range using a straightforward Monte Carlo analysis. The calculation takes into account the relativistic nuclear recoil for coherent mechanisms (electromagnetic and nuclear amplitudes) plus a time dependent multi-collisional intranuclear cascade approach (MCMC) to describe the transport properties of mesons produced in the surroundings of the nucleon. A detailed analysis of the meson energy spectra for the photoproduction on ^{12}C at 5.5 GeV indicates that both the Coulomb and nuclear coherent events are associated with a small energy transfer to the nucleus ($\lesssim 5$ MeV), while the contribution of the nuclear incoherent mechanism is vanishing small within this kinematical range. The angular distributions are dominated by the Primakoff peak at extreme forward angles, with the nuclear incoherent process being the most important contribution above $\theta_{\pi^0} \gtrsim 2^0$. Such consistent Monte Carlo approach provides a suitable method to clean up nuclear backgrounds in some recent high precision experiments, such as the PrimEx experiment at the Jefferson Laboratory Facility.

Keywords. meson photoproduction in-medium effects intranuclear cascade

PACS. PACS number

1 Introduction

The total π^0 photoproduction cross section from complex nuclei can be evaluated as a sum of the electromagnetic part, known as the Primakoff effect[1], the nuclear coherent, the interference between the nuclear coherent and Primakoff amplitude and the nuclear incoherent contribution. The interference mechanism was neglected in this work since it consists of a genuine quantum process which is out of the context of the present semi-classical cascade calculation. While the electromagnetic part is almost non-sensitive to the Final State Interactions of the produced pions, the nuclear amplitudes are highly distorted by the medium and the complete nuclear ensemble should be taken into account in a realistic calculation.

In this paper we describe a sophisticated Monte Carlo method to address the dynamical evolution of an excited nuclear system. The calculation takes advantage of the Impulse Approximation (IA) to incorporate the interaction of the incoming photon with the nucleus in three different ways:

(1) a coherent electromagnetic interaction, where the pion is produced in the nuclear Coulomb field and, consequently, far away from the range of the nuclear potential which results in low Final State Interaction effects;

(2) a coherent nuclear production mechanism, where the pion is produced in the nuclear field and is susceptible to strong Final State Interactions, and;

(3) an incoherent photoproduction mechanism, where the pion is produced nearby a single nucleon, leaving the nucleus in an excited state and being also very sensitive to strong Final State Interactions.

These three scenarios apply, respectively, to the Primakoff, Nuclear Coherent and Nuclear Incoherent contributions for the total π^0 production cross section and are described in more detail below.

The main focus of this work consists in the determination of the nuclear background in high energy-forward angles π^0 photoproduction from spin zero nuclei, such as ^{12}C and ^{208}Pb. The calculation takes into account multiple scattering processes between the produced pions and the bound nucleons during the cascade stage. The Primakoff angular distribution is forward peaked because the Coulomb amplitude (γ^* exchange) goes as $1/t$, where t is the square of the four momentum transfer in the s channel (photoproduction channel) and also scales with the target charge and the square root of the $\pi^0 \rightarrow \gamma\gamma$ decay width. The measurement of this decay width, related to the $\gamma\gamma^*\pi^0$ coupling constant of the Quantum Chromodynamics (QCD) has been the subject of various experiments, like the one performed at Cornell[2], and is also a major issue in a recent experiment carried out at the Jefferson Laboratory Facility, the PrimEx Collaboration[3]. In this regard, the measurements of the π^0 angular distributions at forward angles, one of the PrimEx Collaboration's priorities, provide an indirect method to determine such an important and fundamental quantity.

2 The Intranuclear Cascade Monte Carlo Method

The calculation outlined on this paper follows some previous work based on the Monte Carlo Multi-Collisional Intranuclear Cascade Model MCMC. Some improvements were incorporated to accommodate the Final State Interactions of the produced pions also for the nuclear coherent mechanisms, as well as the relativistic nuclear recoil for both the nuclear coherent and the electromagnetic (Primakoff) processes. The MCMC model was first developed to study high energy photonuclear reactions [4, 5, 6, 7], being more recently successfully extended to intermediate energies[8, 9]. A special and very sophisticated version of the MCMC code were also proposed in a recent paper[10] in order to study the incoherent π^0 photoproduction both at the $\Delta(1232)$ resonance energy region, as well as in the $4.0 - 6.0$ GeV range and forward angles[10]. Such work showed the accuracy of the cascade model in predicting recent and intriguing results of the π^0 differential cross section for ^{12}C in the Delta region[11]. For details, see ref.[10] and references therein.

The MCMC model is essentially a typical transport calculation with some advantages related with the continuous time dependent structure available in the Monte Carlo analysis. Such an approach does not require a pre-determined time span between two successive collisions, like the well known BUU (Boltzmann-Uehling-Uhlenbeck) transport model [12, 13]. This special feature of the MCMC is related with the constant updating of the relativistic space-time, energy-momentum particle's four vectors along time.

3 The π^0 Photoproduction Mechanisms

3.1 The electromagnetic π^0 photoproduction mechanism

The differential cross section of the electromagnetic photoproduction of a pseudo-scalar meson P from a spin-zero nuclei may be written as[14]:

$$\frac{d\sigma_P}{d\Omega} = \Gamma_{P \to \gamma\gamma} \frac{8\pi\alpha Z^2}{m_P^3} \frac{\beta^3 k^4}{Q^4} |F_{el}(Q)|^2 \operatorname{sen}^2 \theta_P, \tag{1}$$

where $\Gamma_{P \to \gamma\gamma}, m_P, \beta$ and θ_P are, respectively, the radiative decay width, mass, velocity and polar angle of the pseudo-scalar meson P. Z is the charge of the nucleus, $F_{el}(Q)$ the nuclear electromagnetic form factor, k the incident photon energy and Q the four momentum transfer to the nucleus, with $\alpha = \frac{1}{137}$.

Eq.(1) indicates that the Coulomb production is forward peaked with its maximum at $\theta_{peak} \sim \frac{m_P^2}{2k^2}$. Such kinematical characteristic makes is possible to disentangle different production mechanisms in nuclei, shedding light to determine $\Gamma_{P \to \gamma\gamma}$ via angular distribution measurements.

Since the main subject on this paper is the nuclear contribution of the total π^0 photoproduction, we have neglected the FSI effects in the Coulomb contribution wherever it appears. The relativistic nuclear recoil was taken into account applying the relativistic energy and momentum conservation to the photon — nucleus system, with the assumption that the whole nucleus has absorbed the momentum transfer Q, as it does in a coherent mechanism. With this approach, the complete kinematics of the interaction were determined by the meson polar angle, which was sampled in the Monte Carlo routine in accordance with eq.(1).

3.2 The nuclear coherent π^0 photoproduction mechanism

Following the steps elegantly delineated in ref.[14], we may write the nuclear coherent cross section as:

$$\frac{d\sigma_{NC}}{d\Omega} = L^2 A^2 |F_N(Q)|^2 \operatorname{sen}^2 \theta_P, \tag{2}$$

where L is an angular-independent constant, A the nuclear mass and $F_N(Q)$ the strong nuclear form factor. The constant L can be interpreted as a free parameter and was taken as $L \approx 12.2k$ following the parameterization achieved for the Cornell data[2]. The nuclear coherent cross section peaks at approximately $\theta_{peak} \sim \frac{2}{kR}$, where R is the nuclear radius.

After being produced in the nuclear field, the pions are highly susceptible to FSI effects and may re-scatter in the nuclear matter, bringing the nucleus to an excited state. This process significantly changes the pion energies and angular distributions, attenuating eq.(2). Such a complex mechanism is out of the range of the conventional scattering theories, since various diagrams are relevant with the additional difficulty of dealing with many coupled channels. For this reason, the Monte Carlo method described here provides a powerful technique to address the correlation between coherent and incoherent processes via FSI.

The distortion effects on eq.(2) depend significantly on the nuclear volume and pion production energy. Besides it, nucleon-nucleon correlations due to the Pauli principle and the momentum distribution of the nucleons at the nuclear ground state play a key role for pion-nucleon scatterings at low polar angles (low momentum transfer).

3.3 The nuclear incoherent π^0 photoproduction mechanism

The total incoherent production amplitude is characterized by a sum of single nucleon amplitudes. The photon probes a high density nuclear environment and is supposed to interact with a single nucleon, bringing the nucleus to an excited state. The fundamental property of such mechanism is that no interference occurs between the accessible states and the nuclear cross section simply scales with the nuclear mass. So, the total incoherent cross section may be written as:

$$\frac{d\sigma_{NI}}{d\Omega} = A\xi(k,\theta)f(Q)\frac{d\sigma_N}{d\Omega}(k,\theta)\left[1+\lambda(k,\theta)\right], \tag{3}$$

where ξ is the π^0 nuclear absorption, f is the Pauli-blocking suppression factor for small Q, $\frac{d\sigma_N}{d\Omega}$ is the photoproduction from the nucleon and λ is a factor accounting for the π^0 secondary production and re-scattering. The elementary photoproduction cross section $\frac{d\sigma_N}{d\Omega}$ can be written in terms of the helicity amplitudes, which are calculated in the context of the Regge theory under the assumption that the production process is largely dominated by vector meson (ρ and ω) exchange for low t. Such elementary operator was calculated in ref.[10], reproducing quite succesfully the proton data at few GeV[15, 16]. For details on the calculations of the helicity amplitudes see ref.[10] and references therein.

The nucleon-nucleon correlations and the momentum distribution of the ground state determine the Pauli-blocking factor f. Indeed, this suppression function significantly changes eq.(3) for low momentum transfer and is the crucial step in reproducing the structures on the π^0 differential cross section for ^{12}C near the Delta resonance[10, 11].

4 The transport of pions through the nuclear matter

To access the Final State Interactions between the produced pions on their way out of the nucleus it is important to take into account the most relevant scattering processes for a given πN channel. This analysis was already performed in a previous work[10], where we have included the following πN channels for $E_\pi \leqslant 6.0$ GeV: i) elastic scattering; ii) resonance production and decay; iii) charge exchange processes; iv) pion re-absorption by a nucleon pair, and v) multiple pion production processes (up to seven particles in the final state). Such multiple scattering scenario dictates how the initial π^0 flow is distributed among the available channels. In order to illustrate the rigorous Monte Carlo method to describe multiple pion-nucleon scatterings, we show in figure 1 the time derivative of the average number of pion-nucleon scatterings $\langle n_{int} \rangle$ with and without the Pauli-blocking for a typical cascade at 6.0 GeV for ^{12}C and ^{208}Pb. The dependence of $\frac{d\langle n_{int}\rangle}{dt}$ on the nuclear volume is evident, showing that secondary scatterings are not relevant in light nuclei. The result for ^{12}C exhibits a peak at the earliest cascade stages falling rapidly to zero around 5 fm/c. This gives a nice estimate ($\sim$ 5 fm) of the nuclear dimensions for a pion traveling at the speed of light. For the ^{208}Pb, however, we obtained a much broader distribution with a maximum shifted towards $\sim$ 5 fm/c, indicating that secondary scatterings with multiple pion productions are likely to occur for heavier systems.

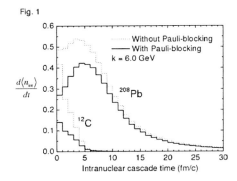

Figure 1 Time derivative of the average number of pion-nucleon interactions during the cascade stage for ^{12}C and ^{208}Pb at 6 GeV. The dotted/solid histogram represents all the collisions without/with the Pauli exclusion principle.

5 Results

5.1 π^0 angular distributions for ^{12}C at the GeV range and forward angles

With the prescriptions outlined above, we can calculate the angular distributions of neutral pions in the GeV range for the Primakoff, Nuclear Coherent and Nuclear Incoherent mechanisms. The results for ^{12}C at k = 5.5 GeV in the laboratory frame are shown in figures 2, 3 and 4. Kinematical cuts on the total π^0 energy were applied in order to estimate the nuclear background at different kinematical domains. The quantity ΔE represents the energy difference between the incoming photon and the outgoing pion: $\Delta E = k - E_{\pi^0}$. While for the coherent mechanisms (electromagnetic/nuclear) most of the pions have the total energy very close to the incident photon energy ($\Delta E \lesssim 3$ MeV), for the incoherent process (figure 4) there is a significant lowering on the pion energies caused by the requirement of a higher momentum transfer to excite the nucleus.

The Primakoff cross section (figure 2) exhibits a peak around $\frac{m_\pi^2}{2k^2} \sim 0.017$ degrees and falls rapidly to zero because of the $\frac{1}{Q^4}$ dependence. It is interesting to note that almost all the pions have total energy between k and k − 1 MeV. This result reflects the forward peak constraint imposed by eq.(1).

The nuclear coherent cross section (figure 3) peaks at approximately $\frac{2}{kR} \sim 1.49$ degrees, falling to zero for $\theta_{\pi^0} \gtrsim 3.5$ degrees. Most of the pions produced via the nuclear coherent mechanism have total energy between k and k − 3 MeV, in accordance with the kinematical characteristic inherent to eq.(2). The solid line of figure 3 represents the Plane Wave Impulse Approximation (PWIA), while the dashed line also includes the FSI effects of the produced pions.

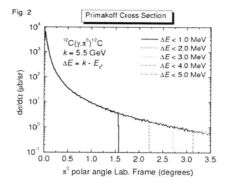

Figure 2 Primakoff π^0 photoproduction cross section from ^{12}C at 5.5 GeV in the PWIA. The kinematical cuts are as follows: $\Delta E < 1$ MeV (solid), $\Delta E < 2$ MeV (dashed), $\Delta E < 3$ MeV (dotted), $\Delta E < 4$ MeV (dashed-dotted) and $\Delta E < 5$ MeV (dashed-dotted-dotted).

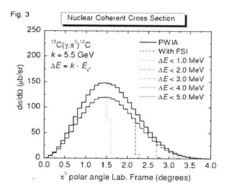

Figure 3 Nuclear coherent π^0 photoproduction cross section from ^{12}C at 5.5 GeV. The solid histogram is the PWIA and the dashed histogram includes the FSI effects of the produced pions. The kinematical cuts are as follows: $\Delta E < 1$ MeV (dotted), $\Delta E < 2$ MeV (dashed-dotted), $\Delta E < 3$ MeV (dashed-dotted-dotted), $\Delta E < 4$ MeV (short-dashed) and $\Delta E < 5$ MeV (short-dashed-dotted).

The nuclear incoherent cross section is shown in figure 4. The PWIA is the solid line and the dashed line represents the calculations including the Pauli exclusion principle. The full calculation is the dotted line and takes into account the FSI effects. The structures of the cross section for low polar angles are attributed to the kinematics of the photoproduction channel, which takes into account the Pauli principle and a realistic momentum distribution of the nucleons at the ground state based on the shell model. For details on the momentum distributions see ref.[10].

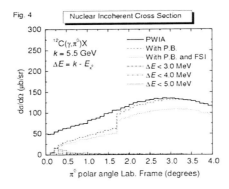

Figure 4 Nuclear incoherent π^0 photoproduction cross section from ^{12}C at 5.5 GeV. The solid histogram is the PWIA, the dashed line takes into account the Pauli exclusion principle and the dotted line represents the full calculation with FSI. The kinematical cuts are as follows: $\Delta E < 3$ MeV (dashed-dotted), $\Delta E < 4$ MeV (dashed-dotted-dotted) and $\Delta E < 5$ MeV (short-dashed).

The kinematical cuts on the incoherent cross section are very effective, showing that only few events can survive after a typical cut $\Delta E < 5$ MeV.

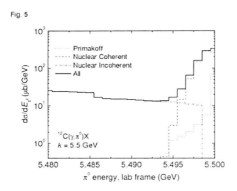

Figure 5 π^0 energy spectra for ^{12}C at 5.5 GeV and its Primakoff (dotted), nuclear coherent (dashed) and nuclear incoherent (dashed-dotted-dotted) contributions.

5.2 π^0 energy spectra for ^{12}C at the GeV range and forward angles

The neutral pion energy spectra can be combined with the angular distributions to disentangle the different production processes in nuclei. The structure of the MCMC model permits a continuous evaluation of particle's energies along time, propitiating an accurate method to access information of an excited nuclear system.

Figure 5 shows the normalized π^0 energy spectra for ^{12}C at k = 5.5 GeV for the three different production channels. The Primakoff contribution (dotted line) exhibits a remarkable peak at $\Delta E < 1$ MeV, as suggested by the angular distribution results. The nuclear coherent processes (dashed line) is also concentrated at the higher energy region, typically $\Delta E < 5$ MeV, and dominates the photoproduction mechanism due to the small fraction of Primakoff events on the total (integrated) cross section. Both of the coherent processes are not relevant above $\Delta E \sim 6$ MeV, since the respective form factors vanish at this kinematical domain ("high"momentum transfer region). The nuclear incoherent π^0 energy spectrum (dashed-dotted-dotted line) is spread almost uniformly for lower pion energies. This contribution vanishes below the Primakoff peak as it demands a much higher momentum transfer in order to excite the nucleus. It is worth noting, however, that both the angular distributions and the pion spectra do not uniquely determine the type of photoproduction mechanism as different processes can take place at the same angular/energy interval. In fact, the nuclear incoherent contribution for high energy pions ($\Delta E \lesssim 5$ MeV) is substantially attenuated (figure 4) and the investigation of the pion spectra at these energies may serve as an additional constraint to clean up the incoherent contribution.

6 Conclusions

A Monte Carlo method has been applied to study both coherent (Primakoff/nuclear) and incoherent π^0 photoproduction mechanisms at the GeV range and forward angles from spin zero nuclei. The Coulomb process was analyzed taking into account the exact relativistic recoil of the nucleus and neglecting FSI effects of the produced pions (PWIA). The other processes, namely, the nuclear coherent and incoherent contributions were analyzed in the context of the MCMC intranuclear cascade model including full FSI effects.

The differential cross sections for ^{12}C at 5.5 GeV are largely dominated by the Primakoff peak up to $\theta_{\pi^0} \lesssim 0.5^0$, with the nuclear coherent mechanism being the most relevant contribution in the range $0.5^0 \lesssim \theta_{\pi^0} \lesssim 2.0^0$. The nuclear incoherent part is Pauli-suppressed at lower angles and becomes the largest contribution for higher angles $\theta_{\pi^0} \gtrsim 2.0^0$. The energy spectra of the pions indicate that both the Coulomb and nuclear coherent processes are concentrated within a narrow energy range ($\Delta E \lesssim 5$ MeV) in contrast to the nuclear incoherent part, which is greatly inhibited in this energy gap due to a much larger momentum transfer to the nucleus.

In conclusion, a careful kinematical analysis using the MCMC model was performed to establish an efficient method of background subtraction for future reference in high energy π^0 photoproduction experiments, such as the PrimEx Collaboration at the Jefferson Laboratory.

The authors thank the Brazilian agencies FAPESP and CNPq, and the Latin-American Physics Center (CLAF) for partial support to this work.

Bibliography

[1] H. Primakoff, Phys. Rev. 81, 899 (1951).

[2] A. Browman, J. De Wire, B. Gittelman, K. M. Hanson, D. Larson, E. Loh and R. Lewis, Phys. Rev. Lett. 33, 1400 (1974).

[3] http://www.jlab.org/primex/primex_notes/PR99-014.ps (PrimEx Proposal, accessed in Aug-2006)

[4] M. G. Gonçalves, S. de Pina, D. A. Lima, W. Milomen, E. L. Medeiros and S. B. Duarte, Phys. Lett. B 406, 1 (1997).

[5] M. G. Gonçalves, E. L. Medeiros and S. B. Duarte, Phys. Rev. C 55, 2625 (1997).

[6] S. de Pina, E. C. Oliveira, E. L. Medeiros, S. B. Duarte and M. Gonçalves, Phys. Lett. B 434, 1 (1998).

[7] S. B. Duarte (private communication).

[8] T. E. Rodrigues, J. D. T. Arruda-Neto, A. Deppman, V. P. Likhachev, J. Mesa, C. Garcia, K. Shtejer, G. Silva, S. B. Duarte, O. A. P. Tavares, Phys. Rev. C 69, 064611 (2004).

[9] T. E. Rodrigues, PhD thesis, University of São Paulo (2005).

[10] T. E. Rodrigues, J. D. T. Arruda-Neto, J. Mesa, C. Garcia, K. Shtejer, D. Dale and I. Nakagawa, Phys. Rev C **71**, 051603(R) (2005).

[11] B. Krusche et al., Eur. Phys. J. A **22**, 277 (2004).

[12] M. Effenberger, E. L. Bratkovskaya and U. Mosel, Phys. Rev. C60, 044614 (1999).

[13] J. Lehr, M. Effenberger, U. Mosel, Nucl. Phys. A671, 503 (2000).

[14] C. A. Engelbrecht, Phys. Rev. 133, B988 (1964).

[15] M. Braunschweig, W. Braunschweig, D. Husmann, K. Lübelsmeyer and D. Schmitz, Nucl. Phys. B 20, 191 (1970).

[16] R. L. Anderson, D. B. Gustavson, J. R. Johnson, I. D. Overman, D. M. Ritson, B. H. Wiik and D. Worcester, Phys. Rev. D 4, 1937 (1971).

Study of Different Configurations for a Subcritical Ensemble Using MCNP-4C Code

A. O. Casanova[1], N. López-Pino[1,2], M. V. Manso Guevara[1],
K D'Alessandro[1], E. Herrera Peraza[1], A. Gelen[1], O. Díaz[1] and
M. N. Martins[2]

[1] Instituto Superior de Tecnología y Ciencias Aplicadas (InSTEC)
Ave. Salvador Allende y Luaces. Quinta de los Molinos. Habana 10600. A.P. 6163,
InSTEC,
La Habana, Cuba. Fax: (537) 204 1188

[2] Laboratório do Acelerador Linear, Instituto de Física da USP
Rua do Matão, Travessa R, 187, 05508-900, SP, Brasil
e-mailneivy@if.usp.br, *neivy@fctn.isctn.edu.cu*

Abstract. Different configurations for a subcritical ensemble (CS-InSTEC) located at the Nuclear Physics Department of InSTEC (Havana, Cuba) are studied, in order to get the maximum flux for thermal and epithermal neutrons. MCNP-4C Monte Carlo code is used to simulate several configurations of the multiplier ensemble, which is composed of natural uranium rods, light water as moderator, and several radioactive neutron sources (Pu-Be and Am-B). Different configurations are achieved combining the three hexagonal geometries (grid step with 4.5, 5.0 or 6.0 cm) with different materials around the irradiation position, which is located at the geometrical center of CS-InSTEC. The neutron fluxes (thermal, epithermal and fast) were sampled in irradiation position using the MCNP tallies 4 and 5. The thermal to epithermal ratio (f) and the shape parameter (α) of the epithermal spectrum for each configuration are obtained from the Monte Carlo simulations. Finally the results obtained for different configurations were compared in order to choose the maximum flux configuration.

Keywords. Monte Carlo, Neutron, Flux, Subcritical, MCNP

---

1 Introduction

During many years the Subcritical ensemble (CS-InSTEC), placed at Nuclear Physics Department of the Instituto Superior de Tecnología y Ciencias Aplicadas (InSTEC) in Havana, has been used basically for teaching purposes, due to its low neutron density flux. Combining fuel rods and several neutron sources, the

CS-InSTEC neutron fluxes can be enhanced [1, 2], allowing its use as irradiator of big samples in Instrumental Neutron Activation Analysis (INAA) [3]. In this sense, Monte Carlo simulations are carried out analyzing different combinations of ensemble components. Several configurations of fuel rods, moderator and neutron sources, were studied to achieve the maximum neutron fluxes.

Usually to apply INAA, the neutron spectrum is divided in three regions named: Thermal (0 – 0.35 eV), Epithermal (0.35 eV – 0.5 MeV) and Fast neutrons (> 0.5 MeV). The thermal - epithermal neutron spectrum is described by two parameters: the thermal to epithermal ratio $f = \phi_{Thermal}/\phi_{Epitermal}$ [4, 5] and the shape parameter (α) of the epithermal spectrum [4 - 6]. α-parameter describes the epithermal flux as function of neutron energy according to $\phi_{Epithermal} \propto 1/E^{1+\alpha}$ behavior, and it is a measure of epithermal flux deviation from the ideal case ($\alpha = 0$) [4-7]. Experimentally, thermal and epithermal components are separated using a Cadmium cover of 1 mm thickness.

The neutron fluxes according to above criteria were sampling in irradiation position for each configuration simulated by MCNP-4C code [8], and then was determined the parameters f and α. Finally the results obtained for different configurations were compared to choose the maximum flux configuration.

2 InSTEC Subcritical Ensemble (CS-InSTEC)

The CS-InSTEC consists of an aluminum tank of $\emptyset$ 120 cm $\times$ 125 cm, in which are placed the moderator (light water) and the fuel rods. The tank is located on the top side of graphite prism (120 $\times$ 120 $\times$ 60 cm). Each fuel rod is composed

Table 1 Neutron Sources Specifications

Neutron Sources	Source Type	Diameter(mm)	Height (mm)	Neutron Yield (ns^{-1})
A	Pu-Be	38.0 ± 0.2	40.0 ± 1.5	(5.1 ± 2.0) × 10^6
B	Am-B	20.0 ± 0.1	60.0 ± 1.5	(2.9 ± 1.1) × 10^6
C	Pu-Be	18.0 ± 0.8	22.0 ± 0.4	(1.0 ± 0.4) × 10^7
D	Am-B	38.0 ± 0.2	40.0 ± 1.5	(5.0 ± 2.0) × 10^6
E	Pu-Be	50.0 ± 1.2	52.0 ± 1.4	(1.4 ± 0.2) × 10^7
F	Pu-Be	45.0 ± 1.2	50.0 ± 1.4	(2.6 ± 1.0) × 10^7
G	Pu-Be	22.5 ± 1.2	48.7 ± 1.2	(8.6 ± 3.5) × 10^6
H	Pu-Be	17.8 ± 1.2	22.0 ± 1.2	(1.6 ± 0.7) × 10^7
I	Pu-Be	28.8 ± 1.2	35.0 ± 1.2	(7.8 ± 3.1) × 10^6
J	Pu-Be	14.0 ± 0.1	37.0 ± 1.2	(8.5 ± 3.4) × 10^7
TOTAL				(1.80 ± 0.68) × 10^8

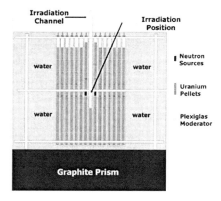

Figure 1 Schematics of InSTEC Subcritical Ensemble. Longitudinal Section

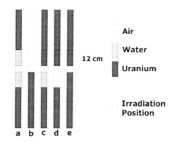

Figure 2 Central Channel Configurations

by eight pellets of U_2O_3 and 1.7 kg of weight. Uranium isotope fractions are equal to natural uranium. The fuel rods can be put in three hexagonal geometries, with grid step equal to 4.5, 5.0 or 6.0 cm. The net is composed by 164 fuel rods, and the central hole is selected as irradiation channel. Around irradiation position (IP) are placed the neutron sources (see Table 1), as is shown in Figure 1.

Five configurations for central channel, varying the materials around the irradiation position, were simulated: (a) Water – Water, (b) Uranium – Air, (c) Water – Uranium, (d) Air – Uranium, (e) Uranium – Uranium (see Figure 2). Combining these central channel configurations and the three Hexagonal geometries, fifteen combinations were analyzed.

3 MCNP Simulations

The CS-InSTEC configurations were simulated using the MCNP-4C Monte Carlo code [8]. Details of Subcritical Ensemble were reproduced, using the "repeated structure geometry" feature of the MCNP code. Aluminum covers of the uranium rods, uranium isotope fractions, moderators (water and graphite) and their thermal treatment, as well as the neutron sources were spelled out in the MCNP input file. Neutron sources inside the lattice were specified as volumetric sources, including the Pu-Be or Am-B mixture as material cell. Neutron spectra for both types of sources were specified according to reported in the IAEA Technical Reports Series No. 318 [9].

Variance reduction techniques in Monte Carlo calculations reduce the computer time required to obtain results of sufficient precision [8]. In order to optimize the simulations, one of these techniques was used: the particles importance was increased in the direction of flux decrease.

Neutron fluxes were estimated using the tallies F4 (Track length estimate of cell flux) and F5 (Flux at a point or ring detector). For F5 tally a "sphere detector" of radius equal to 1.0 cm was placed at the irradiation position. For each configuration the runtime was approximately 24 hours in a PC with 256 MB of RAM, and a Microprocessor Celeron (R) 1.7 GHz, which assured relative errors less than 10%.

4 Results and Discussion

Neutron fluxes were estimated along the central channel, to locate the maximum (see Figure 3) and place the Irradiation Position. After that, a carefully estimation of energies neutron spectra was carried out for each combination. Table 2 and 3 show the Thermal Fluxes for each configuration using the tallies F4 and F5, respectively. There are little differences between both estimation (F4 or F5), but the trend are very similar, showing a maximum thermal flux for the hexagonal geometry with grid step equal to 6.0 cm, and the water-water configuration at central channel. However, the differences between the combinations are not significant. The fluxes range from 1.6×10^5 to 5.2×10^5 n cm^{-2} s^{-1}.

Table 2 Thermal Flux (neutrons cm^{-2} s^{-1}), Tally F4, (Relative errors less 1.0%)

Central Channel Configuration	Grid Step		
	4.5 cm	5.0 cm	6.0 cm
Water-Water	3.97×10^5	4.86×10^5	5.13×10^5
Uranium-Air	2.33×10^5	3.10×10^5	3.58×10^5
Water-Uranium	2.48×10^5	3.07×10^5	3.35×10^5
Air-Uranium	2.22×10^5	2.84×10^5	3.18×10^5
Uranium-Uranium	1.73×10^5	2.21×10^5	2.43×10^5

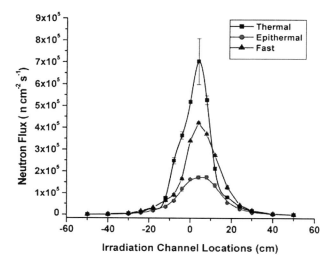

Figure 3 Neutron Fluxes behavior in central channel (Water-Uranium Configuration, Grid Step 6.0 cm, Tally 4)

Table 3 Thermal Flux (neutrons cm^{-2} s^{-1}), Tally F5, (Relative errors less 1.0%)

Central Channel Configuration	Grid Step		
	4.5 cm	5.0 cm	6.0 cm
Water-Water	4.14×10^5	4.95×10^5	5.22×10^5
Uranium-Air	2.28×10^5	3.05×10^5	3.51×10^5
Water-Uranium	2.41×10^5	2.96×10^5	3.20×10^5
Air-Uranium	2.16×10^5	2.77×10^5	3.11×10^5
Uranium-Uranium	1.62×10^5	2.07×10^5	2.28×10^5

Table 4 and 5 show the thermal to epithermal ratio (f) values estimated, using Tally F4 and F5, respectively. There are not great variations on f values. The greater values were obtained for the hexagonal geometry with grid step equal to 6.0 cm, and the Water-Water configuration at central channel. It is due to the increment of moderator material (water) between fuel rods, neutrons sources and irradiation position. However, the f values are substantially less than reported for critical reactors [4, 10-12], because the main contribution to flux at irradiation position is due to neutrons from isotope sources, which are more energetic than fission neutrons.

Table 4 Thermal to epithermal ratio (f), Tally F4, (Relative errors less 1.3 %)

Central Channel Configuration	Grid Step		
	4.5 cm	5.0 cm	6.0 cm
Water-Water	1.44	2.12	3.12
Uranium-Air	0.88	1.38	2.16
Water-Uranium	0.91	1.33	1.97
Air-Uranium	0.87	1.31	1.99
Uranium-Uranium	0.65	0.97	1.43

Table 5 Thermal to epithermal ratio (f), Tally F5, (Relative errors less 1.7 %)

Central Channel Configuration	Grid Step		
	4.5 cm	5.0 cm	6.0 cm
Water-Water	1.49	2.18	3.14
Uranium-Air	0.87	1.37	2.14
Water-Uranium	0.89	1.30	1.91
Air-Uranium	0.87	1.31	2.02
Uranium-Uranium	0.63	0.94	1.39

The Shape parameter (α) of the epithermal spectrum is determined from the slope of Epithermal Flux vs. Energy log-log plot, like displayed in Figure 4. Table 6 and 7 report α-values estimated for Tally F4 and F5 respectively. The behavior of α-values are similar for both tallies, however some differences between Tally F4 and F5 are found. Two reasons could explain its: Firstly the estimation of neutron spectrum near or inside moderator are not recommendable using tally F5 [8]; and secondly, the sphere corresponding to tally F5 is not exactly equal to the cell where was applied the F4 tally. These reasons affect the shape of neutron spectrum, but don't change the total amount of thermal or epithermal neutrons considerably.

$\alpha-$parameter are between $-0.019 < \alpha < -0.14$, showing a hard spectra of epithermal flux, as compared with the ideal one. Both positive and negative α-values have been reported in the literature according to $\alpha < ?0.2?$ [4, 10-13]. In our case, the hardest epithermal spectrum is obtained for the hexagonal geometry with grid step equal to 4.5 cm, and the Uranium-Uranium central channel configuration. On the other hand, the softest spectrum is found for 6.0 cm and Water-Water configuration.

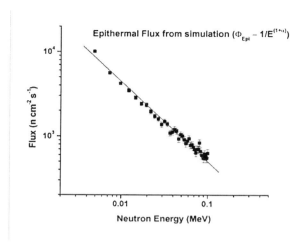

Figure 4 Epithermal Flux from MCNP simulation (Water-Uranium Configuration, Grid Step 6.0 cm, Tally F4)

Table 6 Shape parameter (α) of the epithermal spectrum, Tally F4, (Relative errors less 1.7 %)

Central Channel Configuration	Grid Step		
	4.5 cm	5.0 cm	6.0 cm
Water-Water	-0.068	-0.052	-0.021
Uranium-Air	-0.084	-0.066	-0.067
Water-Uranium	-0.075	-0.064	-0.054
Air-Uranium	-0.097	-0.092	-0.043
Uranium-Uranium	-0.119	-0.111	-0.115

Table 7 Shape parameter (α) of the epithermal spectrum, Tally F5, (Relative errors less 1.2 %)

Central Channel Configuration	Grid Step		
	4.5 cm	5.0 cm	6.0 cm
Water-Water	-0.034	-0.034	-0.019
Uranium-Air	-0.088	-0.082	-0.085
Water-Uranium	-0.078	-0.078	-0.064
Air-Uranium	-0.098	-0.087	-0.068
Uranium-Uranium	-0.138	-0.134	-0.130

5 Conclusions

Fifteen combinations of Hexagonal Geometries, Central Channel Configurations and 10 Radioisotope Neutron sources for a Subritical Ensemble (CS-InSTEC) were simulated, by means of MCNP code. The neutron spectra at Irradiation Position showed a hard epithermal behavior for all combinations. Maximum thermal flux obtained was around 5×10^5 n cm^{-2} s^{-1}, suitable to use for INAA in big samples and major elements.

Acknowledgements

Authors would like to thanks the financial support of Centro Latinoamericano de Física (CLAF).

Bibliography

[1] E. F Herrera Peraza, O. Díaz Rizo, M. E. Montero, et all. J. Radioanalytical Nuclear Chemistry Vol. 240, No.2, 437-443, (1999)

[2] O. Díaz Rizo, E. F. Herrera Peraza, M. E. Montero, et all. J. Radioanalytical. Nuclear Chemistry Vol. 240, No.2, 445-450, (1999)

[3] R. M. W. Overwater, The Physics of big sample Instrumental Neutron Activation Analysis, Delf University Press, ISBN 90-407-1048-1 (1994)

[4] F. De Corte, "The ko Standardization Method", PhD. Thesis, Gent University, Belgium (1987)

[5] F. De Corte, A. Simonits, Vade Mecum for k$_o$-users, DSM Research, Gelen (NL) R29/11492, December, (1994)

[6] T. B Ryves, E. B. Paul, J. Nucl. Energ. 22/12, 759-775 (1968)

[7] S. Jovanovic, P. Vukotic, B. Smodi, R. Jasimovic, N. Mihaljevic, P. Stegnar, J. Radioanalytical. Nuclear Chemistry, Vol. 129, No. 2, 343 (1989)

[8] J. F. Briesmeister, (editor). MCNP 4C General Monte Carlo N-Particle Transport Code, Version 4A. Report LA-12625-M, Los Alamos National Laboratory (2000)

[9] R. V. Griffith, J. Palfalvi, U. Madhvanath. Compendium of neutron spectra and detector responses for radiation protection purposes. Technical Reports Series No. 318, IAEA, Vienna, (1990)

[10] O.Diaz, A. M. G Figueiredo, C. A. Nogueira, N. Lopez, H. González, M. V. Manso Guevara, M. Saiki, M. B .A. Vasconcellos, J. Radioanalytical and Nuclear Chemistry. Vol. 266, No. 1 (2005)

[11] O. Díaz, E. Herrera, M. C. López, I. Alvarez, M. V. Manso, M. Ixquiac. J. Radioanalitycal Nuclear Chemical Letter Vol. 220, No. 1 (1997)

[12] O. Díaz, M. V. Manso, E. Herrera. J. Radioanalytical Nuclear Chemistry Letter Vol. 221, No. 2 (1997)

[13] F. De Corte, S. Jovanovic, A. Simonit, L. Moens, J. Hoste, Kernenergie-Kerntechnik, Suppl. Vol. 44, 385 (1984)

Anais da XXVIII Reunião de Trabalho sobre Física Nuclear no Brasil
SP, Brasil, 2005

Analysis of the Internal Structure in Bone From 3D Microtomography Using Microfocus X-Ray Source

Inaya C. B. Lima[1,], Ricardo T. Lopes[1] and Mônica Rocha[2]*

[1] Laboratório de Instrumentação Nuclear - COPPE/UFRJ
C.P. 68509, Ilha do Fundão
21941-972 Rio de Janeiro, RJ, Brazil

[2] Instituto de Biofísica, Departamento de Farmacologia - CCS, UFRJ/RJ, Brazil

Abstract. Tomographic evaluations of biological specimens, small industrial parts and small animals can be accomplished with X-rays. The 3D micro computed tomography (3D μCT) is a non-invasive technique that can produce a map of the internal structure of the inspected sample, in microns order of the space resolution. The goal of this study is to characterize the internal structure of bone samples and investigate the influence of ethanol on those samples using 3D μCT. The results show that this technique is capable of analyzing these kinds of structure especially on rat bone, once structures in rats (trabecular diameter) are thinner than in humans' bones.

1 Introduction

The development of computed tomography technologies is one of the most important advances in the field of radiology [1]. The 3D micro computed tomography (3D μCT) is a kind of non-invasive imaging technique that provides x-ray attenuation maps of the specimen up to a few millimeters in size, with a resolution typically in order of several microns (from 1 to 100 microns) [2].

The 3D μCT is becoming an important probe and diagnostic tool for investigations in medicine and biology [3,4]. An example of this kind of study, through this procedure, is the analysis of the bone structure, which allows the characterization of the effects of diseases processes on trabecular bone architecture in human bone and a diversity of animal models. In biological studies, the principal advantage of the technique is its ability to quantify the mineral content of hard tissue in three dimensions at a microscopic level. One important application is the research of the

---

* inaya@lin.ufrj.br

relationship between the internal structure (trabecular part) of the bone and the intake of ethanol. Excessive alcohol intake is well-recognized cause of secondary osteoporosis [5], which is a systemic disease characterized by low bone mass and microarchitectural deterioration of bone tissue, with a consequent increase in bone fragility and susceptibility to fracture [6]. A best knowledge of the quality of the bone structure can help a best prevention of the fracture risk. The concept of how ethanol affects bone metabolism is not still fully elucidated.

Osteoporosis occurs due to an imbalance between bone synthesis and resorption. Ethanol affects profoundly mineral metabolism [7,8] and it seems that it also inhibits osteoblastic function [9,10], thus leading to impaired bone synthesis.

The aim of this research is through 3D µCT to investigate the effects of the ethanol treatment on bone samples to provide insights into the contributions of bones three-dimensional morphology to bone quality.

2 Background Informations

2.1 Tomography Technique

The 3D µCT is a miniature version of medical computed tomography (CT) and has a great advantage to the space resolution that is in the microns scale, giving a better quality of image. This technique is used to reconstruct an image of the interior of an object after collecting projection data from different directions at its exterior. Applications of x-ray procedure generally assume that attenuation is governed by equation 1, where I_o corresponds to the incident intensity, I to the attenuated intensity, d to the thickness of the sample and α is the sample of representative linear attenuation coefficient, which depends on the electron density of the material, the energy of the radiation, and the bulk density of the sampling material.

$$I = I_o \exp(-\alpha x) \tag{1}$$

To carry out a tomogram, a series of radiographs is collected at different viewing angles by rotating the specimen, obtaining the projection data. These set of radiographs are processed with a reconstruction algorithm and a three dimensional image of the entire object is acquired, when these data are included in a 3D visualization software. A hundred of radiographic images and each slice represent the cross-section of the sample at a different vertical position, making a complete 3D tomographic data set. The data acquired for one CT can be displayed before the reconstruction. This type of display is called a sinogram. In other words, the sinogram is an image of the raw data acquired by a CT scanner before the reconstruction. The horizontal axis corresponds to the different rays in each projection and the vertical axis represents each projection angle. These procedures are summarized in figure 1.

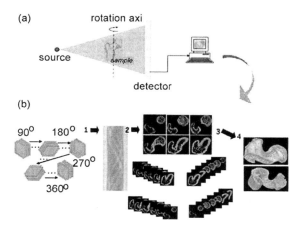

Figure 1 A CT procedure. (a) Illustration of the CT protocol (acquisition data), (b) Steps required to obtain the 3D images: 1 represents the plane rotations until the 360o degrees needed; 2 represents the sinogram; 3 are some examples of the reconstructed images and 4 are examples of the 3D visualization of all scanned sample.

As x-rays from a point source and through a spherical specimen form a cone, this approach is commonly called cone-beam tomography. Because of this the specimens were reconstructed using the cone-beam reconstruction algorithm, which is based on the Feldkamp approach [11]. After the acquisition and reconstruction processes, the data are ready to be analyzed.

2.2 Bone and its Analyses

There are two primary classes of the bone that compose the skeleton:cortical (80%) and trabecular (20%) (figure 2). The cortical bone is a solid structure forming the outer casing of the bones, while the trabecular part consists of plates, arches and struts with marrow occupying the spaces between these structures. Trabecular bone has a higher metabolic rate of volume per unit than cortical bone. Rates of change in bone density may be greater at sites that are predominantly trabecular in composition, compared to sites that are predominantly cortical [12], making the trabecular region a better region to be analyzed when it is interesting investigate the changes occurring by any bone disorder.

The parameters used by μCT to characterize the bones are similar to those used in static histomorphometry (HST) [13]. By applying the theorems of stereology [14] on histological sections (in the case of μCT considers reconstructed slices), it is possible to obtain structural parameters such as the ratio between the bone volume and the total volume (BV/TV), the ratio between bone surface and bone volume, the trabecular thickness (TbTh), the trabecular separation (TbSp) and

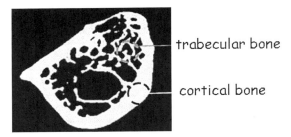

Figure 2 Histological types of bone.

the trabecular number (TbN). The best advantage in extracting those parameters from µCT instead of HST is beacuse the first method is non destructive and the investigation can be made in the entire specimen without extensive work.

To obtain the refereed parameters, the image must be segmented (figure 3). In the first moment, the original image is binarized: a cut-off value (threshold) is chosen to separate bone and non-bone, which appears white and black, respectively. The threshold value is very critical and depends so much on the quality of the image [15]. The second step is to generate the outlines on the image (identify the edges) to allow the surface identification.

An overall characterization of bone quantify will certainly lead to a more knowledge about this kind of structure.

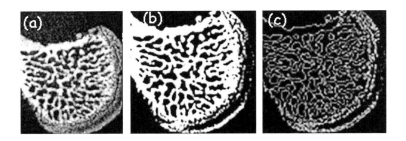

Figure 3 Segmented Process: (a) original image, (b) binary image, (c) edges of the image.

3 Experimental Issues

3.1 Ethanol Protocol and Sample Preparation

Nine animals were chosen randomly in which five of them were submitted to a liquid diet with the ethanol and water treatment (two males and three females) and four animals participated to the control group. In the beginning of the treatment, all the males were twenty-one months old and all the females were nine months old. In the first week, the ethanol group was submitted to 10 % of ethanol plus water. In the second week, the ethanol dose was increased to 20 % and this protocol continued until the week ten, completing three months of treatment. All the animals were fed with normal food over the liquid diet.

The femoral bones were removed randomly, in accordance with the Ethic Commits of the University, gently cleaned with a normal saw and washed in deionized water in order to remove the cutting fluid.

3.2 The System and Experimental Conditions

As in conventional CT units, µCT incorporates an x-ray source, a specimen support and an image detector (figure 4). The system used on this research is a FEINFOCUS x-ray FXE 160 unit, with a microfocus x-ray tube, a tungsten target and characterized by a maximum anode current and voltage of 1mA and 160 kV, respectively. Unavoidable mechanical tolerances in manufacture and assembly of the tube components, including the filament, indicate that the electron beam is generally at a slight angle to the axis of the tube. The skew would produce a travelling focal spot on the surface of the target during focusing, and imaging errors would be aggravated. A magnetic device built into the tube for centering the beam in two perpendicular planes is used for correction purpose. The specimen support consists in a mechanical arm with several degrees of freedom (3 of translation, 1 of rotation and 1 of tilt) making the alignment easier and decreasing the artifacts on the image.

To register the x-ray transmitted, it is used an image intensifier with a couple charge device camera that has a noise suppression. Both devices connected to a microcomputer permit a real time inspection. The scans were performed in nine specimens, four males and five females, with 40 kV of voltage and 0.1mA of current.

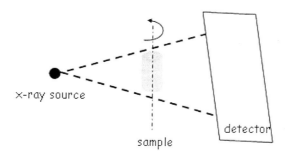

Figure 4 A typical CT geometry

The Pharmacology Laboratory of the Federal University of Rio de Janeiro donated all samples. The magnification factor was 13.2, the projection numbers were 600 and the picture number was 32. The photoelement size on the detector was 0.143 mm for all groups, the pixel size on the reconstruction process was 0.011 mm, and the resolution of the experiment was about 11m.

To quantify the structures, a region of interest (ROI) was chosen. It corresponds to the maximum possible volume of trabecular bone in the femoral head. The process was standardized for all specimens and the same volume was manually chosen. The better threshold value was obtained in accordance with the [16]. A statistical inference was carried out applying the Anova (single factor) and Bonferoni tests, with a confidence interval equal to 95 %.

4 Results and Discussion

Figure 5 illustrates the results of the quantification process for all samples, while figure 6 shows one example of the reconstructed image and the 3D visualization of all the analyzed specimens. Table 1 shows the results of the Anova test. From figure 7 it can be observed the influence of the threshold value on the bone structure. The choice of this parameter is very delicate because a wrong value can lead to a false information of the inspect image: lower values can contribute to enlarge the trabecular region and superestimate the trabecular thickness and subestimate the trabecular separation, for example. On the other hand, high threshold values make the trabeculae thinner than they really are. So, the best threshold must be chosen in a criterion manner. Until now, no standard method has been appeared and the issue is still opened. The results can be improved increasing the samples number, which would be enhanced the statistical count and they show that a control group has a largest bone percentage than the ethanol group, especially in the males' specimens (all of them were statistically different, $p<0.001$). This fact corroborates the hypothesis of this research in respect of the negative influence in bone by the ethanol intake. The Alcohol consumption leads to an inadequate nutrition and/or a vitamin deficit. And another important fact that must be taken into account is that the ethanol can produce bone resorption and can block the simulation of bone new formation. The alcohol decreases the plasmatic testosterone level that decreases the osteoblastics activity and conducts a reduction in the osteocalcine series. In consequence, it falls down the calcium absorption by the intestine, occurring an increase of the PTH (parahormone). Male gender is also prone to osteoporosis with a considerable loss of trabecular bone, although there is a low incidence with them. The major influence of this negative effect in the male group can be occur because of this information.

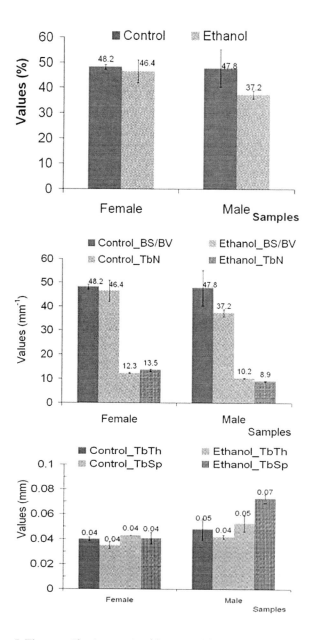

Figure 5 The quantification results: (a) BV/TV (b) TbTh and TbSp, (c) BS/BV, TbN.

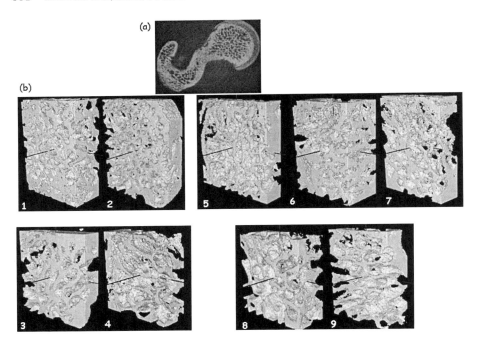

Figure 6 Examples of (a) a reconstructed image, and the (b) 3D visualization of all of them: 1,2- female control group; 3,4 - male control group; 5,6,7 - female ethanol, group; 8,9 - male ethanol group.

Anova: Single Factor

SUMMARY				
Groups	Count	Sum	Average	Variance
Control_female_no1	8	143.0	17.9	456.0
Control_female_no2	8	146.4	18.3	478.3
Ethanol_female_no1	8	140.8	17.6	549.8
Control_female_no2	8	151.6	19.0	521.2
Control_female_no3	8	131.4	16.4	454.2
Control_male_no1	8	114.0	14.3	331.8
Control_male_no2	8	112.5	14.1	322.5
Ethanol_male_no1	8	113.0	14.1	306.7
Ethanol_male_no2	8	112.7	14.1	322.3

ANOVA						
Source of Variation	SS	df	MS	F	P-value	F crit
Between Groups	271.3	8	33.9	0.08	0.9996	2.09
Within Groups	26199.5	63	415.9			
Total	26470.7	71				

Table 1 Anova results.

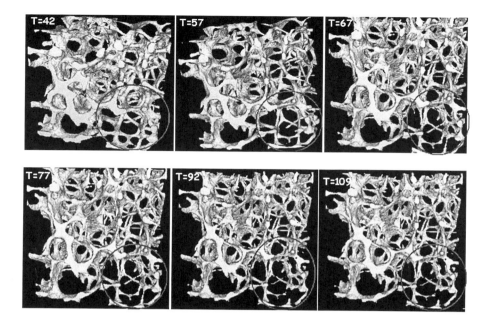

Figure 7 The influence of the threshold parameter on the 3D structures.

5 Conclusions

Through the 3D μCT, the results were capable of showing, that the internal bones structures can be examine and better than this, the ethanol intake can be investigated, making this technique a powerful and an alternative method to looking for images changes on those kind of structures. The three dimensional reconstruction of the trabecular structures enable the quantification of morphological parameters such as trabecular volume and surface area, supporting the knowledge about that kind of bone, especially in microstructural level.

Acknowledgements

This work was partially supported by Conselho Nacional de Desenvolvimento Científico e Tecnológico (CNPq) and FAPERJ.

Bibliography

[1] A. Hessenbruch. Endeavour. **26(4)**, 137 (2002).

[2] S. Robinson, A. Suomalainen and M. Kortesniemi. European Journal of Radiology. **56**, 185 (2005).

[3] M. Ding, A. Odgaard, F. Linde and I. Hvid. Journal of Orthopaedic Research. **20**, 615 (2002).

[4] V. Patel et al. Journal of Orthopaedic Research. **21**, 6 (2003).

[5] J. A. Kanis et al. Osteoporos Int. **16**, 737 (2005).

[6] J. A. Kanis P. Delmas, P. Bruckhardt, C. Cooper, D. Torgerson. Osteoporos. Osteoporos Int. **7**, 390 (1997).

[7] F. H. Wezeman, D. Juknelis, N. Frost and J.J. Callaci. Alcohol. **31**, 87 (2003).

[8] P. Angeli, A. Gatta, L. Caregaro, G. Luisetto, F. Menon, C. Merkel M. Bolognesi and A. Rulo. Gastroenterol. **100**, 502 (1991).

[9] T. Diamond, D. Stíel, M. Lunzer, M. Wilkinson and S. Posen. Am. J. Med. **86**, 282 (1989).

[10] N. R. H. El-Maraghi, B.S. Platt and R.J.C. Stewart. Brit.. J. Nutr. **19**, 491 (1965).

[11] L.A. Feldkamp, L.C. Davis, J.W. Kress. J. Orpt. Soc. Am. A. **1**, 612 (1984).

[12] M. Ding and I. Hvid. Bone. **26(3)**, 291 (2000).

[13] Recker *et al.* Bone Histomorphometry: Techniques and Interpretation. (CRC Press, USA 1990).

[14] Underwood *et al. Quantitative Stereology.* (Assison-Wesley Publishing Company, USA, 1970).

[15] T. Hara *et al.* Bone, **31(1)**, 107 (2002).

[16] A. Barbier *et al.* Bone Mineral Metabolism. **17**, 37 (1999).

Anais da XXVIII Reunião de Trabalho sobre Física Nuclear no Brasil
SP, Brasil, 2005

Evaluation of entrance surface dose to chest x-rays in pediatric radiography performed in a University Hospital in Curitiba, Brazil

N. A. Lunelli[1], H. R. Schelin[1], S. Paschuk[2], V. Denyak[1], J. G. Tilly Jr[3], E. Rogacheski[3], H. J. Khoury[4] and A. C. P. de Azevedo[5]

[1] Programa de Pós-Graduação em Engenharia Elétrica e Informática Industrial – CPGEI/UTFPR, Av. Sete de Setembro, 3165, CEP 80230-901, Curitiba, PR, Brazil

[2] Programa de Pós-Graduação em Engenharia Mecânica e de Materiais – PPGEM/UTFPR, Curitiba, PR, Brazil

[3] Serviço de Radiologia Médica - Hospital de Clínicas da UFPR, Curitiba, PR, Brazil

[4] Laboratório de Metrologia e Radiações Ionizantes – DEN - UFPE, Brazil

[5] FIOCRUZ – Escola Nacional de Saúde Pública – CESTEH - RJ, Brazil

Abstract. The aim of this work is to evaluate the Entrance Surface Dose (ESD) in pediatric chest x-ray. A study of 176 x-ray examinations in three common projections [anterior-posterior (AP), posterior-anterior (PA) and lateral (LAT)] was carried out. The patients were divided into age intervals of 0-1 year, 1-5 years, 5-10 years and 10-15 years. The examinations were made in the Pediatric Room at the Clinical Hospital of the Federal University of Paraná. The ESD was determined with the software DoseCal that calculates the dose for each patient. The program uses the following data as input: patient mass, age, skin-focus distance, parameters of radiographic technique (x-ray tube applied potential, tube current, exposure time), and the radiation output measurement made with a calibrated ionization chamber. The results obtained are compared with the reference dose values established by the European Community.

Keywords. Dosimetry, TLD, X-Ray

PACS. 73.61.Ph;

1 Introduction

There is a growing concern about the amount of absorbed dose by patients submitted to radiographic examinations [1], [2], [3]. This concern is greater in the case of children who have a longer life expectancy than adults and are more sensitive to ionizing radiation [4]. For radiodiagnostic purposes, the image obtained must have the largest amount of information and should be obtained with the lowest possible dose. It is crucial to assure the minimum requirements of radiological protection to the patients, the staff and the general public. The ALARA principle must be applied so that no practice be performed unless it produces enough benefit for society in order to compensate the detriment that can be caused, optimizing the dose. Reference dose values in diagnostic radiology have been developed to optimize patient imaging and to support educational programs. The purpose of the present work is to estimate the entrance surface dose (ESD) of chest x-rays in pediatric examinations performed in AP/PA and LAT projections at the Clinical Hospital of the Federal University of Paraná, Brazil. The chest radiography is the most frequent pediatric x-ray examination. It is important to measure the ESD during the examination of the patient, because all factors which affect the dose to the patient are included. Such factors are the size of the patient, experience of the staff, equipment performance, etc. The doses were evaluated using the DoseCal software developed by the Radiological Protection Center of Saint George's Hospital, in London.

2 Methodology

In this study 176 expositions distributed among patients were made. These patients were divided into age groups of 0-1 year, 1-5 years, 5-10 years and 10-15 years. For each examination, having used radiological technique (kV, exposure time, etc.), the medical-morphological data of the patients were collected.

In Phase 1, 100 x-ray examinations were evaluated. After noticing that the exposure parameters did not attend the European Guidelines on Quality Criteria for Diagnostic Radiographic Images in Pediatrics, a revision was made in the procedures, reducing the mAs and increasing the kV (kV is the tube potential (in kV), mAs is the product of the tube current (in mA) and the exposure time in s). In this Phase 2, 76 examinations were evaluated.

The amount of samples was different for each age group, mainly due to the limited number of examinations by group performed at the hospital in study. The 0-1 year group had less expositions compared to the other groups. Most of the patients were male 66% while 34% were female.

The x-ray machine used was a GE MEDICAL SYSTEMS, model AL01F-PROTEUS XR/A, type 2261765, with 3mm Al total beam filtration.

In order to obtain the ESD with the DoseCal software, the radiation output of the x-ray tube was measured using the ionization chamber RADCAL COR-PORATION, model 10X5-6, 6cm^3 sensitive volume. The ionization chamber was connected to the electrometer and was placed at 25 cm from the table top, in order to prevent inaccurate readings caused by the backscattering, and at 100 cm from the x-ray focus. Exposures were made and the output recorded for a certain mAs value throughout the kV range used for the examinations. Measurements have been made with 5 kV increments, in the range from 46 to 74 kV. The fixed exposure parameters are the same values used in the examinations performed during this study. The yield of the x-ray equipment was used as input parameters of the DoseCal software. The program performs a simulation of the dose received by the patient, taking into account the variation between these parameters provided by the ionization chamber and the values of kV, mAs and the focus-skin distance used in each exam and also the backscattering factor.

The software uses the following equation to calculate the ESD:

$$ESD = \frac{output}{mAs_{cal}} \cdot mAs_{util} \cdot \left(\frac{kV_{cal}}{kV_{util}}\right)^2 \cdot \left(\frac{d_{cal}}{d_{util}}\right)^2 \cdot (BSF + 1) \qquad (1)$$

here output is the kerma in air value measured with the ionization chamber at the kV_{cal} and mAs_{cal} and d_{cal} is equal to 100 cm, mAs_{util}, kV_{util} and d_{util} are the parameters of radiographic technique. The BSF is calculated automatically by the DoseCal software after all input data are entered manually in the program.

The ESD values obtained were compared with the reference values of the NRD for pediatric patients published by the European Community.

3 Results and Discussion

Table 1 shows the exposure parameters used for the chest radiography, the age and weight of the patients, distributed by age group. It is possible to observe that in Phase 1 (before changes) the mean value of the kV is 56.15, lower than the value recommended by the European Guidelines on Quality Criteria for Diagnostic Radiographic Images in Pediatrics [5]. The kV value recommended by the European Commission is in the range of 60 to 80 kV. The results of Table 1 show that the mean kV value in Phase 2 is in this interval. The mAs in Phase 2 is approximately half the value of Phase 1. The higher kV and lower mAs contributed to the reduction of the ESD values.

Tables 2 and 3 show the results of the ESD obtained for each patient age group with the chest examinations in AP/PA and LAT projections. The minimum, maximum, median, standard deviation values are presented along with the size of the sample for AP/PA projections.

Table 1 Average of the patient data and radiographic parameters in Phase 1 and in Phase 2 before and after the changes in the radiographic techniques were made in the chest x-ray examinations.

Age groups (years)	Age (months)	Mass (kg)	kV		mAs	
			Phase 1	Phase 2	Phase 1	Phase 2
AP/PA examinations						
0 - 1	6.7	7.76	51.9	55.0	3.20	1.51
1 - 5	33.7	13.56	53.6	62.5	3.20	1.63
5 - 10	83.5	21.81	59.4	65.6	3.20	1.63
10 - 15	139.4	37.19	59.7	71.7	3.20	1.60
LAT examinations						
0 - 1	6.7	7.76	56.1	59.8	3.20	1.70
1 - 5	34.9	13.56	60.1	71.1	3.20	1.64
5 - 10	84.7	22.77	70.1	75.1	3.20	1.64
10 - 15	139.6	35.78	71.5	83.9	3.20	1.60

Table 2 ESD for AP/PA projection in Phase 1 and Phase 2 before and after the changes in radiographic technique.

ESD (μGy)	0 - 1 year		1 -5 years		5 -10 years		10 -15 years	
	Phase 1	Phase 2	Phase 1	Phase 2	Phase 1	Phase 2	Phase 1	Phase 2
Average	83	48	66	35	41	30	42	38
Standard deviation	12	17	28	15	9	5	7	6
Minimum	69	25	30	19	31	21	34	30
Maximum	102	62	110	63	56	36	54	45
Median	81	53	77	30	37	31	41	39
Sample	8	4	18	20	16	8	14	6

Table 3 ESD for LAT projection in Phase 1 and Phase 2 before and after the changes in radiographic technique.

ESD (μGy)	0 - 1 year		1 -5 years		5 -10 years		10 -15 years	
	Phase 1	Phase 2	Phase 1	Phase 2	Phase 1	Phase 2	Phase 1	Phase 2
Average	105	66	86	52	61	45	66	55
Standard deviation	19	13	34	17	8	9	15	4
Minimum	78	48	45	34	48	31	48	50
Maximum	136	79	143	83	73	55	95	61
Median	101	68	90	48	59	48	65	55
Sample	8	4	10	19	13	8	13	7

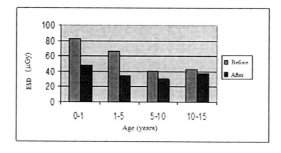

Figure 1 ESD comparison before and after the changes in the radiographic techniques.

Figure 1 shows the mean ESD values obtained per patient group in the two phases, before and after the changes in the radiographic techniques.

The significant difference in ESD before and after the change in radiological technique was obtained for the 0-1 years group (more than 2 standard deviations). For the other three age groups, the conclusions concerning the difference in ESD before and after the change in the radiological technique could not be assumed since the standard deviation of these measurements is very big.

It is possible to observe that when the exposure parameters used are in the range of the parameters established by the European Commission, the ESD has a significant reduction, specially pronounced for the 0-1 year group.

According to NRPB 1999 [6], the reference values for ESD for AP and PA projections in chest radiographies for children are: 50 µGy for newborn babies and 1-year olds, 70 µGy for 5-year olds and 120 µGy for 10-year olds.

It could be seen that after the changes in the radiographic techniques have been introduced (Phase 2) for all four age groups of patients, the mean ESD values were obtained within the limits established by the NRPB 1999.

The minimum and maximum of the data presented in Tables 2 and 3, whose difference reached a factor bigger than 3, show the large range in dose in the examinations, mainly in the 1-5 year group. For this group the standard deviation is also very big (> 30%). The reason of such result is due to the rather big distinction between the size of the body of 1 and 5 year-old children.

The proximity of the average with the median and lower standard deviation in the 10-15 year group show a smaller change in the dose showing that the employed technique is more stable in this range.

On the other hand, the mean doses of patients from the 0-1 year group are higher than the obtained with the other age groups. These results show that the exposure techniques are not optimized. The inappropriate use of antiscatter grid may be a factor for the increase of the ESD because it requires higher radiation intensity. According the European Guidelines, a grid is not needed for chest radiography until 8 years of age because there is no significant scattering. The other factor is the radiographic voltage that can be increased.

4 Conclusions

The increase of the tube potential with corresponding decrease of its current leads to significant dose reduction for children up to 1 year old. This effect is much smaller for older children.

The large range of entrance surface doses found is due to several factors, such as variation in patient thickness, field size, potential of the x-ray tube, etc. The results obtained provide useful guidance on dose levels and optimization strategies in Curitiba.

The results of this study emphasize the necessity to fix and follow the regulatory guidelines for a better staff training and in the standardization of the employed techniques. There is a great possibility to reduce the dose at low costs.

Acknowledgements

The authors are very thankful to the Radiological Protection Center of Saint George's Hospital, in London, for the DoseCal software and CAPES and CNPq for the financial suport.

This research work was authorized by the Ethical Committee of the Hospital de Clinicas (UFPR), under the certificate number 2004015706 from 30/11/2004.

Bibliography

[1] AZEVEDO, A. C. P.; MOHAMMEDAN, K. E. M.; ROSA, L. A. R.; GREBE, M. R. N.; BOBCAT, M. C. B. Dose measurements Using Thermo luminescent Dosimeters and Dose Cal Software at Two Pediatric Hospitals in Rio de Janeiro. Apelei Radiation and Isotopes 59, April 2003.

[2] MOHAMADAIN, K. E. M.; ROSA, L. A. R.; AZEVEDO, A. C. P.; GUEBEL, M. R. N.; BOECHAT, M. C. B.; HABANI, F. Dose Evaluation for Paediatric Chest X-Ray Examinations in Brazil and Sudan: Low Doses and Reliable Examinations Can Be Achieved in Developing Countries. Physics in Medicine and Biology, 49 March 2004 - UK.

[3] MONTGOMERY, A.; MARTIN, C. J. A. Study of the Application of Paediatric Reference Levels. The British Journal of Radiology 73 (2000) 1083-1090.

[4] NRPB - National Radiological Protection Board Occupational Public and Medical Exposute. Documents of NRPB, vol. 4 (1993) UK National Radiological Protection Board, Chilton.

[5] European Guidelines on Quality Criteria for Diagnostic Radiographic Images in Paediatrics. Office for Official Publications of the European Communities, 1996.

[6] NRPB 1999 Guidelines on Patient Dose to Promote the Optimization of Protection for Diagnostic Medical Exposures vol. 10 (1999) UK National Radiological Protection Board, Chilton.

Anais da XXVIII Reunião de Trabalho sobre Física Nuclear no Brasil
SP, Brasil, 2005
Artigo originalmente publicado no Brazilian Journal of Physics, Vol 36 – nº 4B,
Special Issue: XXVIII Workshop on Nuclear Physics in Brazil, pp.1371–1374 (2006).

High-K Band in ^{140}Gd

J. R. B. Oliveira[1], F. Falla-Sotelo[1], M. N. Rao[1],
J. A. Alcántara–Núñez[1], E. W. Cybulska[1], N. H. Medina[1], R. V. Ribas[1],
M. A. Rizzutto[1], W. A. Seale[1], D. Bazzacco[2], F. Brandolini[2],
S. Lunardi[2], C. Rossi-Alvarez[2], C. Petrache[2], Zs. Podolyák[2], C. A. Ur[2],
D. de Acuña[3], G.de Angelis[3], E. Farnea[3], D. Foltescu[3], A. Gadea[3],
D. R. Napoli[3], M. de Poli[3], P. Spolaore[3], M. Ionescu-Bujor[4] and
A. Iordachescu[4].

[1] Instituto de Física, DFN, Universidade de São Paulo, São Paulo, Brazil

[2] Dipartimento di Física and INFN, Sezione di Padova, Padova, Italy

[3] INFN, Laboratori Nazionali di Legnaro, Legnaro, Italy

[4] Institute of Physics and Nuclear Engineering, Bucharest, Romania.

Abstract. High-spin states in the neutron-deficient^{140}Gd nucleus have been studied with the ^{92}Mo(^{54}Fe,α2p) reaction at a beam energy of 240 MeV. The level scheme of ^{140}Gd was considerably extended from what was previously known. In this work we concentrate on one of the 9 bands observed, which presents relatively strong M1 transitions and negligible signature splitting, and has an isomeric band-head, indicating a strongly coupled or high-K configuration. We compare this band to the $K^\pi = 8^-$ bands observed in N = 74 isotones and propose a similar configuration assignment, but with a somewhat larger deformation.

1 Introduction

Several high-K states are known to exist in the mass 130-140 region. For the N = 74 even-even isotones, $K^\pi = 8^-$ isomers, with lifetimes ranging from ns to ms, are known in ^{128}Xe[1], ^{130}Ba[2], ^{132}Ce[3], ^{134}Nd[4], ^{136}Sm[5], and ^{138}Gd[6]. The systematics of the $K^\pi = 8^-$ isomeric states in N = 74 isotones has been studied by A.M. Bruce *et al.* [7]. These states decay towards the K = 0 ground state band, and the transitions are K-forbidden. The variation of the strengths of the transitions, however, is not clearly understood. The explanation might involve a variation of the structure of the isomer or of the ground state band. In the present work have extended the level scheme of ^{140}Gd from what was previously known[8], and, in particular, we have observed for the first time a band also based on an $I^\pi = 8^-$ state.

343

This could be the first case of a $K^\pi = 8^-$ state observed in an $N = 76$ even-even nucleus. The ^{140}Gd case presents strong similarities but also some significant differences with relation to the $N = 74$ isotones. The interpretation of this band is quite difficult, in particular because the addition of the 2 neutrons is expected, in principle, to modify the configuration with respect to the one observed in the $N = 74$ cases, fully occupying the orbitals involved and decreasing the deformation as the $N = 82$ closed shell is approached. We propose that the deformation is larger in ^{140}Gd and a pair of neutrons from the $N = 5$ shell coupled to angular momentum zero is incorporated, from which the $7/2[404]^{-1} \otimes 9/2[514](K^\pi = 8^-)$ neutron excitation (in particle-hole terminology) is built.

2 Experimental Procedure and Analysis

The ^{92}Mo$(^{54}$Fe,α2p) reaction at 240-MeV was employed to populate high-spin states in ^{140}Gd. The incident beam was provided by the tandem XTU accelerator of the Legnaro National Laboratories. A self–supporting and isotopically enriched ($>97\,\%$) ^{92}Mo target of approximately 1.0 mg/cm^2 thickness was used. The GASP array [9] was used for obtaining gamma–ray double and triple coincidence spectra. The photopeak efficiency of the array is approximately 0.5 % for 1.3 MeV. The charged-particle detector array (ISIS) [10] enabled the selection of evaporated charged particles in coincidence with the γ–rays. The recoil mass spectrometer (Camel) [11] allowed for mass identification. Events were collected on tape when at least two HPGe detectors fired in coincidence. The data were taken for a period of 136 hours. A total of 85×10^6 Compton-suppressed events were collected. The data were sorted into *charged particle fold*–γ–γ, A–γ–γ and γ–γ–γ cubes, from which specific matrices for the various exit channes were extracted, and were analyzed using the VPAK [12] and RADWARE [13] spectrum analysis codes. The detectors were energy and efficiency calibrated with standard ^{152}Eu and ^{133}Ba radioactive sources.

Figure 1 presents the background subtracted sum spectra of the 246 and 287 keV gates. The in-band M1, E2 crossovers (except the 533 keV which is the crossover of the gating transitions) and the ground-state band (gsb) transitions from the levels fed by the band can be clearly seen. Table 1 presents the list of transitions and intensities observed. DCO (Directional Correlation from Oriented states) measurements confirm some of the multipolarity assignments presented [14]. Details will be given in a forthcoming article. The same DCO procedure has been employed in previous analises from the same experiment [15, 16]. A careful re-examination of the efficiency was performed and corrections have been applied due to the variation of the electronic response at low γ–ray energies. As a result, the assignment of the band-head spin is different from that of ref. [14], and is discussed in the next section.

Figure 2 presents the relevant partial level scheme of ^{140}Gd obtained in the present measurement. This level scheme is based on the coincidence relationships, intensity balances on each level and energy sums from different paths using the $1\alpha 2p$–gated matrix.

Table 1 Energies, intensities, and spin assignments to the transitions belonging to the strongly coupled band observed in ^{140}Gd (the first two rows are linking transitions to the ground state band). The energies of the transitions are given in keV. The intensities are normalized (to 100 %) with respect to the 507.6 keV transition ($4^+ \rightarrow 2^+$) of the ground state band.

E_γ (keV)	I_γ	$I_i^{\pi_i} \rightarrow I_f^{\pi_f}$
71.3(2)	0.77(12)	$8^- \rightarrow 8^+$
747.1(3)	0.80(18)	$8^- \rightarrow 6^+$
245.8(1)	2.38(20)	$9^- \rightarrow 8^-$
287.0(1)	2.00(13)	$10^- \rightarrow 9^-$
317.4(1)	1.40(10)	$11^- \rightarrow 10^-$
342.1(1)	1.08(8)	$12^- \rightarrow 11^-$
363.7(1)	0.54(6)	$13^- \rightarrow 12^-$
383.3(1)	0.64(7)	$14^- \rightarrow 13^-$
403.1(3)	0.23(5)	$15^- \rightarrow 14^-$
423.0(15)	0.04(5)	$16^- \rightarrow 14^-$
532.9(10)	0.36(7)	$10^- \rightarrow 8^-$
604.0(3)	0.53(8)	$11^- \rightarrow 9^-$
658.8(3)	1.08(15)	$12^- \rightarrow 10^-$
704.9(3)	0.60(8)	$13^- \rightarrow 11^-$
747.1(8)	0.16(7)	$14^- \rightarrow 12^-$
785.7(3)	0.71(11)	$15^- \rightarrow 13^-$
826.3(3)	0.65(8)	$16^- \rightarrow 14^-$

3 Assignments of Band-Head Spin, Parity, and Lifetime

The spin and parity of the $I^\pi = 8^-$ state was established by the following arguments: Firstly the experimental total electronic conversion coefficient of the 71 keV transition was determined from the ratio of gamma intensities of itself to that of the 676 keV gsb transition, as observed in the 246 and 287 keV gated spectra. The value obtained corresponds to a conversion coefficient of $\alpha_{exp} = 0.72(20)$, consistent with the expected value for a pure E1 ($\alpha_{tco} = 0.736$). The experimental value is clearly in disagreement with other possible assignments like

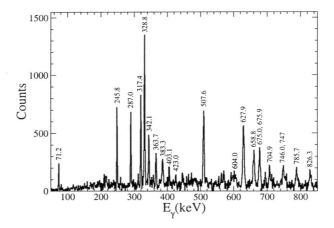

Figure 1 Sum spectrum of the 246 and 287 keV gate transitions from the 1α2p−gated matrix. The energies of the γ−ray lines are indicated in keV (see Sec. 3 with regards to the 71 and 746 keV line energies).

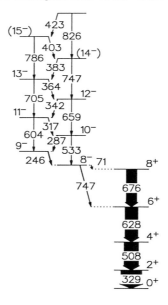

Figure 2 Partial level scheme of ^{140}Gd showing the strongly coupled band discussed in this work and its decay to the gsb. The transition energies are indicated in keV. The transition intensities are proportional to the arrow widths.

M1 or E2 which present very different electronic conversion coefficient values. This establishes that the 71 keV line is of E1 character, and the band has negative parity. Secondly, if the 747 keV transition (from the same state) were of the same multipolarity (E1), it's decay to the gsb should be governed by an intrinsic transition matrix element of the same order of magnitude as that for the 71 keV, assuming not too different structure of the 8^+ and 6^+ gsb states. In this case, the energy factor should favor by three orders of magnitude the 747 keV with relation to the 71 keV transition. However, the experimental intensities of both transitions have been calculated to be of the same order of magnitude. This, therefore, indicates a multipolarity with weaker strength, viz. M2, with $\triangle I = 2$, to the 747 keV transition. In conclusion, the only assignment consistent with these results is of an 8^- for the strongly coupled band head.

The lifetime of the $I^\pi = 8^-$ band-head was estimated to be of about 1.5 ns. This comes from two independent evidences: First, in the 287 keV gate, the intensity of the 328 keV transition $(2^+ \rightarrow 0^+)$ is reduced to about 85 % when compared to that of the 246 keV. This is due to the loss of efficiency when the $\gamma-$decay occurs off-center in the GASP spectrometer. The flight velocity of the recoils after traversing the target is about 1cm/ns. Second, the gsb transitions appear to be shifted to lower energies, when gated in-band from above the isomer, by about -0.15 % relative to the usual Doppler corrected $\gamma-$ray energies. This is again due to the decay off-center, since the transitions become observed by the detectors at larger angles than those originally used in the Doppler corrections of the spectra, which assume decay in the center of the target. Simulations with the GEANT code applied to the specific experimental setup of GASP show that these results are consistent with a lifetime of 1.5 ns for the negative parity band-head, with an uncertainty of about 0.4 ns. The apparent energies of the transitions from the isomeric state are therefore also expected to be slightly shifted. For this reason the 0.15 % correction has been applied to the energy values of the transitions from the 8^- state quoted in table 1. For the 747 keV line, this correction is significant and the peak position is close to 746 keV in the spectrum (Fig.1).

4 Discussion

As stated in the introduction, the systematics of the $K^\pi = 8^-$ isomeric states in $N = 74$ isotones has been discussed by A.M. Bruce et al.[7]. In general, the configuration proposed for the isomers is the $7/2^+[404] \otimes 9/2^-[514]$ two quasi-neutron structure, although other possibilities have been considered such as a two proton oblate configuration in ^{130}Ba, too far from Gd to be of interest here. The lifetimes vary considerably, generally decreasing from 11 ms in ^{130}Ba to 6 µs in ^{138}Gd. The hindrance per degree of K forbideness (f_ν) of the decay is defined as $F_W^{1/\nu}$, where $\nu = \Delta K - \lambda$, λ is the transition multipolarity, and $F_W = t_{1/2}(E\lambda)/t_{1/2}^w(E\lambda)$ is the hindrance factor, the ratio of the experimental partial half-life to that of the Weisskopf single particle estimate. The values of f_ν for the $N = 74$ isotones are around 25 for Nd, Sm and Gd, while for Ba it is 44. This variation can be reasonably well interpreted in terms of admixture of K = 8 in the gsb due to interaction with the s (Stockolm) band near band crossing.

From our data we can estimate a value of $f_v \approx 13$ for the ^{140}Gd 71 keV transition. The partial lifetime for the 747 keV M2 transition results in essentially no hindrance with respect to the Weisskopf estimate.

Total Routhian Surfaces Calculations (TRS), together with the presence of a gamma-vibrational band at low excitation energy, indicate a large degree of triaxiality for ^{140}Gd and a reduced β_2 deformation (from about 0.28 in N = 74 Sm and Gd to 0.21 in ^{140}Gd), near the ground state. This could also favor high-K admixture in the structure of the gsb, and reduce the hindrance factor. The interpretation is not clear, however, because no high-K configuration appears to be near yrast according to the same TRS calculations, since the $7/2^+$[404] orbits become fully occupied and away from the Fermi energy as an effect of the addition of 2 neutrons and reduction of deformation. Besides that, the alignment angular momentum (the projection on the rotational axis) of the strongly coupled band relative to the gsb is much larger in ^{140}Gd (4-5$\hbar$) than in the nearby N = 74 isotones (Gd, Sm and Nd, around 1$\hbar$). This is not expected since in a high-K configuration the quasi-particle angular momentum tends to couple along the deformation axis, perpendicularly to the rotation axis. A 4-quasiparticle configuration would be very unlikely since the excitation energy of the $K^\pi = 8^-$ relative to the ground state is quite small (2.21 MeV), and in fact, quite similar to the one of the N = 74 isotones (2.23 and 2.27 MeV in Gd and Sm respectively). A proton configuration, which could present large alignment, would have a low-K, and should be discarded (see also the electromagnetic properties discussed in the next paragraph). We tentatively propose, therefore, that the strongly coupled band is built on a quite deformed shape core containing 2 paired particles from the N = 5 shell, and in addition the $7/2^+$[404] $\otimes$ $9/2^-$[514] two quasi-neutron excitation coupled to $I^\pi = 8^-$ as in the N = 74 nuclei. The N = 5 single-particle states 1/2[541] (from $f_{7/2}$) and 1/2[660] (from $i_{13/2}$) in the deformed shell model are strongly down-sloping as a function of β_2 and are therefore "β−driving", the excitation energy of the former being lower than the latter. A larger deformation could explain why the alignment is apparently so high in ^{140}Gd. Since the band would have a larger moment of inertia in comparison to the gsb, it develops more collective angular momentum at the same rotational frequency and the concept of quasiparticle alignment becomes inadequate. The difficulty in such explanation resides in the comparatively high strength of the inter-band transition for such a significant shape change and, still, in the small experimental excitation energy. However, this appears to be the most likely interpretation with comparison to the other possibilities. It should be noted that highly deformed bands have been observed at relatively low excitation energy in odd-nuclei of this region, such as ^{137}Sm and ^{139}Gd [17]. These bands are built upon the ν1/2[660] (from $i_{13/2}$) configuration, with large quadrupole moments ($Q_0 = 5 - 7$eb, [18, 19]). In ^{133}Nd [20] the ν1/2[541] band (with an estimated deformation of $\beta_2 = 0.29$) can be found at much lower excitation energy than the ν1/2[660].

Figure 3 presents the experimental ratio of reduced transition probabilities B(M1)/B(E2) in comparison with the nearby nuclei and theoretical estimates. It can be observed that for all nuclei presented the values have the same order of magnitude, being on average lower in ^{140}Gd than in the N = 74 isotones. The dotted line corresponds to a prediction based on the rotational model with an intrinsic quadrupole moment of $Q_0 = 4.5$ eb, the value measured from the lifetime of the first 2^+ state in ^{136}Sm [21], while the dashed line is calculated with $Q_0 = 5.5$ eb. This last value reproduces better the ^{140}Gd data, in agreement with the hypothesis of large deformation.

The absolute value of the difference between valence and collective g-factors $|g_K - g_R|$ for the band states can be obtained from the experimental mixing ratios δ (obtained from the $\Delta I = 2$ to $\Delta I = 1$ branching ratios) according to the procedure of refs.[5, 6], based on the rotational model, assuming a value for the quadrupole moment Q_0. The results are presented in fig. 4. For the N = 74 nuclei the value of $Q_0 = 4.5$ eb was adopted, while for ^{140}Gd $Q_0 = 5.5$ eb was used. Again the same order of magnitude is observed for all the presented nuclei and the values

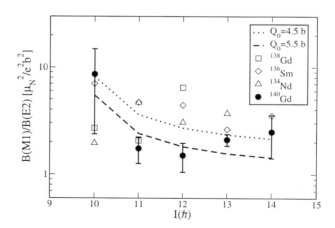

Figure 3 The experimental ratio of reduced transition probabilities B(M1)/B(E2) as a function of the spin of the state in ^{140}Gd. The values for the N = 74 isotones are also shown for comparison (the error-bars are roughly of the same order of magnitude as in ^{140}Gd, but are not shown for clarity of the figure). Theoretical estimates from the pure rotational model for two values of intrinsic electric quadrupole moments are also shown by the lines (considering the value of $|g_K - g_R| = 0.4$).

are consistent with the expected for the $7/2^+[404] \otimes 9/2^-[514]$ quasi-neutron configuration, according to the estimates of ref. [5] (the dotted line at $|g_K - g_R| = 0.4$ in fig.4). For this configuration, the 2 quasiparticle g-factors mutually cancel, remaining the collective value of $g_R = 0.4$. This result also corroborates our assumptions. It should be noted that for 2 quasi-proton configuration with large K the g-factor difference would be much larger and could not be consistent with the present experimental results.

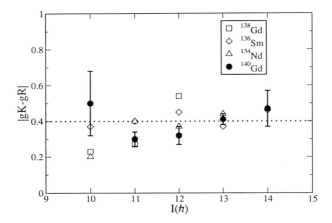

Figure 4 The experimental and theoretical values of $|g_K - g_R|$ as a function of the state spin in ^{140}Gd, and in N = 74 isotones (with values of the quadrupole moment $Q_0 = 5.5$ eb, and $Q_0 = 4.5$ eb, respectively). The value expected for the $\nu 7/2^+[404] \otimes 9/2^-[514]$ ($K^\pi = 8^-$) configuration is shown by the dotted line.

5 Conclusion

A strongly coupled band has been observed for the first time in ^{140}Gd. The bandhead was determined to be an $I^\pi = 8^-$ isomer state, with a lifetime of about 1.5ns. The band presents strong similarities to the high-K bands in neighboring N = 74 isotones, based on the $7/2^+[404] \otimes 9/2^-[514]$ two quasi-neutron configuration coupled to $K^\pi = 8^-$. The apparent alignment, relative to the ground state band is, however, larger than in the lower N isotones, and can perhaps be explained by an increase in deformation. With a significantly deformed shape (say, $\varepsilon > 0.25$), 2 paired particles from the N = 5 shell from the 1/2[541] Nilsson state (from $f_{7/2}$) are incorporated, and the $\nu 7/2^+[404] \otimes 9/2^-[514]$ ($K^\pi = 8^-$) excitation would become the lowest available negative parity configuration, providing the necessary characteristics to explain the presently observed band. This assignment is corroborated by the experimental B(M1) and B(E2) branching and mixing ratio values. It remains to be explained the large reduction of the hindrance factor of the isomer decay to the gsb and the relatively low excitation energy of the structure. Search for additional cases of strongly coupled bands in $N = 76$ isotones would be of great interest, as well as the measurement of in-band transition lifetimes.

Acknowledgements

We acknowledge partial support from FAPESP (Fundação de Amparo à Pesquisa do Estado de S. Paulo) and CNPq (Conselho Nacional de Desenvolvimento Científico e Tecnológico), and thank the LNL staff for the operation of the accelerator.

Bibliography

[1] L. Goettig *et al.,* Nucl. Phys. **A357**, 109 (1981).

[2] H.F. Brinkman *et al.,* Nucl. Phys. **81**, 233 (1966).

[3] D. Ward, Nucl. *et al.,* Phys. **A117**, 309 (1968).

[4] D.G. Parkinson *et al.,* Nucl. Phys. **A194**, 443 (1972).

[5] P.H. Regan *et al.,* Phys. Rev. C **51**, 1745 (1995).

[6] D.M. Cullen *et al.,* Phys. Rev. C **58**, 846 (1998).

[7] A.M. Bruce *et al.,* Phys. Rev. C **55**, 620 (1997).

[8] E. S. Paul *et al.,* Phys. Rev. C **39**, 153 (1989).

[9] D. Bazzacco, in Proc. of the Int. Conf. on Nucl. Struct. at High Ang. Momentum, Ottawa, 1992, Report No. AECL 10613, Vol. 2, p. 376.

[10] E. Farnea *et al.*, Nucl. Instrum. Meth., **A400**, 87 (1997).

[11] P. Spolaore, *et al.*, Nucl. Instr. Meth., **A238**, 381 (1985).

[12] W. T. Milner, Holifield Heavy Ion Research Facility Computer Handbook, Oak Ridge National Laboratory, Oak Ridge, Tennessee, USA (1987).

[13] D. Radford, Nucl. Instr. Meth. Phys. Res. **A361**, 297 (1995).

[14] F. Falla-Sotelo, MSc Thesis, Universidade de S. Paulo, IFUSP, Brazil, (2002).

[15] F.R. Espinoza-Quinones *et al.*, Phys. Rev. C **60**, 054304 (1999).

[16] J.R.B. Oliveira *et al.*, Phys. Rev. C **62**, 064301 (2000).

[17] C.A. Rossi-Alvarez, *et al.*, Nucl. Phys. **A624**, 225 (1997).

[18] E.S. Paul *et al.*, J. Phys. **G 18**,121 (1992).

[19] P.H. Regan *et al.*, J. Phys. **G 18**, 847 (1992).

[20] D. Bazzacco *et al.*, Phys. Rev. C **58**, 2002 (1998).

[21] R. Wadsworth *et al.*, J. Phys. **G 13**, 205 (1987).

Anais da XXVIII Reunião de Trabalho sobre Física Nuclear no Brasil
SP, Brasil, 2005
Artigo originalmente publicado no Brazilian Journal of Physics, Vol 36 – nº 4B,
Special Issue: XXVIII Workshop on Nuclear Physics in Brazil, pp.1379–1382 (2006).

Trajectory Effects in Coulomb Excitation

F. W. Fernandes and B. V. Carlson

Departamento de Física, Instituto Tecnológico de Aeronáutica,
Centro Técnico Aeroespacial, 12228-900 São José dos Campos, São Paulo, Brazil

Abstract. We compare the cross sections for Coulomb excitation of multiple giant dipole resonances in ^{208}Pb + ^{208}Pb scattering using Coulomb trajectories and straight-line trajectories that have the same point of closest approach as the Coulomb one. We find the effects of the Coulomb deflection relative to the straight line trajectory to be small at incident energies above about 500 MeV/nucleon.

1 Introduction

Coulomb excitation has proven itself as an important tool for studying the structure of both stable and exotic nuclei.[1, 2, 3, 4] The phenomenon has been well-studied in a nonrelativistic context[5] and studied perturbatively in a relativistic one[6] by Alder and Winther. More recent experimental studies have considered multiple Coulomb excitation at relativistic energies.[2, 3, 4]

Many calculations of high-energy Coulomb excitation include the effects of Coulomb deflection, but only through the effective impact parameter used in a straight-line trajectory.[7, 8, 9, 10, 11] The curvature of the trajectory is usually not taken into account. The curvature of the Coulomb trajectory increases the distance between the projectile and target, relative to a straight-line trajectory, and is thus expected to decrease Coulomb excitation, which is strongly dependent on the separation. In the following, we derive the semiclassical coupled-channel equations for Coulomb excitation and then use them to compare excitation cross sections obtained using straight-line and Coulomb trajectories.

2 The Semiclassical Coupled Equations

The scattering between two heavy ions can be well-approximated semi-classically when the wavelength of relative motion is much smaller than the length scales on which the interaction varies. We will assume this to be the case here. For a localized wave-packet, we can then determine the expectation values

$$\vec{x}(t) = \langle \vec{x} \rangle \qquad \text{and} \qquad \vec{p}(t) = \langle \vec{p} \rangle$$

directly from the classical equations of motion,

$$\frac{d\vec{x}(t)}{dt} = \frac{\vec{p}(t)}{m} \qquad \text{and}$$

$$\frac{d\vec{p}(t)}{dt} = -\langle \nabla V(\vec{x}) \rangle \approx -\nabla V(\vec{x}(t)).$$

In the following, we reduce the coupled wave equations for Coulomb excitation to coupled equations for the average trajectory and for the occupation amplitudes of the ground and excited states.

We begin with the Schrödinger equation for the relative and internal motion,

$$\left[\frac{\hbar^2 \nabla^2}{2m} + h(\alpha) + U_0(\vec{x}) + V_c(\vec{x}, \alpha) \right] \psi = i\hbar \frac{\partial}{\partial t} \psi, \tag{1}$$

where $h(\alpha)$ is the Hamiltonian of the internal degrees of freedom, represented by α, $U_0(\vec{x})$ is a complex optical potential that depends on the relative coordinate alone and $V_c(\vec{x}, \alpha)$ is the interaction coupling the relative and internal motion. We will assume that the Coulomb-excited states can be expressed as eigenstates of this Hamiltonian,

$$h(\alpha)|n\rangle = \varepsilon_n|n\rangle, \tag{2}$$

with ε_n being the energy of the excited state $|n\rangle$. For uniformity, we represent the ground state as $|0\rangle$ with energy ε_0.

We write the wavefunction of the system as a linear combination of the ground and excited states,

$$\psi(\vec{x}, t) = \exp\left[iS(\vec{x}, t)\right] \sum_n a_n(t) |n\rangle, \tag{3}$$

where $\exp[S(\vec{x}, t)]$ describes the average relative motion and $a_n(t)$ is the occupation amplitude of state $|n\rangle$. Before the collision, as $t \to -\infty$ and $|\vec{x}| \to \infty$, only the ground state is occupied, so that

$$a_n(t \to -\infty) = \delta_{n0}. \tag{4}$$

Substituting the wavefunction $\psi(\vec{x}, t)$ in the Schrödinger equation and taking the matrix element of the state $\langle m|$, we obtain

$$\left[-\frac{\hbar^2}{2m}\nabla^2 + \varepsilon_m\right] a_m(t)\, e^{iS(\vec{x},t)}$$

$$+ \left[\sum_n V_{mn}(\vec{x})a_n(t) + U_0(\vec{x})a_m(t)\right] e^{iS(\vec{x},t)}$$

$$= \left[i\hbar\frac{\partial}{\partial t}a_m(t) + i\hbar\frac{d}{dt}a_m(t)\right] e^{iS(\vec{x},t)}, \quad (5)$$

where we have defined

$$V_{mn}(\vec{x}) = \langle m| V_c(\vec{x}, \alpha)|n\rangle. \quad (6)$$

To isolate the equation of relative motion, we multiply the above expression by $a_m^*(t)$ and sum over m. We obtain

$$\left[-\frac{\hbar^2}{2m}\nabla^2 + \sum_m \varepsilon_m |a_m(t)|^2\right] e^{iS(\vec{x},t)} + \left[U_0(\vec{x})\sum_m |a_m(t)|^2\right] e^{iS(\vec{x},t)}$$

$$+ \left[\sum_{n,m} a_m^*(t) V_{mn}(\vec{x})a_n(t)\right] e^{iS(\vec{x},t)}$$

$$= \left[i\hbar\frac{\partial}{\partial t} + i\hbar\sum_m a_m^*(t)\frac{d}{dt}a_m(t)\right] e^{iS(\vec{x},t)},$$

where we assume that

$$\sum_m |a_m(t)|^2 = 1. \quad (7)$$

We can rewrite this equation as,

$$\left[-\frac{\hbar^2}{2m}\nabla^2 + U_0(\vec{x}) + \tilde{V}_c(\vec{x}, t) = i\hbar\frac{\partial}{\partial t}\right] e^{iS(\vec{x},t)}, \quad (8)$$

by defining the contribution of the coupling to the potential for relative motion as,

$$\tilde{V}_c(\vec{x}, t) = \sum_{n,m} a_m^*(t) V_{mn}(\vec{x}) a_n(t)$$

$$+ \sum_m \varepsilon_m |a_m(t)|^2 + \hbar\,\mathrm{Im}\left[\sum_m a_m^*(t)\frac{d}{dt}a_m(t)\right], \quad (9)$$

where we have used Eq. (7) to obtain

$$\sum_m a_m^*(t)\frac{d}{dt}a_m(t) = -\sum_m \frac{d}{dt}a_m^*(t)\,a_m(t)$$

and thus

$$-i\sum_m a_m^*(t)\frac{d}{dt}a_m(t) = \text{Im}\left[\sum_m a_m^*(t)\frac{d}{dt}a_m(t)\right].$$

We observe that $\tilde{V}_c(\vec{x},t)$ is real and the coupling matrix $V_{mn}(\vec{x})$ is hermitian when the coupling potential $V_c(\vec{x},\alpha)$ is real.

We write the phase $S(\vec{x},t)$ as an expansion in $\hbar$ and retain only the first term,

$$S(\vec{x},t) = \frac{1}{\hbar}\left(S_r(\vec{x},t) + iS_i(\vec{x},t)\right) + \dots.$$

When this is substituted in the equation for the relative motion, Eq.(8), we find, to order 0 in $\hbar$,

$$\frac{1}{2m}(\nabla S_r + i\nabla S_i)^2 + \left[U_0(\vec{x}) + \tilde{V}_c(\vec{x},t)\right] = -\frac{\partial}{\partial t}S_r - i\frac{\partial}{\partial t}S_i$$

We separate this into its real and imaginary parts,

$$\frac{1}{2m}\left((\nabla S_r)^2 - (\nabla S_i)^2\right) + \text{Re}\,(U_0) + +\tilde{V}_c(\vec{x},t) + \frac{\partial}{\partial t}S_r = 0 \qquad (10)$$

and

$$\frac{1}{m}\nabla S_r\cdot\nabla S_i + \text{Im}\,(U_0) = -\frac{\partial}{\partial t}S_i$$

If we neglect the term $(\nabla S_i)^2$ in the first of these, it becomes the Hamilton-Jacobi equation, the solution of which provides the trajectory, $\vec{x}(t)$ and $\vec{p}(t) = \nabla S_r$. Substituting $\vec{p}(t)$ in the second equation, we have

$$\frac{d}{dt}S_i = \frac{\partial}{\partial t}S_i + \frac{\vec{p}(t)}{m}\cdot\nabla S_i = -\text{Im}\left[U_0(\vec{x}(t),t)\right],$$

which yields

$$S_i(t) = -\int_{-\infty}^{t}\text{Im}\left[U_0(\vec{x}(t),t)\right]dt.$$

The factor $S_i(t)$ furnishes absorption along the trajectory due to the imaginary part of the optical potential U_0.

To obtain the coupled equations for the occupation amplitudes, we multiply Eq. (8) by $a_m(t)$ and subtract it from Eq. (5).

The expression that results can be reduced to

$$\left[-\widetilde{V}_c(\vec{x}(t), t)\,(t) + \varepsilon_m \right] a_m(t) + \sum_n V_{mn}(\vec{x}(t))\, a_n(t) = i\hbar \frac{d}{dt} a_m(t).$$

Defining

$$a_m(t) = \exp\left[\frac{i}{\hbar} \int \widetilde{V}_c(\vec{x}(t), t)\,dt \right] \widetilde{a}_m(t), \tag{11}$$

and substituting in the previous equation, we obtain

$$i\hbar \frac{d}{dt} \widetilde{a}_m(t) = \varepsilon_m \widetilde{a}_m(t) + \sum_n V_{mn}(\vec{x}(t))\widetilde{a}_n(t) \tag{12}$$

which is the equation we solve numerically to obtain the occupation amplitudes. Since the cross sections depend on the occupation probabilities rather than the amplitudes, the overall phase factor provided by the real potential $\widetilde{V}_c(\vec{x}(t), t)$ is irrelevant and can be ignored. The trajectory dependence of the occupation amplitudes can be represented as a dependence on the impact parameter $\vec{b}$. We then obtain the cross section for excitation of the state n by integrating the product of the absorption factor due to S_i and the asymptotic occupation probability of state n over the impact parameter,

$$\sigma_n = \int db^2 \exp\left(2\int_{-\infty}^{\infty} \mathrm{Im}\left[U_0(\vec{x}(t), t)\right] dt \right) \times \left| \widetilde{a}_n\left(t \to \infty, \vec{b} \right) \right|^2. \tag{13}$$

3 Comparison Between Linear and Coulomb Trajectories

The Coulomb trajectory for relative motion between a pair of ions of reduced mass μ and charges Z_P and Z_T satisfies

$$\frac{p_r^2}{2\mu} + \frac{L^2}{2\mu r^2} + \frac{Z_P Z_T e^2}{r} = E,$$

where p_r is the radial component of the relative momentum, L the conserved angular momentum and E the center-of-mass energy.

We can rewrite this in terms of the impact parameter b as

$$\frac{p_r^2}{2\mu} + E\left(\frac{b}{r} \right)^2 + \frac{Z_P Z_T e^2}{r} = E.$$

The minimum value of the radial distance between the ions r_{min} occurs where the radial momentum $p_r = 0$. In a head-on collision, for which $b = 0$, this is

$$r_{min}(b = 0) = 2a \qquad \text{where} \qquad a = \frac{Z_P Z_T e^2}{2E}.$$

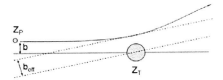

Figure 1 Coulomb and effective straight-line trajectories that are used to calculate Coulomb excitation.

It can be easily shown that for arbitrary values of the impact parameter, the minimum separation is given by

$$r_{min}(b) = a + \sqrt{a^2 + b^2}.$$

We perform straight-line Coulomb excitation calculations using the effective trajectory for which the minimum separation is the same as that of the Coulomb trajectory, as shown in Fig. 1. This is done by using a $b_{eff}(b) = r_{min}(b)$ in the straight-line trajectory calculation of the the damping factor and occupation probabilities in the expression for the cross sections, Eq. (13) (but not in the area d^2b). The additional effects of the Coulomb trajectory will then be due to its curvature relative to the straight line one, as illustrated in Fig. 1.

In the numerical calculations, we use the instantaneous Coulomb interaction and a dipole coupling interaction,

$$V_c(\vec{x}, \alpha) = -\vec{d} \cdot \vec{E}(\vec{x}),$$

where $\vec{E}(\vec{x})$ is the electric field and $\vec{d}$ the isovector dipole operator of the nucleus ^{208}Pb, with the dipole strength given by the energy-weighted sum rule. The excited states were assumed to be harmonic with the first excited state at 13.4 MeV. We used the São Paulo potential[12] for the real part of the optical potential and an imaginary part given by 0.78 times the real part. We note however that the real part of the nuclear potential plays an extremely limited role in the calculations shown here. The important contributions to the scattering come from the Coulomb deflection and the absorption due to the imaginary part of the optical potential. The latter was included in both the Coulomb and straight-line trajectory calculations.

Our results, as a function of the incident energy per nucleon, are shown in Fig. 2. There, we present calculations in which the excited states are stable and others in which they have non-zero widths. In both cases, as expected, we find the excitation cross sections to be smaller when the trajectory is a Coulomb one. The

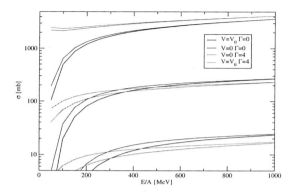

Figure 2 Single (top curves), double (middle curves) and triple (lower curves) giant dipole excitation cross sections in ^{208}Pb + ^{208}Pb as a function of incident energy/nucleon.

single giant dipole resonance (GDR) cross section is almost identical for straight-line and Coulomb trajectories at energies above about 500 MeV/nucleon. In the case of zero width, the GDR cross section for a Coulomb trajectory is about 20 % below its straight-line value at 100 MeV/nucleon and about a factor of two smaller at 50 MeV/nucleon. The deviation between the the two GDR cross sections is much smaller in the case of states with widths.

The discrepancy between the cross sections obtained with Coulomb and straight-line trajectories increases with the number of modes excited. At an incident energy of 500 MeV/nucleon, the difference between the two double giant dipole cross sections is about 5 % and for the triple giant dipole cross sections is about 15 %. These differences are even larger at lower energies and are of about the same order of magnitude for states with and without widths.

In conclusion, we can say that the Coulomb trajectory corrections are negligible for the one and two giant dipole resonance cross sections in the incident energy range in which they have been observed (above about 600 MeV/nucleon).[2, 4] These corrections become more important at lower incident energies and should certainly be taken into account at energies below about 200 MeV/nucleon.

BVC aknowledges partial support from FAPESP and the CNPq.

Bibliography

[1] C. A. Bertulani and G. Baur, Phys. Rep. **163**, 299 (1988).

[2] H. Emling, Progr. Part.Nucl. Phys. **33**, 792 (1994).

[3] Ph. Chomaz and N. Frascaria, Phys. Rep. **252**, 275 (1995).

[4] T. Aumann, P.F. Bortignon and H. Emling, Ann. Rev. Nucl. Sci **48**, 351 (1998).

[5] K. Alder and A. Winther, *Coulomb Excitation*, (Academic Press, New York, 1965).

[6] A. Winther and K. Alder, Nucl. Phys. A **319**, 518 (1979).

[7] C. A. Bertulani, L. F. Canto, M. S. Hussein and A. F. R. de Toledo Piza, Phys. Rev. C **53**, 334 (1997).

[8] B.V. Carlson, L.F. Canto, S. Cruz-Barrios, M.S. Hussein, A.F.R. de Toledo Piza, Phys. Rev. C **59**, 2689 (1999).

[9] B.V. Carlson, L.F. Canto, S. Cruz-Barrios, M.S. Hussein, A.F.R. de Toledo Piza, Ann. Phys.(N.Y.) **276**, 111 (1999).

[10] B.V. Carlson, M.S. Hussein, A.F.R. de Toledo Piza, L.F. Canto, Phys. Rev. C **60**, 014604 (1999).

[11] M.S. Hussein, B.V. Carlson and L.F. Canto, Nucl. Phys. A **731**, 163 (2004).

[12] L.C. Chamon, B.V. Carlson, L.R. Gasques, D. Pereira, C. De Conti, M.A.G. Alvarez, M.S. Hussein, M.A. Cândido Ribeiro, E.S. Rossi Jr. and C.P. Silva, Phys. Rev. C 66, 014610 (2002).

Anais da XXVIII Reunião de Trabalho sobre Física Nuclear no Brasil
SP, Brasil, 2005
Artigo originalmente publicado no Brazilian Journal of Physics, Vol 36 – nº 4B,
Special Issue: XXVIII Workshop on Nuclear Physics in Brazil, pp.1383–1387 (2006).

Fissility of Actinide Nuclei by 60-130 MeV Photons

Viviane Morcelle* and Odilon A. P. Tavares

Centro Brasileiro de Pesquisas Físicas - CBPF/MCT, Rua Dr. Xavier Sigaud 150,
22290-180 Rio de Janeiro - RJ

Abstract. Nuclear fissilities obtained from recent photofission reaction cross section measurements carried out at Saskatchewan Accelerator Laboratory (Saskatoon, Canada) in the energy range 60-130 MeV for ^{232}Th, ^{233}U, ^{235}U, ^{238}U, and ^{237}Np nuclei have been analysed in a systematic way. To this aim, a semiempirical approach has been developed based on the quasi-deuteron nuclear photoabsorption model followed by the process of competition between neutron evaporation and fission for the excited nucleus. The study reproduces satisfactorily well the increasing trend of nuclear fissility with parameter Z^2/A.

1 Introduction

The availability of high-quality monochromatic (or quasi-monochromatic) photon beams generated by different techniques (tagged photons [1,6], and Compton backscattered photons [7,8]) has opened new possibilities for experimental investigation of photonuclear reactions in the energy range $0.02 \lesssim E_\gamma \lesssim 4.0$ GeV. In particular, the development of high-performance parallel-plate avalanche detectors (PPAD) for fission fragments [2, 5, 9, 10] has allowed researchers to obtain photofission cross section data for actinide target nuclei (Th, U, Np) within $\sim 5\%$ of uncertainty, and for pre-actinide (Pb, Bi) within $\sim 12\%$, covering a large incident photon energy-range from about $\sim$30 MeV on.

The experimental data from such photonuclear reactions (cross section measurements and fission probability) have been generally interpreted on the basis of a current nuclear photoreaction model for intermediate- ($30 \lesssim E_\gamma \lesssim 140$ MeV) and high- ($E \gtrsim 140$ MeV) energy photons [2, 11-14]. In brief, during the first reaction step, the incoming photon interacts with a neutron-proton pair ($E_\gamma \lesssim 500$ MeV)

---

* Present address: Instituto de Física, Universidade de São Paulo, C. Postal 66318, 05315-970 São Paulo-SP, Brazil.

and/or individual nucleons ($E_\gamma \gtrsim 140$ MeV), where pions and baryon resonances, and recoiling nucleons initiate a rapid ($\sim 10^{-23}$s) intranuclear cascade process which produces a residual, excited nucleus. After thermodynamic equilibrium was reached, a second, slow process takes place where fission may occur as a result of a mechanism of competition between particle evaporation (neutron, proton, deuteron, triton and alpha particle) and fission experienced by the excited cascade residual.

Nuclear fissility, f, is the quantity which represents the total fission probability of a target nucleus (Z, A) at an incident photon energy, E_γ, and it is defined as the ratio of photofission cross section, σ_f, to photoabsorption cross section, σ_a^T, both quantities being measured at the same energy, i.e.,

$$f(Z, A, E_\gamma) = \frac{\sigma_f(Z, A, E_\gamma)}{\sigma_a^T(Z, A, E_\gamma)} .$$ (1)

Five years ago, a group of researchers reported on results of high-quality photofission cross section measurements for ^{232}Th, ^{233}U, ^{235}U, ^{238}U and ^{237}Np nuclei carried out at Saskatchewan Accelerator Laboratory (Saskatoon, Canada) [5], covering the energy range of 68-264 MeV. The aim of the present work is to perform an analysis of the nuclear fissility data obtained from such measurements in a semiempirical and systematic way by making use of the two-step photoreaction model [12, 14]. The study will cover the range 60-130 MeV, where nuclear photoabsorption has been described by the interaction of the incident photon with neutron-proton pairs (quasi-deuterons), as described for the first time by Levinger [15,16].

2 Quasi-deuteron Nuclear Photoabsorption

In the incident photon energy range here considered, the incoming photon is assumed to be absorbed by a neutron-proton pair as described by Levinger's original and modified quasi-deuteron model [15, 16]. Accordingly, the total nuclear photoabsorption cross section has been evaluated by

$$\sigma_a^T = LZ \left(1 - \frac{Z}{A}\right) \sigma_d(E_\gamma) f_B(E_\gamma) .$$ (2)

In this expression, $\sigma_d(E_\gamma)$ is the total photodisintegration cross section of the free deuteron, the values of which have been taken from the data analysis by Rossi et al. [17] (Fig.1-a). L is the so-called Levinger's constant, and it represents the relative probability of the two nucleons being near each other inside the nucleus as compared to a free deuteron. An evaluation of Levinger's constant for nuclei throughout the Periodic Table by Tavares and Terranova [18] gives L = 6.5 for

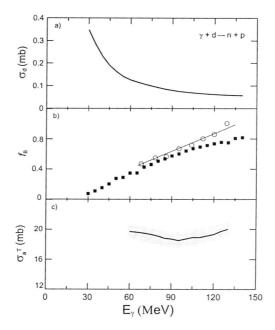

Figure 1 a) Total photodisintegration cross section for the "free"deuteron, σ_d, versus energy[17]. b) Pauli blocking function, $f_B(E_\gamma)$: open circles represent semiempirical values obtained as described in the text; the straight line is the trend of f_B in the range 60-130 MeV (Eq.(4)); full squares are results from a Monte Carlo calculation [19]; c) total nuclear photoabsorption cross section for actinide nuclei as given by Eq.(15); the shaded area indicates the uncertainties (estimated to $\pm 5\%$).

actinide nuclei. Finally, $f_B(E_\gamma)$ is the Pauli block function, which can be established in a semiempirical way, if we observe that, among the actinide nuclei which have been investigated, ^{237}Np is the one which has a low fission barrier (4.63 MeV, therefore, a good chance for fission) and, at the same time, the greater neutron separation energy (6.58 MeV). Thus, we can assume that the photofission cross section of ^{237}Np represents its total photoabsorption cross section, i.e.,

$$\sigma_a^T(E_\gamma)_{Np} = \sigma_f(E_\gamma)_{Np} . \tag{3}$$

This is the same as to assume fissility constant and equal to unity for ^{237}Np. Hence, from Eqs. (2,3), and by making use of the experimental data of σ_f for ^{237}Np (Table 1, 10th column), the Pauli block function in the incident photon energy range 60-130 MeV, can be expressed as

$$f_B(E_\gamma) = KE_\gamma , \; K = 0.00718 \text{ MeV}^{-1} . \tag{4}$$

Table 1 Photofission cross section[a] (σ_f) and nuclear fissility[b] ($\bar{f}$) for actinide nuclei in the quasi-deuteron photoabsorption energy region.

E_γ	^{232}Th		^{233}U		^{235}U		^{238}U		^{237}Np	
(MeV)	σ_f	f	σ_f	f	σ_f	f	σ_f	f	σ_f	f
68	8.8	0.45	15.9	0.82	16.7	0.86	15.6	0.80	19.0	0.98
78	8.8	0.46	15.3	0.80	16.3	0.85	14.4	0.76	19.0	~ 1
86	8.8	0.46	15.9	0.84	16.7	0.88	14.4	0.76	17.7	0.93
95	9.5	0.51	16.4	0.88	16.9	0.91	14.8	0.80	18.5	~ 1
104	9.2	0.49	16.4	0.87	18.8	0.99	15.2	0.80	18.5	0.98
112	9.5	0.50	16.9	0.89	17.3	0.91	15.0	0.79	19.3	~ 1
120	10.6	0.55	17.4	0.90	18.8	0.97	17.0	0.88	19.6	~ 1
129	10.8	0.54	19.6	0.98	19.4	0.97	18.0	0.90	22.0	~ 1

[a] Cross sections (expressed in mb) are experimental data obtained at Saskatchewan Accelerator Laboratory (Saskatoon, Canada) by Sanabria *et al.* [5];

[b] Nuclear fissilies are deduced values from $f = \sigma_f/\sigma_a^T$, where σ_a^T is total nuclear photoabsorption cross section as explained in the text (section 2).

This result is shown in Fig.1-b (straight line) together with the semiempirical values of f_B (circles). For a comparison, the results of a Monte Carlo calculation (full squares in Fig.1-b) obtained by de Pina *et al.* [19] is also shown. Except for 129 MeV, the above results differ from each other by 8 % on the average. Finally, if one adopts for $f_B(E_\gamma)$ the result expressed in (4), the total cross section for nuclear photoabsorption in the range 60-130 MeV for actinide nuclei can be calculated as

$$\sigma_a^T(E_\gamma) = CE_\gamma\sigma_d(E_\gamma)\,, \quad C = LZ\left(1 - \frac{Z}{A}\right)K\,. \tag{5}$$

Since the constant C does not vary significantly for the actinide nuclei ($\bar{C} = 2.60 \pm 0.03$ MeV^{-1}), we can assume the same curve $\sigma_a^T(E_\gamma)$ as depicted in Fig.1-c valid for all actinide here analysed. Therefore, the nuclear fissilities, as defined in (1), can be obtained in this way, and results are listed in Table 1.

3 Nuclear Excitation and Fission Probability

As a consequence of the quasi-deuteron primary photointeraction, $\gamma + (n + p) \rightarrow n^* + p^*$, different residual nuclei can be formed according to one of the three possible modes: 1) The neutron escapes from the nucleus, at the same time that the proton remains retained; 2) the proton escapes and the neutron is retained; 3) both neutron and proton are retained inside the nucleus. A fourth possibility there exists clearly of obtaining residual nuclei, namely, escaping of both the neutron and proton simultaneously from the nucleus. In this case, however, no excitation energy is left to the residual nucleus, with the consequence of null chance for fission from this residual [14]. Let τ_n and τ_p denote, respectively, the probabilities of escaping for neutron and proton from the nucleus (without suffering for any secondary interaction), i.e., the nuclear transparencies to neutron (τ_n) and proton (τ_p). Complete retention (or reabsorption, or non-escaping) of nucleons means $\tau_p = \tau_n = 0$. The probabilities for neutron and proton retention are, respectively, $(1 - \tau_n)$ and $(1 - \tau_p)$. Therefore, the probabilities of residual nuclei formation (excited or not) following one of the four routes mentioned above are given by $p_1 = \tau_n(1 - \tau_p)$, $p_2 = \tau_p(1 - \tau_n)$, $p_3 = (1 - \tau_p)(1 - \tau_n)$ and $p_4 = \tau_p\tau_n$.

Nuclear transparencies depend essentially upon neutron and proton kinetic energies in their final states [12,14]. For photons in the energy range 60-130 MeV interacting with actinides, it is found that for almost $\sim 80\%$ of cases excited residual nuclei are formed ($\overline{E}_1^* \approx \overline{E}_2^* \approx E_\gamma/2$, hence, highly fissionable residuals), and the third mode of formation is the predominant one.

The total fission probability, i.e, the fissility of the target nucleus is thus given by

$$f(E_\gamma) = \sum_{i=1}^{4} p_i P_{fi}(E_i^*), \qquad \sum_{i=1}^{4} p_i = 1, \tag{6}$$

where P_{f_i} represents the fission probability for the residual nucleus formed according to the mode i of formation of residuals. Since $E_4^* = 0$, it follows that $P_{f_4} = 0$, and, therefore

$$f(E_\gamma) = p_1 P_{f_1}(Z, A - 1, \overline{E}_1^*) + p_2 P_{f_2}(Z - 1, A - 1, \overline{E}_2^*) + p_3 P_{f_3}(Z, A, \overline{E}_3^*). \tag{7}$$

The different residual nuclei formed are actinide (like the target ones) and, in this case, the fission barriers are, in general, lower than the respective neutron separation energies ($B_{f_0} < S_n$). Therefore, we can say that the fission probability of residual nuclei formed following the formation modes $i = 1, 2$, and 3 is given approximately by their first-chance fission probability, f_{1_i}, i.e,

$$P_{f_1} \approx f_{1_1}(Z, A - 1, \overline{E}_1^*); \; P_{f_2} \approx f_{1_2}(Z - 1, A - 1, \overline{E}_1^*);$$
$$P_{f_3} \approx f_{1_3}(Z, A, E_\gamma). \tag{8}$$

On the other hand, the first-chance fission probabilities, f_{1_i} ($i = 1, 2, 3$), should not differ significantly for adjacent actinide nuclei, because all of them do exhibit low fission barrier values ($B_f \lesssim 5$ MeV), at the same time that f_{1_i} should not vary significantly with energy too (see Table 1). Calculations of nuclear transparencies for actinides [14] show that ($0.20 \lesssim \tau_n \lesssim 0.50$) and $0.10 \lesssim \tau_p \lesssim 0.15$ for neutron and proton energies considered here, in such a way that the mode $i = 3$ is the predominant one, at the same time that $p_4 \lesssim 8\%$. Thus, to a good approximation, we can write

$$f(Z, A, E_\gamma) \approx f_1(Z, A, E^* = E_\gamma), \quad 60 \lesssim E_\gamma \lesssim 130 \text{ MeV}. \tag{9}$$

This result means that, for actinide nuclei, nuclear photofissility can be described by the first-chance fission probability of the target nucleus with an excitation energy equal to the incident photon energy.

4 Competition between Neutron Evaporation and Fission

The most probable residual nucleus (Z, A, E_γ) formed after the primary quasi-deuteron photoabsorption de-excites by a mechanism of competition between nucleon evaporation and fission. For actinide residuals of low excitation energy ($E^* \approx E_\gamma \lesssim 130$ MeV) the main modes of nuclear de-excitation are neutron evaporation and fission (charged particles, such as proton and alpha particle, have the additional difficult of a coulomb barrier which hinders the emission of these particles). The quantitative description of the process [12,14] is based on the drop model for nuclear fission proposed by Bohr and Wheeler [20], and the statistical theory of nuclear evaporation developed by Weisskopf [21]. Accordingly, the fission probability relative to neutron emission is given by Vandenbosch-Huizenga's equation [22], which reads

$$\frac{\Gamma_f}{\Gamma_n} = F = \frac{15}{4} \frac{\left(\sqrt{4ra_n(E^* - B_f)} - 1\right)}{rA^{\frac{2}{3}}(E^* - S_n)}$$
$$\times \exp\{2\sqrt{a_n}[\sqrt{r}\sqrt{E^* - B_f} - \sqrt{E^* - S_n}]\}. \tag{10}$$

Thus, the first-chance fission probability, i.e., the nuclear photofissility for the nuclei here considered has been calculated as

$$f \approx f_1 = \frac{F}{1 + F}. \tag{11}$$

In expression (10), S_n is neutron separation energy [23], and B_f is the fission barrier height correct for the nuclear excitation,

$$B_f = B_{f_0} \left(1 - \frac{E^*}{B}\right), \tag{12}$$

where B_{f_0} is the ground state fission barrier [24], and B is the total nuclear binding energy [23]. For the level density parameter of residual nucleus after neutron evaporation, a_n, we use a modern expression,

$$a_n = \tilde{a}\left\{1 + [1 - \exp(-0.051E^*)]\frac{\Delta M}{E^*}\right\}, \tag{13}$$

which has been proposed by Iljinov *et al.*[25]. Here, ΔM represents the shell model correction to the nuclear mass [24], and

$$\tilde{a} = 0.114A + 0.098A^{\frac{2}{3}} \quad MeV^{-1} \qquad (14)$$

is the asymptotical value of a_n (a small correction on E^*-values due pairing effects was neglected in (13)). The constants which appear in (13) and (14) are adjustable parameters resulting from the systematic study on level density carried out with hundreds of excited nuclei (for details see [25]).

The quantity $r = a_f/a_n$ (ratio of the level density parameter at the fission saddle point to a_n) is unknown, and it has been determined in a semiempirical way. The experimental fissility data from Table 1 are used together with equations (10-14) to obtain the ratio $r = a_f/a_n$ for each reaction case. Once the r-values were obtained as described above they could be fitted to a general expression of the type

$$r = 1 + a \left(\frac{Z^2}{A} - b \right) , \qquad (15)$$

where a and b are constants to be determined by least-squares analysis. Finally, we insert back into Eq. (11) the semiempirical r-values given by (15) to obtain the nuclear fissility for each target nucleus and the different incident photon energies.

5 Results and Discussion

For the various target nuclei considered in the present nuclear photofissility study we observe, firstly, that the photofission cross sections do not vary remarkably in the range 60-130 MeV ($\sim$ 23% for ^{232}Th and ^{233}U, and $\sim$ 16% for ^{235}U, ^{238}U and ^{237}Np, as it follows from data of Table 1). The same occurs with the total nuclear photoabsorption cross section (a variation not greater than 8%, or practically constant, within the uncertainties, as we can see in Fig. 1-c). Therefore, nuclear fissility f from the measurements results practically constant in the range 60-130 MeV for each target nucleus (Table 1), thus it is sufficient to consider the average value, $\bar{f}$, for each case. However, we remark on an increasing trend of $\bar{f}$ with parameter Z^2/A (Fig.2-a, full circles) except for the ^{233}U isotope. Since this nuclide exhibits the lowest fission barrier height as calculated from the droplet model (4.60 MeV [24]), one should expect for a value of nuclear fissility greater than that for ^{235}U, which has a higher fission barrier (4.77 MeV). On the contrary, this was not the case observed experimentally (Fig. 2-a).

The values for ratio $r = a_f/a_n$ determined semiempiracally for each target nucleus (section 4) result rather independent of energy, and their average value, $\bar{r}$, calculated in the range 60-130 MeV is plotted versus Z^2/A, where the value for ^{233}U isotope appears displaced somewhat from the general linear trend. By assuming the

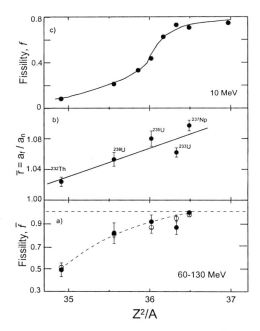

Figure 2 a) Average nuclear fissility, $\bar{f}$, versus parameter Z^2/A: full circles are average values from experimental data listed in Table 1; open circles are calculated values from the present analysis (see text); the curve is drawn by eye. b) ratio $r = a_f/a_n$ versus Z^2/A: points represent average semiempirical values as obtained in this work; the straight line is a least-squares fit to the points (cf. Eq.(15)). c) Experimental nuclear fissility-values at 10-MeV photon energy for various actinides [27,28]; the curve is drawn by eye.

parametrization defined by Eq.(15) as proposed in [26], which is valid only for actinides, we found for the constants a and b the values $a = 0.0389$ and $b = 34.24$. Finally, the $\bar{r}$-values determined in this way are inserted into (10) to obtain the calculated nuclear fissilities. The average values are shown in Fig. 2-a (open circles), where we can see that the relative position of the average measured and calculated fissilities for each nucleus is similar to that shown in Fig.2-b for the quantity $\bar{r}$, between the points and the straight line fitted to them.

Finally, let us consider the unexpected case of $\bar{r}$- and $\bar{f}$- values for ^{233}U target nucleus being slightly lower than the predicted ones ($\sim$ 10% lower in the case of $\bar{f}$ and $\sim$ 3% in the case of $\bar{r}$, as shown in Fig.2-a,b). Such a result contradicts experimental fissility data at low energies [27,28] (Fig. 2-c), as well as the prediction from droplet model for fission [24]. However, serious difficulties there exist when evaluating small (but important) shell effects which surely contribute to the final value of the fission barrier. This latter value, in turn, change the semiempirical r̄-values. In addition, systematic uncertainties associated with σ_f-measurements of isotopes very close to each other, such as ^{235}U and ^{233}U can make difficult a better determination of the experimental σ_f-values for these isotopes (for details see Ref. [5]).

6 Conclusion

A semiempirical approach has been developed based on the current two step-model (quasi-deuteron nuclear photoabsorption followed by a process of competition between neutron evaporation and fission) for photonuclear reactions and used to analyse available experimental data on photofission cross section of actinide nuclei (^{232}Th, ^{233}U, ^{235}U, ^{238}U and ^{237}Np) in the range 60-130 MeV [5]. The adjustable parameter $r = a_f/a_n$ resulted constant for each target nucleus in the incident photon energy range considered, but the average value r̄ shows an increasing linear behavior with parameter Z^2/A (Fig. 2-b). This result is valid only for actinide nuclei ($Z^2/A \gtrsim 34.25$), which is in quite good agreement with previous conclusions by Martins *et al* [26]. The average nuclear fissilities have been quite well reproduced by the present analyses, and have shown also an increasing trend with Z^2/A (Fig. 2-a), although differences not yet completely resolved there exist in the case of ^{233}U isotope.

Acknowledgements

The authors wish to express their gratitude to Silvio C. Conceição, Margaret Q.N. Soares e Silva, and Susana R. de Pina for technical supporting. Special thanks are due to S.B. Duarte for valuable discussions. VMA acknowledges the Brazilian CNPq for partially supporting of this work.

Bibliography

[1] H. Ries, U. Kneissl , G. Mank , H. Ströher, W. Wilke, R. Bergère, P. Bourgeois, P. Carlos, J.L. Fallou, P. Garganne, A. Veyssière, and L.S. Cardman, Phys Letters B**139**, 254 (1984).

[2] A. Leprêtre, R. Bergère, P. Bourgeois, P. Carlos, J. Fagot, J.L. Fallou, P. Garganne, A. Veyssière, H. Ries, R. Gobel, U. Kneissl, G. Mank, H. Ströher, W. Wilke, D. Ryckbosch, and J. Jury, Nucl. Phys. **A472**, 533 (1987).

[3] D. Babusci, V. Bellini, M. Capogni, L. Casano, A. D' Angelo, F. Ghio, B. Girolami, L. Hu, D. Moricciani, and C. Schaerf, Riv. Nuovo Cimento **19** (5), 1 (1996).

[4] M.L. Terranova, G.Ya. Kezerashvili, V.A. Kiselev , A.M. Milov, S.I. Mishnev, I.Ya. Protopopov, V.N. Rotaev, D.N. Shatilov, and O.A.P. Tavares, Journal of Physics G: Nucl. Part. Phys. **22**, 1661 (1996).

[5] J.C. Sanabria, B.L. Berman, C. Cetina, P.L. Cole, G. Feldman, N.R. Kolb, R.E. Pywell, J.M. Vogt, V.G. Nedorezov, A.S. Sudov , and G.Ya. Kezerashvili, Phys. Rev. C **61**, 034604 (2000).

[6] C. Cetina, P. Heimberg, B.L. Berman, W.J. Briscoe, G. Feldman, L.Y. Murphy, H. Crannell, A. Longhi, D.I. Sober, J.C. Sanabria, and G.Ya. Kezerashvili, Phys. Rev. C **65**, 044622 (2002).

[7] D. Babusci, L. Casano, A. D'Angelo, P. Picozza, C. Schaerf, and B. Girolami, Prog. Part. Nucl. Phys. **24**, 119 (1990).

[8] M.L. Terranova, G.Ya. Kezerashvili, A.M. Milov, S.I. Mishnev, N.Y. Muchnoi, A.I. Naumenkov, I.Ya. Protopopov, E.A. Simonov, D.N. Shatilov, O.A.P. Tavares, E. de Paiva and E.L. Moreira, J. Phys. G: Nucl. Part. Phys. **24**, 205 (1998).

[9] N. Bianchi, A. Deppman, E. De Sanctis, A. Fantoni, P. Levy Sandri, V. Lucherini, V. Muccifora, E. Polli, A.R. Reolon, P. Rossi, A.S. Iljinov, M.V. Mebel, J.D.T. Arruda-Neto, M. Anghinolfi, P. Corvisiero, G. Gervino, L. Mazzaschi, V. Mokeev, G. Ricco, M. Ripani, M. Sanzone, M. Taiuti, A. Zucchiatti, R. Bergère, P. Carlos, P. Garganne, and A. Leprêtre, Phys. Rev. C **48**, 1785 (1993).

[10] N. Bianchi, A. Deppman, E. De Sanctis, A. Fantoni, P. Levy Sandri, V. Lucherini, V. Muccifora, E. Polli, A.R. Reolon, P. Rossi, M. Anghinolfi, P. Corvisiero, G. Gervino, L. Mazzaschi, V. Mokeev, G. Ricco, M. Ripani, M. Sanzone, M. Taiuti, A. Zucchiatti, R. Bergère, P. Carlos, P. Garganne, and A. Leprêtre, Phys. Letters **B299**, 219 (1993).

[11] A.S. Iljinov , D.I. Ivanov, M.V. Mebel , V.G. Nedorezov, A.S. Sudov, and G.Ya. Kezerashvili: Nucl. Phys. **A539**, 263 (1992).

[12] E. de Paiva, O.A.P. Tavares, and M.L. Terranova: J. Phys. G: Nucl. Part. Phys. **27**, 1435 (2001).

[13] A. Deppman, O.A.P. Tavares, S.B. Duarte, E.C. de Oliveira, J.D.T. Arruda-Neto, S.R. de Pina, V.P. Likhachev, O. Rodriguez, J. Mesa, and M. Gonçalves: Phys. Rev. Letters **87**, 182701 (2001)

[14] O.A.P. Tavares and M.L. Terranova: Z. Phys. A: Hadrons and Nuclei **343**, 407 (1992).

[15] J.S. Levinger: Phys. Rev. **84**, 43 (1951).

[16] J.S. Levinger: Phys. Letters **B82**, 181 (1979).

[17] P. Rossi , E. De Sanctis, P.Levy Sandri, N. Bianchi, C. Guaraldo, V. Lucherini, V. Muccifora, E. Polli, A.R. Reolon, and G.M. Urciuoli: Phys. Rev. C **40**, 2412 (1989).

[18] O.A.P. Tavares and M.L. Terranova: J. Phys. G: Nucl. Part. Phys. **18**, 521 (1992).

[19] S.R. de Pina, J. Mesa, A. Deppman, J.D.T. Arruda-Neto, S.B. Duarte, E.C. de Oliveira, O.A.P. Tavares, E.L. Medeiros, M. Gonçalves, and E. de Paiva, J. Phys. G: Nucl. Part. Phys. **28**, 2259 (2002).

[20] N. Bohr and J.A. Wheeler: Phys. Rev. **56**, 426 (1939).

[21] V. Weisskopf: Phys. Rev. **52**, 295 (1937).

[22] R. Vandenbosch and J.R. Huizenga: Nuclear Fission, 1st edition, p. 227 (N. York, Academic Press, 1973).

[23] G. Audi and A.H. Wapstra: Nucl. Phys. **A565**, 1 (1993).

[24] W.D. Myers: Droplet model of atomic nuclei, 1st edition, pp.35 (N. York, Plenum Press), 1977.

[25] A.S. Iljinov , M.V. Mebel, N. Bianchi, E. De Sanctis, C. Guaraldo, V. Lucherini, V. Muccifora, E. Polli, A.R. Reolon, and P. Rossi: Nucl. Phys. **A543**, 517 (1992).

[26] J.B. Martins, E.L. Moreira , O.A.P. Tavares, J.L. Vieira, L. Casano, A. D'Angelo, C. Schaerf, M.L. Terranova, D. Babusci, and B. Girolami: Phys. Rev. C **44**, 354 (1991).

[27] B.L. Berman, J.T. Caldwell, E.J. Dowdy, S.S. Dietrich, P. Meyer, and R.A. Alvarez: Phys. Rev. C **34**, 2201 (1986).

[28] J.T. Caldwell, E.J. Dowdy, B.L. Berman, R.A. Alvarez, and P. Meyer: Phys. Rev. C **21**, 1215 (1980).

Anais da XXVIII Reunião de Trabalho sobre Física Nuclear no Brasil
SP, Brasil, 2005

Analysis of Multi-Parametric Coincidence Measurements with n Detectors

Frederico Antonio Genezini[1], Guilherme Soares Zahn[2],
Marcus Paulo Raele[2], Cibele Bugno Zamboni[2] and
Manoel Tiago Freitas da Cruz[3]

[1] Centro Regional de Ciências Nucleares - CRCN-CNEN/PE

[2] Instituto de Pesquisas Energéticas e Nucleares, Caixa Postal 11049, 05422-970, São Paulo, SP, Brazil

[3] Instituto de Física da Universidade de São Paulo, Caixa Postal 66318, 05315-970, São Paulo, SP, Brazil

Abstract. This work describes a method for calculating transition intensities from coincidence measurements. In order to use all the available statistics, the spectra obtained for all the detector pairs are summed up, and the resulting bi-dimensional spectrum is used to calculate the transition intensities. This method is then tested in the calculation of intensities for well-known transitions from both ^{152}Eu and ^{133}Ba standard sources.

Keywords. intensity calculation, multi-detector array, gamma-gamma coincidence

1 Introduction

The use of multi-detector setups for data acquisition in coincidence measurements of γ-ray spectroscopy has become very frequent. While most multi-detector experiments aim to analyze angular correlation parameters, those facilities are useful for non-angular coincidence measurements as well, as in the verification of gamma-gamma coincidence relations and the determination of transition intensities. In these cases, the use of a multi-parametric coincidence system, where n detectors are associated as $L = n(n-1)/2$ independent pairs, represents an improvement, due to the reduction of the acquisition time, when compared to setups with only two detectors.

2 Theoretical Basis

Assume that there are two photons, γ_a and γ_b, emitted by a nucleus in a rapid succession (it is said that they belong to a *cascade*, $\gamma_a\gamma_b$), forming the angle ϕ_{ab} between their directions of emission. Two infinitesimal detectors, 1 and 2, subtending the solid-angle elements $d\Omega_a$ and $d\Omega_b$ respectively, are positioned at the direction of emission of the photons, at some distance from the radioactive source. The radioactive source is point-like and is positioned at the origin of the coordinate system. The number of detected coincidences is given by

$$dN(\phi_{ab}) = Sy_{ab}\left[\frac{d\Omega_a}{4\pi}\frac{d\Omega_b}{4\pi}W(\phi_{ab})\right]\epsilon^1(E_a,\Omega_a)\,\epsilon^2(E_b,\Omega_b) \tag{1}$$

where ϕ_{ab} is the angle between the solid-angle elements $d\Omega_a \equiv (d\vartheta_a, d\varphi_a)$ and $d\Omega_b \equiv (d\vartheta_b, d\varphi_b)$ with ϑ, φ representing angles in spherical coordinates. The term between brackets is the emission probability of the photons γ_a and γ_b inside the corresponding solid-angle elements. ϵ^1 and ϵ^2 are the intrinsic photo-peak efficiencies of detectors 1 and 2, respectively. Finally, S is the number of disintegrations occurred during the measurement time and y_{ab} is the emission probability of the $\gamma_a\gamma_b$ cascade (per disintegration). $W(\phi_{ab})$ is expressed as

$$W(\phi_{ab}) = 1 + A_{22}P_2(\cos\phi_{ab}) + A_{44}P_4(\cos\phi_{ab}) \tag{2}$$

where P_k are the Legendre polynomials of order k, with coefficients A_{kk}. It is seen that this is the angular correlation function, truncated due to the experimental sensitivity, normalized and not affected by solid angle effects.

Passing now to detectors of finite size, the number of detected coincidences becomes

$$N_{ab}(\theta^{12}) = \int dN(\phi_{ab}) = \iint \frac{d\Omega_a}{4\pi}\frac{d\Omega_b}{4\pi}W(\phi_{ab})\epsilon^1(E_a,\Omega_a)\epsilon^2(E_b,\Omega_b) \tag{3}$$

Here, θ^{12} is a reference angle, between the symmetry axes of detectors 1 and 2. Substituting $W(\phi_{ab})$ by (2), we have:

$$N_{ab}(\theta^{12}) = Sy_{ab}\sum_{k\ even=0}^{4}A_{kk}$$
$$\times \iint \frac{d\Omega_a}{4\pi}\frac{d\Omega_b}{4\pi}\epsilon^1(E_a,\Omega_a)\epsilon^2(E_b,\Omega_b)P_k(\cos\phi_{ab}) \tag{4}$$

with $A_{00} = 1$. Factoring the zeroth-order term, we produce:

$$N_{ab}(\theta^{12}) = \alpha\left[1 + \sum_{k\ even=2}^{4}A_{kk}F_k(\theta^{12})\right] = \alpha w_{ab}(\theta^{12}) \tag{5}$$

with

$$\begin{aligned}\alpha &= Sy_{ab}\int\frac{d\Omega_a}{4\pi}\epsilon^1(E_a,\Omega_a)\int\frac{d\Omega_b}{4\pi}\epsilon^2(E_b,\Omega_b)\\ &= Sy_{ab}\epsilon^1(E_a)\epsilon^2(E_b)\end{aligned} \qquad (6)$$

and

$$\omega_{ab}(\theta) = 1 + A_{22}Q_{22}P_2(\cos\theta) + A_{44}Q_{44}P_4(\cos\theta) \qquad (7)$$

where one should note that now $\epsilon^{i(j)}(E_{a(b)})$ is the absolute photo-peak efficiency of the i(j)-th detector at the energy $E_{a(b)}$. Q_{kk} are the solid angle correction coefficients that account for the attenuation of the angular correlation function. From equations (5)-(7), we can see that the number of detected coincidences is proportional to the product of the absolute efficiencies times the angular correlation function. Generalizing for several detectors and rewriting y_{ab}, the total number of coincidences for cascade $\gamma_a\gamma_b$ may be written as:

$$N_{ab}^{tot} = \sum_{(ij)=1}^{L} N_{ab}(\theta^{ij}) = SI_a br_b \sum_{(ij)=1}^{L} \epsilon^i(E_a)\epsilon^j(E_b)\omega_{ab}(\theta^{ij}) \qquad (8)$$

where ij is the detector pair, θ^{ij} is the angle between the symmetry axes of the detectors, I_a is the emission probability of the first photon of the cascade, γ_a, and br_b is the normalized branching ratio of the second photon of the cascade, γ_b.

In order to eliminate the influence on the intensity calculations from both the source activity and the real time of acquisition (which is hard to determine in multi-parametric measurements), two $\gamma\gamma$ cascades with a common transition were used. The ratio of the total numbers of detected coincidences for these cascades, by all detector pairs, is:

$$\frac{N_{ab}^{tot}}{N_{cb}^{tot}} = \frac{SI_a br_b \sum_{(ij)=1}^{L} \epsilon^i(E_a)\epsilon^j(E_b)\omega_{ab}(\theta^{ij})}{SI_c br_b \sum_{(ij)=1}^{L} \epsilon^i(E_c)\epsilon^j(E_b)\omega_{cb}(\theta^{ij})} \qquad (9)$$

Now, in order to allow the analysis of very weak cascades, it would be interesting to use the spectrum summed over all angles. For that purpose, the following approximation shall be made, which relies on two basic assumptions: **a.** that the detector efficiencies are not very dissimilar from each other; and **b.** that the sum of $\omega(\theta)$ over the six detector pairs is approximately constant for all cascades:

$$\frac{N_{ab}^{tot}}{N_{cb}^{tot}} = \frac{I_a \sum_{(ij)=1}^{L} \epsilon^i(E_a)\epsilon^j(E_b)\omega_{ab}(\theta^{ij})}{I_c \sum_{(ij)=1}^{L} \epsilon^i(E_c)\epsilon^j(E_b)\omega_{cb}(\theta^{ij})} \approx \frac{I_a \sum_{(ij)=1}^{L} \epsilon^i(E_a)\epsilon^j(E_b)}{I_c \sum_{(ij)=1}^{L} \epsilon^i(E_c)\epsilon^j(E_b)} \qquad (10)$$

and now the intensity of the transition γ_a may be written as:

$$I_a = I_c \frac{N_{ab}^{tot} \sum_{(ij)=1}^{L} \varepsilon^i(E_c)\varepsilon^j(E_b)}{N_{cb}^{tot} \sum_{(ij)=1}^{L} \varepsilon^i(E_a)\varepsilon^j(E_b)} \qquad (11)$$

3 Experimental Validation

In order to give experimental support for the approximation made in eq.10, the intensities of both the 779- and 1090-keV transitions from ^{152}Eu, as well as the 356-keV transition from ^{133}Ba, were calculated using the forementioned methodology, and the results compared to the values found in reference [2].

3.1 Data Acquisition

The multi-parametric system used in this test, located at the Linear Accelerator Laboratory of the Instituto de Física da Universidade de São Paulo, is composed of four Ge detectors with active volumes ranging from 50 to 190 cm^3. The signals from these detectors are inserted into a usual fast-slow electronics, where the decisory functions are performed by a custom CAMAC module. If all logic conditions are met, a CAMAC controller, directly coupled to the bus of an IBM-compatible personal computer (PC) through an ISA-bus lengthener, manages the data transfer to the PC. Both devices, the controller and the lengthener, were designed and built in-house [1], and allow a high data throughput. The data were recorded on an event-by-event mode on hard disk. The whole system can be seen in Figure 1.

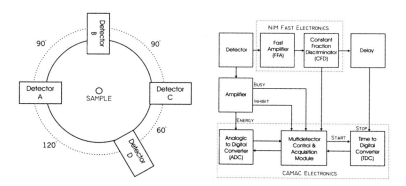

Figure 1 Experimental setups used in this work.
(left) Placement of the four detectors around the sample;
(right) Simplified scheme of the electronic setup.

In the present measurement, the system was exposed to the radiation from a 3.6μCi ^{152}Eu standard source for a total of 60 h; a standard ^{133}Ba source of approximately the same activity was also counted for the same time.

3.2 Data Reduction

The summation procedure mentioned above must be carefully done, and two steps are required. In the first step, a data set taken with a particular detector pair is time-gated in order to produce a pair of two-dimensional spectra, one containing accidental coincidences only (AC) and other with the total (true + accidental) coincidences (TC). In the second step, the AC spectra from all pairs are relocated in energy, in order to correct for differences in the energy calibrations of the detectors, and then summed together to produce a single two-dimensional AC spectrum; the same procedure is applied to the TC spectra, and the two "all-in-one" two-dimensional spectra are analyzed as if they were obtained from a single detector pair.

4 Results and Discussion

For this method to be applicable, the decay scheme of the nucleus under study must contain pairs of cascades with a common transition. Four cascades from ^{152}Eu have been used in this test: 344.3-778.9 keV and 344.3-411.1 keV, to determine the intensity of the 778.9 keV transition, and 344.3-1089.8 keV and 344.3-1299.2 keV to determine the intensity of the 1089.8 keV transition. Both decay schemes are shown in Figure 2. The results are presented in Table 1.

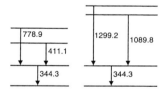

Figure 2 Simplified decay schemes showing the transitions from ^{152}Eu used in this work.

In addition to those, the 81.0-356.0 keV and the 81.0-302.9 keV cascades from ^{133}Ba were also used to determine the intensity of the 356.0 keV transition; the decay scheme for these transitions is shown in Figure 3, and the results are shown in Table 2.

As shown in Table 1, the intensity values obtained for the ^{152}Eu transitions by this method are statistically compatible with the corresponding values from the literature, while Table 2 shows that the result obtained for the 356 keV ^{133}Ba transition is roughly within 2 standard deviations from the value found in [2]. These results show that the proposed method is suitable for the obtention of transition intensities from multi-detector coincidence experiments.

Table 1 Comparison of the intensity results obtained for the ^{152}Eu transitions by the present method to the values found in [2].

Transition	778.9 keV	1089.8 keV
Peak Area [cascade 1]	$2.385(7) \times 10^6$ [344.3 - 778.9] keV	$1.973(22) \times 10^5$ [344.3 - 1089.8] keV
Peak Area [cascade 2]	$6.91(4) \times 10^5$ [344.3 - 411.1] keV	$1.669(19) \times 10^5$ [344.3 - 1299.2] keV
I_γ (this work) I_γ ref [2]	13.1 (3) 12.98 (25)	1.69 (3) 1.711 (9)

Figure 3 Simplified decay scheme showing the transitions from ^{133}Ba used in this work.

Table 2 Comparison of the intensity results obtained for the ^{133}Ba transition by the present method to the value found in [2].

Transition	356.0 keV
Peak Area [cascade 1]	$3.448(10) \times 10^6$ [81.0 - 356.0] keV
Peak Area [cascade 2]	$1.179(6) \times 10^6$ [81.0 - 302.9] keV
I_γ (this work) I_γ ref [2]	0.605 (5) 0.6205 (19)

Bibliography

[1] D. Bärg-Filho, R. C. Neves, and V. R. Vanin. Proceedings of the XX Brazilian Workshop on Nuclear Physics, S. R. Souza et al. eds. (World Scientific, Singapore, 1998), p. 424.

[2] R. B. Firestone and V. S. Shirley, **Table of Isotopes** John Wiley & Sons, New York, 1996.

Anais da XXVIII Reunião de Trabalho sobre Física Nuclear no Brasil
SP, Brasil, 2005

Compton Scattering and Absolute Measurement of Tagged Photon Energies

W. R. Carvalho Jr., Z. O. Guimarães-Filho and A. Deppman

Instituto de Física da Universidade de São Paulo, C.P. 66318, 05315-970, São Paulo, SP, Brazil.

Abstract. A simulation based on Monte Carlo methods was developed in order to evaluate the potentiality of using Compton scattering at high energies ($\sim$ GeV) to obtain high precision ($\lesssim 0.1\%$) absolute measurements of tagged photon energies. This method is based on angular measurements of the scattering products to reconstruct the incident photon energy using the kinematics of Compton scattering. Through simulation of Jlab's *Primex* experiment[1, 2, 3] that uses this method for energy calibration of a tagged photon beam, it was possible to identify two sources of systematic errors. Analysis methods that minimize one of these systematic errors and creates corrections for the energy measurement were developed. Our results show that, at least in the studied experimental setup, it is possible to obtain energy measurements with a precision of about 0.07 %.

1 Introduction and Experimental Considerations

Compton scattering can be used to get a higher energy resolution than that available from the photon tagger. An absolute energy measurement can be obtained based only on Compton scattering kinematics and on measurements of the scattering angles of the final particles, an electron and a photon.

From the kinematics of Compton scattering one can derive [4] an expression that relates the incoming photon energy E_γ to the scattering angles θ_γ and θ_e (eq. 1, $\hbar = c = 1$), which is the basis of this method:

$$E_\gamma(\theta_\gamma, \theta_e) = m_e \left(\cot \frac{\theta_\gamma}{2} \cot \theta_e - 1 \right) \tag{1}$$

Fig. 1 shows the simplified experimental setup for the Compton method. It consists of a high energy photon beam aligned with the detector symmetry axis, a thin target and a high spatial resolution detector. Some photons from the beam are Compton scattered by atomic electrons in the target, thus generating an ejected electron and a scattered photon. This $e' - \gamma'$ pair hits the detector, giving rise to the angular measurements θ_γ and θ_e. These angles are then fed into Eq. 1, reconstructing the energy of the incident photon that generated this specific $e' - \gamma'$ pair.

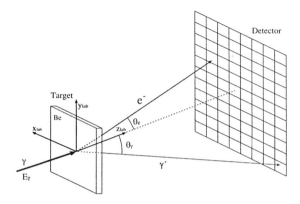

Figure 1 Simplified experimental setup.

However Eq. 1 is only valid in the case of free electrons at rest. So, several experimental and physical effects will produce a dispersion on Eq.1. The most important contributions, that are studied in this work, are:

1. The target electrons are bound and have an internal atomic momentum.

2. The ejected electron is subjected to multiple scattering until leaving the target material, changing its original direction.

3. Several geometrical effects, such as detector angular acceptances, beam intensity distribution (shape) and beam-detector misalignments.

4. Detector spatial resolution. This is the most important contribution.

In order to take these effects into account and estimate their impact on the energy measurements, a Monte Carlo simulation was developed. Some advantages of the Compton method are: it's an absolute method, i.e. no calibration is needed; and it can be used to enhance the photon tagger energy resolution of an experiment with minimum impact on the original experimental setup, or in some cases even with simultaneous acquisition with this main experiment.

2 Description of the Monte Carlo code

An original Monte Carlo simulation was developed in FORTRAN-77, responsible for Compton scattering simulation and particle transport. This code creates histories for incident photons that are Compton scattered in the target. Each photon generates a single $e' - \gamma'$ pair compatible with the detector angular acceptance. The simulation is based on impulse approximation kinematics [4] which takes into account the electron atomic momentum[5]; a single step condensed history technique to simulate electron multiple scattering ($Highland$[6, 7, 8]); and the $HyCal$[2] detector spatial resolution to simulate particle detection. It also takes into account several geometrical parameters such as translations and distortions of the illuminated target area due to beam-detector misalignments.

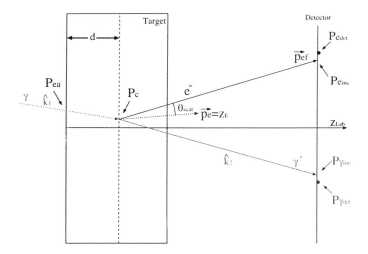

Figure 2 Geometrical schema of the steps that make up an incident photon history.

3 Analysis

For each simulated $e' - \gamma'$ pair the incident photon energy is reconstructed using Eq. 1. These energies are put into a histogram, creating a spectrum, which is then analyzed to reconstruct the average energy of the beam. Since Eq. 1 is not linear, the spectrum is asymmetric. Figure 3 shows a Gaussian fit to a very large spectrum. This spectrum asymmetry will create a systematic error to higher energies, if a simple Gaussian fit is used. So we developed two other methods: The corrected Gaussian fit and the energy expression fit. This was accomplished by developing programs in C++ using the ROOT system[9].

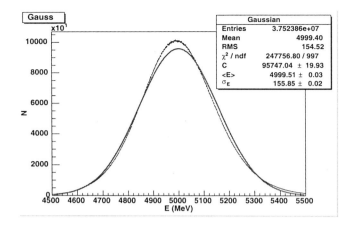

Figure 3 Gaussian fit to a very large spectrum to ilustrate spectrum asymmetry.

The form of the corrected Gaussian fit is given by:

$$F(E) = P_5(E) \cdot C \cdot \exp \frac{-(E - G_m)^2}{2\sigma^2} \quad , \tag{2}$$

where $P_5(E)$ is a fifth degree polynomial that represents the form of the spectrum. It is obtained from a large number of simulated events and then applied to "experiment sized" i.e. spectra with a number of events compatible with the acquisition time of the experiment, ~ 15000 events. This number of events is estimated in [2] and calculated using a simulation in [4]. Figure 4 shows a corrected Gaussian fit to a large spectrum.

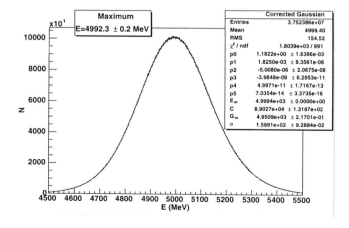

Figure 4 Corrected Gaussian fit to a very large spectrum.

Another way to eliminate the systematic error linked to the spectrum asymmetry is not to use the spectrum at all. This is the energy expression fit, which fits the θ_γ-θ_e pairs to equation 1 directly. Figure 5 shows an energy expression fit to a set of 15000 θ_γ-θ_e pairs.

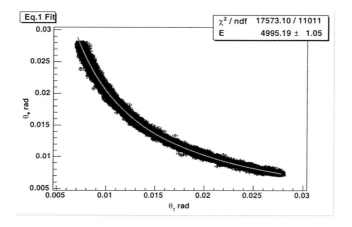

Figure 5 Energy expression fit. θ_γ-θ_e pairs are fited directly to Eq. 1.

The energy measurement is also sensitive to beam-detector misalignments. To minimize this effect, L. Gan et al[2] proposed a kinematical cut $0.4° < \theta_{\gamma,e} < 1.6°$. This is the region where the influence of misalignments is minimal[2, 4]. The effects of this cut were also studied.

A systematic error correction factor for the experiment can be obtained by using the simulation to create a large number of "experiment sized" spectra. Each spectrum is analyzed, and an energy measurement E is obtained. Then the error $\epsilon = E - E_0$ is calculated, where E_0 is the input beam energy of the simulation. The systematic error correction factor $C = -<\epsilon>$ is then calculated for all analyzed spectra. The final corrected energy measurement for each spectra then becomes $E_f = E + C$.

4 Conclusion

By using the above described analysis methods, we discovered another systematic error, this time to lower energies. This effect is opposite to the effect of spectrum asymmetry, which leads to higher measured energies, if a simple Gaussian fit is used. This new effect is related to a systematic error in angular measurements. The scattering angles θ_γ and θ_e are reconstructed from particle detection points (x, y) on the detector plane, which are subject to detector uncertainties σ_x, σ_y. Due to purely geometrical factors, these will create asymmetrical angular uncertainties that will lead, in average, to angular measurements θ greater than the "real" scattering angle θ_0.

Fig. 6(a) shows the center of the detector (+), a circle that represents the "real" scattering angle θ_0 (dotted circle) and a circle of radius σ (full circle and enlarged circle) centered at the incidence point (x_0, y_0) (**x**). This circle represents the region where one expects to find ∼ 68% of the detected incidence points for the "real" incidence point (**x**). The $\theta = \theta_0$ "reference line" divides the circle into two areas A and B. Since A > B the probability to get a reading (x, y) that will lead to an angle $\theta > \theta_0$ is greater. Since $\partial E_\gamma / \partial \theta_{\gamma,e} < 0$ (Eq. 1) this will lead to lower measured energies. As θ_0 gets larger, i.e. an incidence point further from the detector center, the area difference gets smaller and this effect is less pronounced (Fig.6(b)).

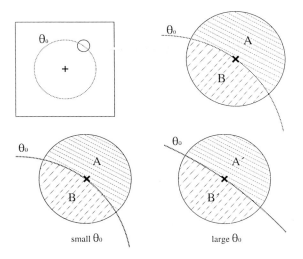

Figure 6 Geometrical schemas of incidence point detection.

Table 1 shows the results for each analysis method with and without applying the kinematical cut. 2500 "experiment sized" data sets for an input energy $E_0 = 5\text{GeV}$ with the same experimental geometry were used in each case. $< \epsilon > = -C$ is the systematic error correction factor and σ_ϵ^2 is the variance of the error distribution of the reconstructed beam energy for 2500 "experiment sized" spectra.

When the kinematical cut is used, $< \epsilon >$ is shifted to higher values. This is because the kinematical cut discards the central part of the detector, where the systematic error on angular measurements is more pronounced. This is the leading systematic error, except for the case of the Gaussian fit with cut. The Gaussian fit is the only method subject to the spectrum asymmetry systematic error to higher

Table 1 Result comparison between the several analysis methods (all energies in MeV).

Method		$< \epsilon >= -C$	σ_ϵ
Gaussian	no cut	−0.50(3)	1.29(2)
	w/ cut	2.57(3)	1.34(2)
corr. Gaussian	no cut	−7.67(3)	1.33(2)
	w/ cut	−4.75(3)	1.35(2)
energy expr.	no cut	−7.41(4)	1.21(3)
	w/ cut	−3.18(4)	1.32(3)

energies, which dominates if the kinematical cut is used. Another effect of the kinematical cut is a smaller number of accepted events. This leads to a slightly higher uncertainty in all cases, even though it diminishes the errors related to beam-detector misalignments.

Table 1 assumes a monocromatic beam. In the case of a 5GeV beam with a Gaussian incident photon energy distribution $\sigma_{k_1} = 25\text{MeV} \cong 1\%$ FWHM, there is less than 1% increase in σ_ϵ if the Gaussian or corrected Gaussian fit is used, and $\sim 4\%$ if the energy expression fit is used (see Table 2).

Table 2 Results for a non-monocromatic 5GeV beam with a Gaussian incident photon energy distribution $\sigma_{k_1} = 25\text{MeV}$ and 50MeV. All energies in MeV.

width σ_{k_1}		$\sigma_{k_1} = 25\text{MeV}$		$\sigma_{k_1} = 50\text{MeV}$	
Method		$< \epsilon >$	σ_ϵ	$< \epsilon >$	σ_ϵ
Gaussian	s/ corte	−0.34(3)	1.30(2)	−0.24(3)	1.37(2)
	c/ corte	2.67(3)	1.35(2)	2.86(3)	1.42(2)
corr. Gaus.	s/ corte	−7.81(3)	1.33(2)	−8.31(3)	1.40(2)
	c/ corte	−4.79(3)	1.36(2)	−4.87(3)	1.44(2)
Energy Expr.	s/ corte	−7.50(4)	1.25(3)	−7.71(4)	1.29(3)
	c/ corte	−3.24(4)	1.37(3)	−3.29(4)	1.41(3)

The results show that it is possible to obtain, at least in the studied experimental setup, an energy measurement with a relative unceartainty $\frac{\sigma_E}{E}$ on the order of 0.03%, which is equivalent to a FWHM precision $\frac{\Delta E}{E}$ on the order of 0.07%, if the systematic error correction factor is used.

Acknowledgements

The authors would like to dedicate this work to V.P. Likhachev, who was Carvalho's supervisor until his death in 2004. This work was made possible by the financial support from FAPESP.

Bibliography

[1] *The Primakoff Experiment at Jefferson Lab.*, http://www.jlab.org/primex/

[2] L. Gan et al., *Forward Electron Compton Scattering and Absolute Energy Calibration of the Tagged Photon Beam in Hall B (Letter of Intent)*, http://www.jlab.org/primex/documents/2002_05_05/ compton_loi.ps, (2002)

[3] Readiness Review Documentation for E-02-103 - A Precision Measurement of the Neutral Pion Lifetime via the Primakoff Effect: *http://www.jlab.org/primex/documents/2003_11_20/ readiness_report.pdf*

[4] W.R. Carvalho Jr., *Espalhamento Compton e medida absoluta da energia de fótons marcados - Uma simulação Monte Carlo* Masters thesis - Physics Institute - University of São Paulo, USP/IF/SBI-019/2005, 2005.

[5] Y.F. Chen, C.M. Kwei and C.J. Tung, *Analytic functions for atomic momentum-density distributions and Compton profiles of K and L shells*, Phys. Rev. A 47, 4502-4505 (1993)

[6] V.L. Highland, *Some practical remarks on multiple scattering* Nucl. Instr. and Meth. *129* (1975) 497-499

[7] G.R. Lynch and O.I. Dahl, *Approximations to multiple Coulomb scattering*, Nucl. Instr. and Meth. in Phys. Res. *B58* (1991) 6-10

[8] S. Eidelman et al., *Review of particle physics*, Phys. Lett. *B592* (2004) 1

[9] *ROOT System - An Object-Oriented Data Analysis Framework*, http://root.cern.ch/

Evaluation of Trace Elements in Whole Blood and its Possible use as Reference Values

C. B. Zamboni[1], L. C. Oliveira[1], M. R. Azevedo[2], E. C. Vilela[3], F. L. Farias[3], P. Loureiro[4], R. Mello[4], J.Mesa[5] and L. Dalaqua Jr[6].

[1] Instituto de Pesquisas Energéticas e Nucleares - IPEN/CNEN-SP

[2] Universidade de Santo Amaro - UNISA-SP

[3] Centro Regional de Ciências Nucleares - CRCN/CNEN- PE

[4] HEMOPE - Recife -PE

[5] Universidade de São Paulo - IFUSP-SP

[6] Promon Engenharia, SP

Abstract. Trace elements in whole blood of human have been determined through Neutron Activation Analysis (NAA), using Au as flux monitor. The estimation for reference values of Br, Cl, K, Mg and Na were evaluated in the 95 % confidence interval for normal population.

Keywords. NAA, reference value, whole blood, gamma spectrometry

1 Introduction

The availability of accurate reference values for trace elements in human tissues represents an important indicator of the health status of the general population. In addition, such values provide the scientific basis for biomedical research in trace elements related diseases, forming the basis of setting legal limits of exposure for public protection and occupational health. In Brazil there is a lack of such data for whole blood. Therefore, the project: "Evaluation of Trace Element in Body Fluids for the Brazilian population using nuclear methodology: Determination of reference values", have been proposed in collaboration with Research Centers, Blood Banks and Universities from several Brazilian regions, aiming to establish reference values for trace elements in human blood for clinical diagnosis.

2 Nuclear Analysis

The basic principle of the NAA technique is the irradiation of the biological material with neutrons followed by the measurement of γ-ray activities induced in the biological sample, where the elements can be identified by its nuclear properties. In order to determine the concentration of the trace elements in whole blood each biological sample is sealed into individual polyethylene bags, together with the flux monitor (small metallic Au foil $\sim$ 1mg) and irradiated in a pneumatic station in the nuclear reactor (IEA-R1, 2-4MW, pool type) at IPEN, allowing the simultaneous activation of these materials [1]. Using this procedure, the γ-ray activity induced in the Au detector as well as in the biological sample was obtained under the exact same irradiation conditions. After the irradiation, the activated materials (blood and Au detectors) were γ-counted using an HPGe spectrometer and the areas of the peaks, corresponding to γ transitions related to the nuclides of interest, were evaluated. The energy spectra analysis was performed using the IDF code [2] and the concentration were calculated using the software developed by Medeiros *et al.* [3]. The samples were irradiated for 2 minutes. The counting period was of 1 minute for Au detector and 10 minutes for each biological sample and the background radiation.

3 Collection and Preparation of the Biological Samples

The biological samples came from several Brazilian Blood Banks. Ethical approval for this study was obtained from the Ethics Committee of the Blank Blood authority. The whole blood samples were collected from healthy group (male and female blood donators), with ages varying from 25 to 60 years, weighing from 50 to 85 kg, select from Blood Banks following a procedure developed for the nuclear performance.

About 2ml of whole blood were collected in a vacuum plastic tubing attached to the donor's arm and immediately after the collection an amount of 100 µl of whole blood, using a calibrated pipette, was transferred to the filter paper. After that, the sample was dried for a few minutes exposed to an infrared lamp.

4 Results

The reference values of Br, Cl, K, Mg and Na were evaluated for whole blood in the 95 % confidence interval for normal population ($\pm$ 2SD). A total of 170 samples (collected in replicate) were analyzed. The mean values and the basic statistical treatment results are shown in Table 1.

Table 1 Mean values and the basic statistical treatment results for the trace elements Br, Cl, K, Mg and Na in whole blood by using the NAA technique.

Element Concentration (g l $^{-1}$)	Br	Cl	K	Mg	Na
Mean	0.0118	2.87	1.57	0.057	1.64
1 SD (67%)	0.0079	0.46	0.27	0.017	0.21
Minimum Value	0.0028	2.10	1.07	0.046	1.24
Maximum Value	0.0362	3.94	2.27	0.069	2.10
2 SD (95%)	0.0158	0.92	0.54	0.034	0.42

5 Conclusion

Using NAA it was possible to perform several clinical analysis simultaneously in whole blood in an agile, fast and economic way. Of course, larger scale studies and the participation of other Institutions, associated with Hospitals, Analysis Laboratories and also with Health Public Services, must be incorporated to the project providing the scientific basis for research in more detailed reference values and to study the common deficiencies in Brazilian population helping its diagnostics.

Acknowledgements

The authors thank the clinical staff at Blood Banks for technical assistance given during the blood collection and CNPq and CNEN by the financial support.

Bibliography

[1] L. C. Oliveira, C. B. Zamboni, F. A. Genezini, A. M. G. Figueiredo, G. S. Zahn. J. Radioanal. Nucl. Chem., 263 (2005) 783.

[2] P. Gouffon, Manual do programa IDEFIX. (Instituto de Física da Universidade de São Paulo, Laboratório do Acelerador Linear, São Paulo, 1987)

[3] J. A. G. Medeiros, C. B. Zamboni, G. S. Zahn,L. C. Oliveira, L. Dalaqua Jr. Software para realização de análises hematológicas utilizando processo radioanalítico Anais do 39° Congresso Brasileiro de Patologia Clinica (SP, Brasil, 2005).

Anais da XXVIII Reunião de Trabalho sobre Física Nuclear no Brasil
SP, Brasil, 2005

Active Photomultiplier Bases for the Saci Perere Spectrometer

T. F. Triumpho, D. L. Toufen, J. R. B. Oliveira, C. M. Figueiredo,
R. Linares, K. T. Wiedemann, N. H. Medina, R. V. Ribas

Instituto de Física da Universidade de São Paulo, São Paulo, SP, Brazil

Abstract. Degradation of output signal in scintillator detectors of the ancillary system Saci of the Perere gamma ray spectrometer (LAFN–IFUSP) is seen at high detection rates. Pulse degradation on scintillator detectors occurs when an elevated current in the photomultiplier causes a tension drop on the passive bases which work as the voltage divider. A new base prototype has been implemented with an association of passive and active electronic elements (transistors). Tests with the ^{16}O + ^{27}Al reaction at 56 MeV were carried on with the 0° detector of the Saci. The tests showed no degradation of the active base output pulse up to 17000 counts/s of α particles (energy $E_\alpha \approx 10 - 30$ MeV) and protons (energy $E_P \approx 5 - 20$ MeV).

1 Introduction

Scintillator detectors are widely used in the measurement and detection of charged particle radiation. Basically, a scintillator is a material capable of converting part of the kinetic energy of an incident charged particle into measurable light [1]. This material has a set of important characteristics such as scintillation efficiency, linear conversion, transparency, decay time and other optical criteria, all of them, properly set up for a specific use. A Phoswich detector consists of a combination of two kinds of scintillator material optically attached to a photomultiplier tube (PMT). A thick long-decay-time plastic scintillator is covered by a very thin plastic scintillator layer of fast decay time – arrangement also known as a ΔE–E detector. This allows discerning between different types of particle radiations and their respective energy. This kind of arrangement produces a typical two-dimensional histogram seen in Figure 1. Different particles appear in different regions of the histogram due to their mass, charge and energy.

This kind of charged particle detection is particularly useful in gamma spectroscopy techniques for observing specific residual nuclei proceeding from a nuclear reaction. Gamma rays detected in temporal coincidences with charged particle radiation (γ–γ–particle coincidence) is a commonly used procedure. The spectroscopy group of the LAFN–IFUSP has been using this technique (see refs. [2, 3]) with the gamma spectrometer *Saci–Perere* [4, 5]. It consists of 4 GeHP detectors with Compton suppression and an ancillary system of 11 Phoswich detectors with a detection solid angle of about 4π.

In using this procedure, it is essential to identify different particles emitted from the residual nuclei. Experience has shown that this is not possible when one uses a high detection rate or high particle energy. In these situations a degradation of the output signal from the Phoswich detectors is observed. Particle regions in the ΔE-E histogram appear superposed thus impeding analysis. This causes a limitation of the data acquisition rate or an increase of the acquisition time.

The degradation of the detector signal was attributed to a gain loss in the stages of electron multiplication of the PMT. The purpose of this work has been to develop and test a new active base in order to expand the charged particle detector's capacity. This allows working with higher rates of particle incidence on the detector without losing data due to equipment limitation.

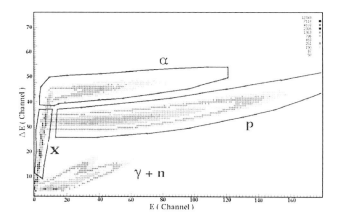

Figure 1 A ΔE-E two-dimensional histogram. Selected regions on the histogram are attributed to different particles. α particles transfer more energy to the fast scintillator (ΔE) than protons, thus, α region appears above the proton one. The X region indicates charged particles with energy less than the necessary to cross ΔE scintillator; mostly α particles and protons in the case of this histogram.

Saci's photomultipliers bases are originally built with passive elements (a Hamamatsu [6] D-type voltage divider part E2924-01). This means that they are built with a set of electrical resistances disposed in series. In the final stages capacitors are connected to the resistances to improve their fast pulse response. Because of that, the passive electronic base does not show a steady behavior under a variation of the counting rate. A rate increase implies an increase of the mean current through the dynodes, especially in the final multiplication stages. When the current on those dynodes becomes comparable to the current that passes through the resistances, the tension tends to drop, thus causing the PMT gain to decrease.

To reduce this problem, one can use an active base [7, 8]. In the case of the Saci detectors, the circuit shown in Figure 2 was used. In this circuit, a set of electrical resistances are used for the first stages, as in the passive base. In the last 4 stages, however, transistors configured as emitter followers, with the bases connected to the resistors, are arranged between the dynodes. The collector–emitter voltage is provided by the resistances. As transistors in the configuration of an emitter follower have low output impedance, a rise in the dynode current produces a much smaller alteration in the voltages in the last stages (compared to the passive base). Thus, it is expected that the response of the active PMT should be more steady.

To illustrate the advantages of the active base, the voltage between the last

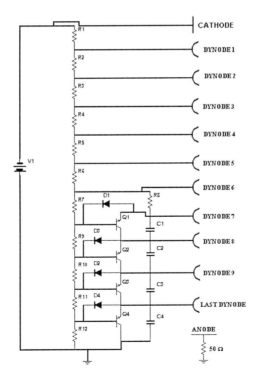

Figure 2 PMT active base circuit. A list of electronic elements used in this base can be seen in Table 1.

dynode and the anode was calculated as a function of the current in the anode. The calculations were made for both the passive and active bases (assuming ideal transistors with hfe = 100). The results can be seen in Figure 3. On the active circuit, diodes were also arranged between the emitter and the base to protect the transistor from electronic overcharge.

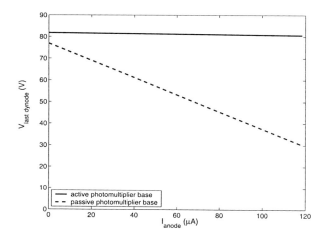

Figure 3 Tension on the last dynode versus current in the anode. It can be noticed that the voltage drop in the dynodes in the active base is very low compared to the one in the passive base.

Table 1 Table of the electronic elements used in the active PMT base.

Electronic Component	Nominal Value
R1	820kΩ
R2 – R6	270kΩ
R7 – R12	560kΩ
Q1 – Q4	MPSU10
C1 – C4	100nF (polyester)
D1 – D4	1N4148
V1	1000V

2 Experimental Arrangements

Preliminary tests showed that the two bases present similar results in low counting rate situations. The detectors were exposed to α radioactive sources of ^{241}Am ($E_\alpha \approx$ 5,5 MeV) with a detection rate of about 1500 counts/s. This radiation was not enough to produce an elevated current in the dynodes.

An in-beam test was carried out at the Pelletron Accelerator of the Universidade de São Paulo. A fusion-evaporation reaction $^{16}O + {}^{27}Al$ ($E_b = 56,0$ MeV) was measured with the Saci ancillary system. A schematic figure of the complete Saci–Perere gamma spectrometer can be seen in Figure 4. The 0° detector of the Saci was used for the measurements.

Particles were detected for several different counting rates twice for the same detector, first using the passive base and then with the active circuit. The detection rate was modified by altering the beam current and maintaining its energy.

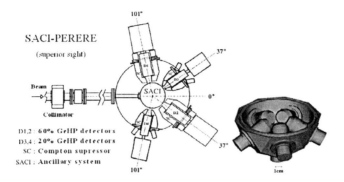

Figure 4 Scheme of the complete Saci-Perere system with the 4 GeHP gamma detectors (Perere) and the Saci ancillary system with 11 Phoswich particle detectors. One half of the Saci is shown in the detail.

3 Experimental Results

Measurements were taken for different beam currents. The $^{16}O + {}^{27}Al$ reaction yielded a continuum spectra of α particles ($E_\alpha \approx 10$ to 30 MeV), protons ($E_P \approx 5$ to 20 MeV) and other heavier nuclei (typically ^{6}Li). An example of a ΔE-E histogram obtained from this reaction can be seen in Figure 5.

With the accelerator, it was possible to reach higher currents in the active base up to the detector's limit. The maximum incidence rates were at about 17000 counts/s.

It was noticed that for beam currents around 10 nA, the signal from the passive base detector was degraded. The output base current depends not only on the rate of the incident particles, but also on their energy. The reaction, therefore, was sufficient to grant the combination (of incidence rate and particle energy) necessary for observing pulse degradation. Close to beam currents of 10 nA, the histogram regions for α particles and protons collapsed into one low energy region (see Figure 6). In this situation it is not possible to set a gate in either of those particles separately (for γ-particle coincidence).

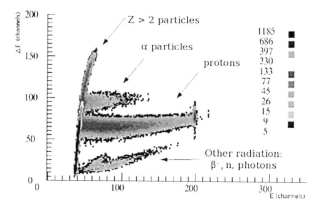

Figure 5 Normal two-dimensional histogram obtained from the ^{16}O + ^{27}Al 1 nA beam current at 56 MeV reaction measured with the active base detector. For this beam current, the spectrum yielded from the passive base detector is the same.

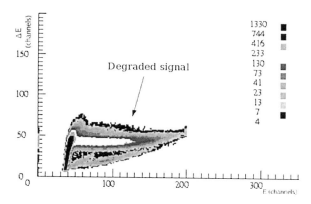

Figure 6 Degraded two-dimensional histogram obtained from the ^{16}O + ^{27}Al reaction with a 12 nA beam current at 56 MeV, measured by the passive base detector.

4 Data analysis

Projections of the ΔE-E histograms onto the ΔE axis were analyzed. Peak shifts of α particles and protons were attributed to a gain variation of the PMT. Ideally, the peak centroids should not change with the variation of the beam current because the beam energy was kept constant. The projections clearly showed a progressive degradation of the passive base detection signal (see Figure 7). This appears to be in agreement with the calculations presented in Figure 3, which predicted a voltage drop in the last and most critical dynodes. Calculations also predicted that the voltage drop should be very small in the new active base. In fact, Figure 8 shows no expressive shifts for the α and proton peaks up to 13 nA.

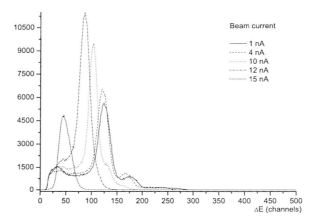

Figure 7 Histograms of ΔE projections for different beam currents of the ^{16}O + ^{27}Al (E_b = 56,0 MeV) reaction measured with the 0° Saci Phoswich detector with the passive base.

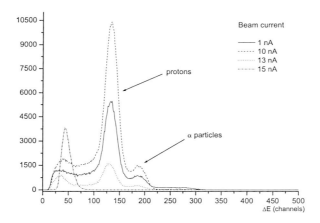

Figure 8 Histograms of ΔE projections for different beam currents of the ^{16}O + ^{27}Al (E_b = 56,0 MeV) reaction measured with the 0° Saci Phoswich detector with the active base.

The spectra show that the signal gain change in the passive base detector can be considered significant for beam currents of about 4 nA in the reaction.

5 Conclusion

With the results presented here it is possible to infer that the projected circuit of an active photomultiplier electronic base is efficient to stabilize the voltage drop in the last multiplication stages of the PMT. The detector's performance was acceptable in testing. For low base currents, the active base response to particle radiation was satisfactory, yielding the same spectra as conventional passive bases. Nevertheless, more energetic and intense radiation could be measured by active detectors without jeopardizing data integrity.

This represents an instrumentation advance for Saci ancillary system that allows one to enhance experimental statistics. Several advantages may be considered. For instance, some low cross section reaction that would require a long data acquisition time can have its experiment time reduced due to considerable rise in experimental statistics. It was estimated that the active base can tolerate 3 to 4 times more current than passive bases without any appreciable gain changes. It is important that the detector is not sensitive to small and momentary variations in the beam current. Stability on the signal yield more reliable results and also opens the possibility of energy calibration.

Bibliography

[1] G.F. Knoll, *Radiation Detection and Measurement*, second ed. (J. Wiley & Sons, 1988).

[2] J.A. Alcántara-Nuñez et al., Phys. Rev. C 69, 024317 (2004).

[3] J.A. Alcántara-Nuñez et al., Phys. Rev. C 71, 054315 (2005).

[4] J.A. Alcántara-Núñez, J. R. B. Oliveira, et al. Nucl. Inst. Meth. A 497, 429 (2003).

[5] J.A. Alcántara-Núñez, Dissertação de Mestrado, IFUSP (1999).

[6] Copyright © Hamamatsu Photonics K.K. All Rights Reserved. Site: < http://www.hamamatsu.com/ > last access in: Feb./2006.

[7] R.D. Hiebert, H.A. Thiessen, A.W. Obst, Nucl. Inst. Meth. 142, 467 (1977).

[8] C. R. Kerns, IEEE Trans. Nucl. Sci. NS-24(1), 353 (1977).

Anais da XXVIII Reunião de Trabalho sobre Física Nuclear no Brasil
SP, Brasil, 2005

Nuclear methodology for Al quantification in patients submitted to dialysis

N. Santos[1], C. B. Zamboni[2], F. F. Lima[1], E. C. Vilela[1]

[1] Centro regional de Ciências Nucleares

[2] Instituto de Pesquisas Energéticas e Nucleares

Abstract. Neutron Activation Analysis was applied to determine Al concentration in whole blood of patients submitted to a dialysis using small quantities of biological material. The results from this nuclear analysis suggesting its application for this clinical investigation. The advantages as well as the limitations of using this nuclear procedure were discussed.

Keywords. Aluminum, NAA, whole blood, dialysis

1 Introduction

In recent years we have applied the Neutron Activation Analysis (NAA) with success to investigate some element concentrations in whole blood, serum, urine, resulting in an agile and economic way to perform clinical investigations [1-7]. More recently, this methodology was used to predict iron level in whole blood of human being to diagnosis anemia (iron deficiency) [8] , that have high prevalence in Brazilian population (60 % of all anemias) [9], also confirming that this procedure can be one alternative technique for clinical investigations.

The basic principle of this nuclear methodology is the irradiation of the biological material with neutrons followed by the measurement of the γ-ray activities induced in the biological sample, where the elements can be identified by the nuclear properties of γ-rays.

In the present study we intend to use this methodology to quantify Al, using whole blood, in patients submitted a dialysis aiming its application, in the future, for checking the Al levels in blood , in fast and economic way . This study is part of a big project entitled: "Determination of reference values for concentrations of trace elements in human whole blood using nuclear methodology", nowadays in development at IPEN (Instituto de Pesquisas Energéticas e Nucleares) in collaboration with Blood Banks and Hematological Laboratories from different regions of Brazil.

For the development of the present investigation Al concentration was first analyzed in healthy group (male and female blood donors), age between 25 and 60 years at 50 and 85 kg, select from Blood Banks, to obtain an indicative interval for their reference values. The necessity to perform measurements in whole blood is related to the fact that conventional analyses are performed using serum so, there are no previous reference value established in whole blood for Brazilian population. After that, this methodology was applied to quantify Al in blood of patients submitted to a dialysis for checking the reference levels in them. This control is important to check disease in metabolism that could resulting in lost of quality of life blood of patients with CRI (chronicle renal insufficiency).

2 Experimental Procedure

To obtain the Al concentration in whole blood, each biological sample (0.1ml of whole blood fixed in filter paper) is sealed into individual polyethylene bag, together with the Au detectors (small metallic foils) used for measurement of the flux distribution [10], and irradiated for 5 minutes in a pneumatic station in the nuclear reactor (IEA-R1, 2-4MW, pool type) at IPEN, allowing the simultaneous activation of these materials. Using this procedure the γ-ray activity induced in the Au detectors as well as in the biological sample are obtained under the exact same irradiation conditions. After the irradiation, the activated materials (biological sample and Au) are gamma-counted using a HPGe Spectrometer of High Energy Resolution and the areas of the peaks, corresponding to gamma transitions related to the nuclides of interest, are evaluated. The gamma spectra analysis evaluation is performed using the IDF computer code [11] and the calculation of the concentration for each element can be obtained from software developed by Medeiros et al. [12] To check the reproducibility of this methodology, a similar experimental procedure was used in the MB-01 nuclear reactor also at IPEN facilities but, due the low flux ($\sim 10^9 \mathrm{n.cm}^{-2}.\mathrm{s}^{-1}$) comparatively from IEA-R1 ($\sim 10^{12} \mathrm{n.cm}^{-2}.\mathrm{s}^{-1}$) the irradiation time was fixed in 20 minutes using 0.5ml of biological material .

3 Results

The estimation for reference values for Al was evaluated as na interval encompassing 95 % of normal population ($\pm$ 2SD). The mean values as well as the results related to the basic statistical treatment of the data are shown in Table 1. In Fig. 1 the concentration results in whole blood are shown and the indicative interval defined by the mean value considering one and two standard deviations (SD). In addition, in Fig. 2 the frequency distribution of Al concentration is shown, with class intervals defined as 0.005, as well as the fitted normal distributions. According to these figures we can notice that the maximum of Gaussian curve distributions is in agreement with the frequency interval of the calculated arithmetic mean value (see Table 1).

Table 1 Indicative interval for the reference values of the element Al in whole blood by using the ANAA technique.

Element concentration	Al (mg/l)
Mean	0.029
1 SD (68%)	0.008
Minimum Value	0.017
Maximum Value	0.042
2 SD (95%)	0.016

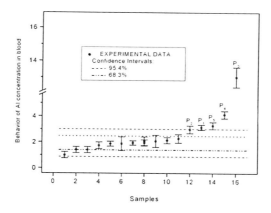

Figure 1 Concentration of Al in whole blood samples.

The Al concentration have also been determined in a group of five patients (denominated P_1, P_2, P_3, P_4 e P_5) submitted to dialysis during 5 years; these results were included in Fig. 1. A significantly variation in Al levels were obtained compared the samples from patients submitted a dialysis with those of the control group, suggesting that NAA could an alternative method for determining Al levels in blood.

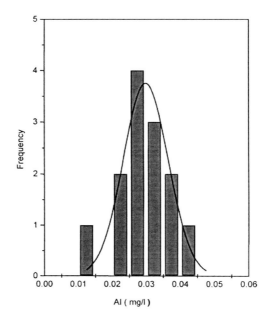

Figure 2 Histogram and Gaussian fit of Al concentration in whole blood samples

4 Conclusions

In this work a method for determination of Al, in whole blood of human being, have been proposed based on neutron activation analysis (NAA). Considering that its application for clinical analyses involve the measurements of hundreds of biological samples, this methodology could be considered fast. Of course, more systematic and large scale studies are needed to establish reference value with high precision aiming its application in for the clinical practical using whole blood helping the diagnostic of patients with CRI.

Acknowledgements

The authors thank the clinical staff at Paulista Blood Bank, staff the UFPE Clinical Hospital and the CRCN for technical assistance given during the blood collection.

Bibliography

[1] C. B. ZAMBONI, I. M. M. A. MEDEIROS, F. A. GENEZINI, A. C. CESTARI, J. T. ARRUDA-NETO, Nuclear Methodology to Study Kidney Anomalies, Proc. in: INAC - V ENAN, Brazil, 2000.

[2] L. C. OLIVEIRA, C. B. ZAMBONI, A. M. G. FIGUEIREDO, A. C. CESTARI, J. T. ARRUDA-NETO, Nuclear methodology to perform clinical examination of urine, Proc. in: III ENBN, Brazil, 2001.

[3] L. C. OLIVEIRA, C. B. ZAMBONI, G. S. ZAHN, A. C. CESTARI, L. DALAQUA JR., M. V. G. MANSO, A. M. G. FIGUEIREDO, J. T. ARRUDA-NETO, Rev. Bras. Pesq. Des., 4 (2002) 1035.

[4] L. C. OLIVEIRA, C. B. ZAMBONI, G. S. ZAHN, M. P. RAELE, M. A. MASCHIO, Braz. J. Phys., 34 (2004) 811.

[5] J. D. T. ARRUDA-NETO, M. V. M. GUEVARA, G. P. NOGUEIRA, I. D. TARICANO, M. SAIKI, C. B. ZAMBONI, J.MESA, Int. J. Rad. Biol., 250 (2004) 10.

[6] L. C. OLIVEIRA, C. B. ZAMBONI, F. A. GENEZINI, A. M. G. FIGUEIREDO, G. S. ZAHN. Use of Thermal Neutrons to Perform Clinical Analyses in Blood and Urine Samples. J. Radioanal. Nucl. Chem. 263 (2005) 783.

[7] C. B. ZAMBONI, L. C. OLIVEIRA, L. DALAQUA JR., Diagnostic Application of Absolute Neutron Activation Analysis in Hematology, Proc. in: ANES, EUA, 2004.

[8] P. S. LINS, Padronização de metodologia nuclear para determinação da concentração de Ferro em amostras de sangue total de humanos, Dissertação de Mestrado, Universidade Santo Amaro - UNISA, São Paulo, 2004.

[9] S. S. QUEIROZ, J. Pediatr., 76(2000)298.

[10] L. C. OLIVEIRA, Desenvolvimento de metodologia nuclear para análises clínicas, Dissertação de Mestrado, Instituto de Pesquisas Energéticas e Nucleares – IPEN, São Paulo, 2003.

[11] P. GOUFFON, Manual do programa IDF, Instituto de Física da Universidade de São Paulo, Laboratório do Acelerador Linear, São Paulo, 1982.

[12] J. A. G. MEDEIROS, C. B. ZAMBONI, G. S. ZAHN,L. C. OLIVEIRA, L. DALAQUA JR., Software para realização de análises hematológicas utilizando processo radioanalítico, Proc. in: 39° CBPC, Brazil, 2005.

Author Index

Aburaya, J. H. 9
Added, N. 9, 25
Alcántara–Núñez, J. A. 343
Anjos, R. M. 127
Appoloni C. R. 171, 179, 187, 199
Arruda-Neto, J. D. T. 289, 305
Azevedo, M. R. 387
Baptista, M. S. 209
Barbosa, M. D. L. 9, 37
Bastos, J. 127
Bazzacco, D. 343
Benevides, C. A. 57
Bernardes, S. 37
Bittencourt-Oliveira, M. C. 289
Borello-Lewin, T. 299
Brandolini, F. 343
Carbonari, A. W. 209
Carlin, N. 209
Carlson, B. V. 17, 353
Carmignotto, M. A. P. 25
Carvalho, C. 127
Carvalho Jr., W. R. 379
Casanova, A. O. 315
César Barbero 263
Coelho, P. R. P. 255
Cybulska, E. W. 343
Dalaqua Jr., L. 387
Dale, D. 305
D'Alessandro, K. 63
da Rocha, C. A. 1
da Silva, A. A. R. 25
de Acuña, D. 343

de Almeida, G. L.	227, 239
de André, J. P. A. M.	299
de Angelis, G.	343
de Azevedo, A. C. P.	335
De Conti, C.	17
Delgado, A. O.	25
de Lima, G. F.	45
Denyak, V.	335
de Oliveira Filho, F. J.	209
de Oliveira Jr., J. M.	45
de Poli, M.	343
Deppman, A.	255, 379
de Souza, J. C.	209
Díaz, O.	315
dos Anjos, M. J.	89
dos Santos, P. R.	9
Duarte, J. L. M.	299
Duarte, S. B.	251
Eichlt, J. R.	187
Fachini, P.	135
Falla-Sotelo, F.	343
Farias, F. L.	387
Farnea, E.	343
Felix Mas	255
Fernandes, F. W.	353
Fernandes, J. S.	199
Figueiredo, C. M.	391
Fiolhais, M.	75
Foltescu, D.	343
Frederico, T.	17
Freitas da Cruz, M. T.	221, 373
Friedberg, E. C.	289
Furieri, R. C. A. A.	227, 239
Gadea, A.	343
Galeão, A. P.	263
Garcia, C.	289, 305
Garcia, F.	289
Gelen, A.	315
Genezini, F. A.	221, 373
Gonçalves, M. J.	227, 239

Gouveia, A. N. 255
Gual, M. R. 255
Zahn, G. S. 221, 373
Guimarães-Filho, Z. O. 273, 281, 379
Helene, O. 273, 281
Hereman, T. C. 289
Herrera Peraza, E. 315
Higa, R. 1
Horodynski-Matsushigue, L. B. 299
Hoyos, O. R. 255
Hussein, M. S. 161
Ionescu-Bujor, M. 343
Iordachescu, A. 343
Javier Ccallata, H. 205
K D'Alessandro 315
Khoury, H. J. 57, 335
Krmpotić, F. 151, 161, 263
Lima, A. R. 25
Lima, F. F. 57, 399
Lima, I. C. B. 325
Linares, L. P. 75
Linares, R. 391
Longacre, R. 135
Lopes, F. 171, 179
Lopes, R. T. 89, 227, 239, 325
López-Pino, N. 63, 315
Loureiro, P. 387
Lunardi, S. 343
Lunelli, N. A. 335
Macario, K. 127
Malheiro, M. 75
Manso-Guevara, M. V. 63, 315
Marestoni, L. D. 187, 199
Martins, A. C. G. 45
Martins, M. N. 315
Matheus, R. D. 99
Medeiros, E. L. 251
Medina, N. H. 343, 391
Mello, R. 387
Melquiades, F. L. 171, 179

Mesa, J.	289, 305, 387
Mesa-Hormaza, J.	221
Morcelle, V.	361
Mosquera, B.	127
Nakagawa, I.	305
Napoli, D. R.	343
Navarra, F. S.	99
Nielsen, M.	99
Oliveira, J. R. B.	343, 391
Oliveira, L. C.	387
Oliveira, L. F.	89
Oliveira, M. C. C.	289
Paizani, E. C.	45
Parreira, P. S.	171, 179
Paschuk, S.	335
Pereira, G. R.	89
Pérez, C. A.	89
Petrache, C.	343
Podolyák, Zs.	343
Raele, M. P.	373
Rao, M. N.	343
Ribas, R. V.	343, 391
Rizzutto, M. A.	9, 25, 37, 343
Robilotta, M. R.	1
Rocha, H. S.	89
Rocha, M.	325
Rodrigues, C. L.	299
Rodrigues, M. M. N.	251
Rodrigues, M. R. D.	299
Rodrigues, T. E.	289, 305
Rogacheski, E.	335
Rogers, J. D.	227
Rossi-Alvarez, C.	343
Samana, A. R.	151, 161
Sammarco, F.	187
Sanches, N.	127
Santos, N.	399
Schelin, H. R.	335
Schenberg, A. C.	255
Schenberg, A. C. G.	289

AUTHOR INDEX

Seale, W. A. 343
Shtejer, K. 289, 305
Silva, E. C. 289
Silvani, M. I. 227, 239
Silveira, R. 57
Spolaore, P. 343
Stabin, M. G. 57
Szanto, E. M. 209
Tabacniks, M. H. 25, 37
Tarutina, T. 161
Taurines, A. R. 75
Tavares, O. A. P. 251, 361
Tilly Jr, J. G. 335
Toufen, D. L. 391
Triumpho, T. F. 391
Ukita, G. M. 299
Ur, C. A. 343
Vanin, V. R. 281
Varona-Ayala, J. N. 63
Veiga, R. 127
Vilela, E. C. 387, 399
Wiedemann, K. T. 391
Xu, Z. 135
Yoshimura, E. M. 205
Zaldívar-Montero, B. 63
Zahn, G. S. 221, 373
Zamboni, C. B. 221, 373, 387, 399
Zhang, H. 135

Projeto gráfico: Casa Editorial Maluhy & Co.
www.casamaluhy.com

Tamanho da página: 160 mm x 230 mm.

O texto desta obra está composto em Esprit 10/14.1 pt.
Os títulos de capítulos e seções em Syntax.